PRAISE FOR *FARMING WHILE BLACK*

"*Farming While Black* helps us remember why land cultivation is such a significant part of the fight for freedom for Black people. Reading this book provides practical tools along with a beautiful visionary template for practicing land development that is rooted in healing and transformation. Thank you, Leah, for your work and for your vision."

—PATRISSE KHAN-CULLORS, author of *When They Call You a Terrorist*; co-founder of Black Lives Matter

"*Farming While Black* makes an important contribution to the growing body of literature on Black farming and foodways. Small farms that grow food using sustainable, regenerative practices... are necessary if humanity is to survive. *Farming While Black* provides ideas and best practices to move us in that direction. It should be read by both new and experienced rural and urban farmers, and by all wanting to participate in creating a just, equitable, earth-friendly food system."

—MALIK YAKINI, executive director, Detroit Black Community Food Security Network

"Nothing is more important than the increasingly visible and energetic role of Black people in moving toward creating and building a food system that actually works for people—one that provides nourishing food and provides a fair standard of living for workers while stewarding the land. *Farming While Black* is a brilliant guide to moving in that direction, regardless of your skin color."

—MARK BITTMAN

"At the heart of the movement for liberation is the opportunity to heal intergenerational trauma. The most authentic way to do so is to cultivate the earth, eat the foods of your ancestors, reweave yourself back into the story that been sprouting from the village hearth since time immemorial. With these teachings of resilience, channeled from her countless generations of wise ancestors, she has watered seeds of hope that will nourish many beyond our time."

—ROWEN WHITE, Mohawk farmer; Indigenous Seed Keepers Network

"*Farming While Black* is such an incredible gift to our movement. From Black history to soil health to movement building to land preservation, this book is incredibly generous in offering a roadmap for Black people to return to our rich, land-based heritage. Calling all farmers, organizers, and lovers of freedom to pick up this book, read, share, study, and build together."

—DARA COOPER, National Black Food and Justice Alliance

"*Farming While Black* is a beautiful and timely work that manages to live at once as a stunning memoir of the extraordinary life of Leah Penniman and her Soul Fire Farm; a methodical and innovative instruction manual for a sustainable farm practice; and a clear-eyed manifesto that uses the rich history of the Black farming legacy as the guiding ethos for an effective modern day resistance movement."

—THERESE NELSON, chef, writer; founder of blackculinaryhistory.com

"*Farming While Black* offers up a bounty of hope and inspiration, not just for farmers of color—but for all of us. A practical and visionary book that challenges us to change how we farm, how we live, and how we treat each other."

—ERIC HOLT-GIMÉNEZ, executive director, Food First/Institute for Food and Development Policy

"Equal parts practical farm instruction and spiritual reflection on mind, body, spirit, and land, *Farming While Black* honors Black folks' connections to land and agriculture while recognizing structural constraints that have ruptured those connections. *Farming While Black* is an important text that (re)centers Blackness and Black people in a conversation about being growers and responsible stewards of land."

—ASHANTÉ REESE, PhD, assistant professor of anthropology; co-director of the Food Studies Program, Spelman College

"*Farming While Black* is a rich and culturally relevant how-to manual for Black and Brown farmers. Filled with uplifting stories of Black contributions to agriculture and the ongoing work at Soul Fire Farm to build an anti-racist and just food system, this is the most inspiring book I have read in years."

—IRA WALLACE, owner of Southern Exposure Seed Exchange; author of *The Timber Press Guide to Vegetable Gardening in the Southeast*

"Leah Penniman's *Farming While Black* is a remarkably thorough—and beautiful!—handbook for successful farming... But this book is not only that. *Farming While Black* shows us how we might *repair* our relationships with the land, which, given as we are the land, means repairing our relationships to ourselves. And each other. It can feel difficult to believe in the possibility of such repair, but this book gives me faith."

—ROSS GAY, poet; author of *Catalog of Unabashed Gratitude*

"Leah Penniman's powerful, informative, and lyrical work reflects her profound love for Black people and her unwavering commitment toward achieving Black land justice—a Black land classic!"

—OWUSU BANDELE, PhD, co-founder of Southeastern African American Farmers Organic Network

"Soul Fire Farm is not just a farm; it is a place of refuge. It is where intergenerational, queer, and trans Black and Brown people go to be nourished in their mind, body, and soul. My family and I are blessed to be part of this community. Wherever you are, after you read *Farming While Black*, make a trip to Grafton, New York, and visit this liberated land."

—ROSA CLEMENTE, journalist and scholar-activist; 2008 Green Party vice presidential candidate

"*Farming While Black* is freedom. A true gift from our ancestors, reminding us that they are always our teachers and inspiring us with the work being carried forward today. This book is a long awaited toolbox for Black farmers by Black farmers, rooting us to the land and empowering us to grow in our own skin."

—NATASHA BOWENS, farmer; author of *The Color of Food: Stories of Race, Resilience, and Farming*

"*Farming While Black* is a genuinely beautiful read, providing a critical guide for self-determination and community sustainability presented in a most accessible, tangible, thoughtful, and clear way; and moving us a step closer toward our over all goal of creating a 'just' economy rooted in true democracy and shared resources where all of us, not some of us, have what we need to thrive."

—RUKIA LUMUMBA, founder and executive director of the People's Advocacy Institute

"Leah is a *griot*! I could feel my own ancestors talk to me as I read each chapter. The message is a clear call to arms for what it takes to be of African descent and to be liberated in these times: we must connect with the Earth, we must put our hands in the soil. I thank the ancestors for guiding her heart, protecting her as child, and whispering to her as she penned *Farming While Black*! A'se."

—MATTHEW RAIFORD, farmer and chef; owner of Gilliard Farms, The Farmer & Larder, and Strong Roots Provisions

"As an agricultural attorney and founder of Family Agriculture Resource Management Services, I find Ms. Penniman's convictions regarding advocacy resonate personally with my own work. From the basic definition of soil testing to the technicalities of lending, Ms. Penniman's book thoroughly defines what *Farming While Black* truly means."

—JILLIAN HISHAW, Esq., founder of Family Agriculture Resource Management Services (F.A.R.M.S.)

FARMING WHILE BLACK

Soul Fire Farm's Practical Guide to Liberation on the Land

Leah Penniman

Foreword by Karen Washington

CHELSEA GREEN PUBLISHING
White River Junction, Vermont
London, UK

The painting on the dedication page is *Foresight* by Naima Penniman.

The song "How Could Anyone" on page 66 is from the recordings *How Could Anyone* and *If You See a Dream* by Libby Roderick (Turtle Island Records). Copyright © 1988 Libby Roderick Music. Reprinted with permission. Please visit www.libbyroderick.com, or contact libbyroderick@gmail.com for more information. All rights reserved.

The song "A Sower Went Out to Sow Her Seed" on page 68 is from *Octavia E. Butler's Parable of the Sower: The Opera by Toshi Reagon.* Reprinted with permission.

The poem "Lineage" on page 149 is from *This Is My Century: New and Collected Poems by Margaret Walker*. Copyright © 1989 by Margaret Walker. Reprinted with the permission of University of Georgia Press.

The poem "A Small Needful Fact" by Ross Gay on page 103 is from Split This Rock's *The Quarry: A Social Justice Poetry Database.* www.splitthisrock.org. Copyright © 2015 by Ross Gay. Reprinted with permission.

All author proceeds from this book will be donated to a fund providing land and training for Black farmers.

Project Manager: Patricia Stone
Editor: Michael Metivier
Copy Editor: Laura Jorstad
Proofreader: Nancy A. Crompton
Indexer: Linda Hallinger
Designer: Melissa Jacobson

Printed in the United States of America.
First printing October, 2018.
10 9 8 7 6 5 4 20 21 22

Our Commitment to Green Publishing
Chelsea Green sees publishing as a tool for cultural change and ecological stewardship. We strive to align our book manufacturing practices with our editorial mission and to reduce the impact of our business enterprise in the environment. We print our books and catalogs on chlorine-free recycled paper, using vegetable-based inks whenever possible. This book may cost slightly more because it was printed on paper that contains recycled fiber, and we hope you'll agree that it's worth it. *Farming While Black* was printed on paper supplied by LSC Communications that contains at least 10% postconsumer recycled fiber.

Library of Congress Cataloging-in-Publication Data
Names: Penniman, Leah, author.
Title: Farming while Black : Soul Fire Farm's practical guide to liberation on the land / Leah Penniman ; foreword by Karen Washington.
Description: White River Junction, Vermont : Chelsea Green Publishing, [2018] | Includes bibliographical references and index.
Identifiers: LCCN 2018027631| ISBN 9781603587617 (pbk.) | ISBN 9781603587624 (ebook)
Subjects: LCSH: Farms, Small—United States. | African American farmers—United States.
Classification: LCC HD1476.U6 P46 2018 | DDC 630.68—dc23
LC record available at https://lccn.loc.gov/2018027631

Chelsea Green Publishing
85 North Main Street, Suite 120
White River Junction, VT 05001
(802) 295-6300
www.chelseagreen.com

This book is dedicated to our ancestral grandmothers, who braided seeds in their hair before being forced to board transatlantic slave ships, believing against the odds in a future of sovereignty on land.

CONTENTS

Foreword *ix*
Acknowledgments *xi*

Introduction Black Land Matters 1

Chapter 1 Finding Land and Resources 11
Accessing Land, 12 • Farm Skills Training, 16 • Gathering Material Resources, 20

Chapter 2 Planning Your Farm Business 27
Worker-Owned Cooperative Business Model, 29 • Farm-Share, Community-Supported Agriculture, 38 • Food Hubs, 42 • Communal Labor Practices, 44 • Writing Your Farm Business Plan, 48

Chapter 3 Honoring the Spirits of the Land 53
Sacred Literature, 56 • Offerings to Azaka and Orisa Oko, 59 • Planting and Harvesting Rituals, 62 • Herbal Baths, 64 • Songs and Chants, 66

Chapter 4 Restoring Degraded Land 71
Remediating Soil Contaminated with Lead, 72 • Healing Erosion with Terraces, 78 • Agroforestry for Soil Restoration, 80 • No-Till and Biological Tillage, 84

Chapter 5 Feeding the Soil 87
Soil Tests, 88 • Compost, 94 • Soil Ecology, 96 • Cover Crops, 99

Chapter 6 Crop Planning 103
Annual Crops, 104 • Distant Cousins, 116 • Polycultures, 123 • Farm Layouts with Rotations, 126

Chapter 7 Tools and Technology 129
Bed Preparation, 130 • Propagation, 135 • Transplanting and Direct Seeding, 139 • Irrigation, 140 • Weeding and Crop Maintenance, 142 • Harvest, 145 • Apparel and Gear, 147

Chapter 8 Seed Keeping 149
Why Save Seed?, 150 • The Seed Garden, 152 • The Seed Harvest, 155 • Seed Exchange, 159

Chapter 9 Raising Animals 163

Raising Chickens for Eggs, 164 • Raising Chickens for Meat, 168 • Raising Pigs, 176 • Meat and Sustainability, 179

Chapter 10 Plant Medicine 181

Species Accounts of Cultivated Plant Allies, 182 • Species Accounts of Wildcrafted Plant Allies, 189 • Growing an Herb Garden, 196 • Herbal Preparations, 199

Chapter 11 Urban Farming 205

Laws and Land Access, 207 • Clean Soil, Clean Water, 210 • Growing in Small Spaces, 215 • Community, 222

Chapter 12 Cooking and Preserving 223

African Food Pyramid, 224 • Recipes, 226 • Food Preservation, 236 • No Money, No Time, 241

Chapter 13 Youth on Land 245

Why Youth on Land?, 246 • Best Practices in Youth Programming, 248 • Youth Food Justice Curriculum, 252

Chapter 14 Healing from Trauma 263

Historical Trauma: An Annotated Timeline, 265 • Internalized Racism, 273 • Healing Ourselves, 275

Chapter 15 Movement Building 281

Litigation, 284 • Education, 285 • Direct Action, 287 • Land Defense, 289 • Policy Change, 291 • Consumer Organizing, 295 • Mutual Aid and Survival Programs, 297

Chapter 16 White People Uprooting Racism 299

Reparations, 301 • Forming Interracial Alliances, 304 • Organizational Transformation, 306 • Calling In, 309 • Personal Development, 312

Afterword 317

Closing Poem *Black Gold* 318

Resources *321*

Notes *325*

Index *338*

FOREWORD

I met Leah Penniman back in the summer of 2010 while attending the NOFA (Northeast Organic Farming Association) Summer Conference. It was a typical food and agriculture conference, with many attendees, mostly white, talking about farming and sustainable agriculture. I often wondered how one could talk about sustainable agriculture without mentioning the contributions of people of color (POC). Throughout the years these conferences had all been white-led, with no reference to the contributions made by Black people; at every one I attended, I could count on one hand how many people looked like me.

During that same year I was in the midst of planning an all-Black conference with friends and colleagues, known as the Black Farmers and Urban Gardeners Conference, to be held in New York City. Calling ourselves BUGS (Black Urban Growers), our conference would address the great contributions of the Black farmer. It was to celebrate our history, knowledge, and leadership in Black agriculture. We were to have Black leaders, farmers, educators, and activists addressing our issues and concerns, while also offering resources, networking, and opportunities.

So this time I came prepared to do outreach at NOFA, seeking out as many Black folks as possible to tell them about the conference. That afternoon, however, while walking to a session, a woman of color handed me a piece of paper with a classroom number. I looked in her eyes, and then we gave each other that familiar nod that only Black people can give and proceeded to the room. That woman was Leah Penniman. She had set out and commandeered one of the classrooms to provide a space solely for POC. This was the first time at a conference that there was a space just for us. We sat with pride as we went around the circle introducing ourselves, talking about our frustrations with not being represented at food and farming conferences. I sat in awe as this young Black woman engaged us in a conversation around race and power. Right then and there I knew she was special.

For the past eight years I have watched Leah grow into a powerful leader, starting her own farm, holding antiracism workshops, and developing a justice and leadership program for youth, all while holding down a teacher's job. We continue to work together, breaking down walls of oppression while seeking out opportunities and justice for our people. We have been told time and time again that we stand on the shoulders of greatness. I am deeply honored that I have been asked to write the foreword for this powerful book.

For centuries our ancestors have cried from beyond the voyage of time for us to hear the truth so that their deaths were not in vain. In *Farming While Black*, Leah Penniman has heard them. Given the centuries of falsehoods, misconceptions, and stolen information, Leah has heard their cries for salvation!

Out from the shadows of darkness we are taken on an emancipated journey of truth, power, reclamation, and fortitude with a resounding *Amen*. With humility and respect, she pays homage to our ancestors as she articulates our agricultural history.

Confronting everything from the destruction of the continent's original populace to the tyranny of colonialism, this masterpiece of Afro-indigenous sovereignty sheds light on the richness of Black culture permeating throughout agriculture. Throughout the book we are reminded that we were the pioneers (Harriet Tubman, Booker T. Washington), the inventors (George Washington Carver, Booker T. Whatley), the trailblazers (Fannie Lou Hamer, The Black Panther Party), and visionaries—intertwining and interconnecting our spirituality and farming.

Farming While Black teaches us the fundamental acts of growing food and growing community. While entertaining us with family adventures, Leah shares stories, cultures, chants, recipes, curricula, and much more from her life's work as a farmer and co-founder and co-director of Soul Fire Farm.

We are reminded that as Black farmers we cannot do this work without consulting the wisdom of our sacred literature (Odu Ifa) and receiving spiritual permission from the deities of the Earth and universe (Orisa) to plant our seeds, to grow our food, to share our harvest, and to give back to the Earth.

We are also reminded of both past and present forms of trauma and oppressive behavior, from colonialism, to the present day slavery of mass incarceration, and yet, globally as a people, still we rise!

In the struggle for civil rights and human rights, Black farmers have always been there. We have endured the test of time in our resilience and resolve. Our power is in the soil, the land, the Earth. Our skin hues are a testament to our belonging.

Farming While Black encourages us to reach for greatest and settle for nothing less. Know your history. Share and tell our stories. Pay respect and honor our elders. Pass on the gift of knowledge and fortitude to our youth. Find strength in family and community, but above all love one another, love the Earth, and be true to one's self.

As we move toward alternative ways of farming, living, and being, away from our oppressors, we cannot fall victim to replicating their behavior. Leah has paved the way by addressing our fears, wants, and desires.

Thank you Leah for giving us this gift. *Farming While Black* is the book we will turn to when we need to be grounded, reminded of our heritage, our greatest, our resources, our networks, and the reason why as Black people we farm!

—KAREN WASHINGTON
Rise & Root Farm, Chester, New York

ACKNOWLEDGMENTS

While I sit indoors and pontificate at my computer screen, the Soul Fire Farm team endures cold fingers and sore shoulders to grow food and create habitat. The biggest shout-out goes to Jonah Vitale-Wolff, Larisa Jacobson, Damaris Miller, Amani Olugbala, Lytisha Wyatt, Ceci Pineda, Olive Watkins, Neshima Vitale-Penniman, and Emet Vitale-Penniman for doing the beautiful and often grueling work of stewarding this sacred land. Jonah, Neshima, and Emet, you are the loves of my life and I am so honored to be part of this family.

Beneath my confident and vivacious exterior dwells an insecure and quiet thinker who wonders, *Can I really do this?* I thank my dear soul brother Enroue Halfkenny for being the unwavering, loving support that got me through. My powerful sisters Naima Penniman and Taina Asili never doubted my capacity and kept reminding me, "The world needs this book."

The entire manuscript came together in just five months, and with only two days per week at that. This was possible thanks to a brilliant behind-the-scenes team of researchers and readers who generously shared their time and knowledge. Thank you Larisa Jacobson for being the lead content editor for this book and to Juliet Tarantino, Myles Lennon, Dara Cooper, Ross Gay, Owen Taylor, Justine Williams, Tagan Engel, Enroue Halfkenny, and Elizabeth Henderson for your research and review of the chapters.

Farmers and activists are the busiest people I know, yet so many of you took the time to talk with me and share your stories for *Farming While Black*. A low bow of gratitude to Eugene Cooke, JoVonna Johnson, Dennis Derryck, Matthew Raiford, Demalda Newsome, Rufus Newsome, Forrest Lahens, Terressa Tate, Chris Bolden Newsome, Ben Burkett, Rhyne Cureton, Dijour Carter, Malik Yakini, Mama Hanifa, Karen Washington, Owen Taylor, Chef Njathi Kabui, Mama Isola, Michael Twitty, Yusuf Burgess, Julie Rawson, Wislerson Pierre Louis, Jean Moliere, Mr. Kwabla of Oborpah-Djerkiti, Jun San Yasuda, Xavier Brown, and all the other growers whose work and mentorship inspired the content of this book.

To all the past, present, and future board members, staff, alumni, and volunteers of Soul Fire Farm, I see you, I honor you, I love you. A special thank-you goes to Adaku Utah, Jalal Sabur, Adele Smith-Penniman, Dan Lyles, Kavitha Rao, Abby Lublin, Tagan Engel, Taina Asili, Kristin Reynolds, Elena Rosenbaum, Naima Penniman, Gabrilla Ballard, Adán Martinez, Gail Myers, and Mama Claudia Ford for the countless hours you put into keeping this project alive and thriving.

Big ups to my editor at Chelsea Green, Michael Metivier, for surviving the Ashburnham-Westminster public school system with me, waiting patiently until I found the time to write this compendium, and having the courage to publish a book that directly confronts white supremacy and colonialism in the food system.

I thank my parents, Adele Smith-Penniman and Keith Penniman and my siblings Naima and Allen Penniman for loving me unconditionally. I offer gratitude to my ancestors Samuel Cornelius Smith, Brown Lee McCullough, Anne Elizabeth McTurman, Eural Allen McCullough, Sarah Ann Jackson, Merton Allen Penniman, Winifred Ruth Curtis-Penniman, and all others in the Smith, McCullough, Curtis, Penniman, Bordeleau, and Jackson lineages for having my back at all times. I thank the Mohican people for stewarding this sacred land for generations. I pay homage to Mawu-Lisa, the orisa, and all the Forces of Nature for the abundant blessings of this life.

The Soul Fire Farm staff gather for our winter retreat. *Top row from the left:* Lytisha Wyatt, Larisa Jacobson, and Damaris Miller. *Bottom row from the left:* Leah Penniman, Amani Olugbala, Ceci Pineda, Jonah Vitale-Wolff, and Olive Watkins. Photo by King Aswad.

INTRODUCTION

Black Land Matters

Revolution is based on land. Land is the basis of all independence. Land is the basis of freedom, justice, and equality.

—Malcolm X

As a young person, and one of three mixed-race Black children raised in the rural North mostly by our white father, I found it very difficult to understand who I was. Some of the children in our conservative, almost all-white public school taunted, bullied, and assaulted us, and I was confused and terrified by their malice. But while school was often terrifying, I found solace in the forest. When human beings were too much to bear, the earth consistently held firm under my feet and the solid, sticky trunk of the majestic white pine offered me something stable to grasp. I imagined that I was alone in identifying with Earth as Sacred Mother, having no idea that my African ancestors were transmitting their cosmology to me, whispering across time, "Hold on daughter—we won't let you fall."

I never imagined that I would become a farmer. In my teenage years, as my race consciousness evolved, I got the message loud and clear that Black activists were concerned with gun violence, housing discrimination, and education reform, while white folks were concerned with organic farming and environmental conservation. I felt that I had to choose between "my people" and the Earth, that my dual loyalties were pulling me apart and negating my inherent right to belong. Fortunately, my ancestors had other plans. I passed by a flyer advertising a summer job at The Food Project, in Boston, Massachusetts, that promised applicants the opportunity to grow food and serve the urban community. I was blessed to be accepted into the program, and from the first day, when the scent of freshly harvested cilantro nestled into my finger creases and dirty sweat stung my eyes, I was hooked on farming. Something profound and magical happened to me as I learned to plant, tend, and harvest, and later to prepare and serve that produce in Boston's toughest neighborhoods. I found an

Soil-kissed hands at Soul Fire Farm. Photo by Neshima Vitale-Penniman.

Children lead the opening ceremonies at the 2017 Black Farmers and Urban Gardeners Conference in Atlanta, Georgia. Photo by Warren Cameron.

anchor in the elegant simplicity of working the earth and sharing her bounty. What I was doing was good, right, and unconfused. Shoulder-to-shoulder with my peers of all hues, feet planted firmly in the earth, stewarding life-giving crops for Black community—I was home.

As it turned out, The Food Project was relatively unique in terms of integrating a land ethic and a social justice mission. From there I went on to learn and work at several other rural farms across the Northeast. While I cherished the agricultural expertise imparted by my mentors, I was also keenly aware that I was immersed in a white-dominated landscape. At organic agriculture conferences, all of the speakers were white, all of the technical books sold were authored by white people, and conversations about equity were considered irrelevant. I thought that organic farming was invented by white people and worried that my ancestors who fought and died to break away from the land would roll over in their graves to see me stooping. I struggled with the feeling that a life on land would be a betrayal of my people. I could not have been more wrong.

At the annual gathering of the Northeast Organic Farming Association, I decided to ask the handful of people of color at the event to gather for a conversation, known as a caucus. In that conversation I learned that my struggles as a Black farmer in a white-dominated agricultural community were not unique, and we decided to create another conference to bring together Black and Brown farmers and urban gardeners. In 2010 the National Black Farmers and Urban Gardeners Conference (BUGS), which continues to meet annually, was convened by Karen Washington. Over 500 aspiring and veteran Black farmers gathered for knowledge exchange and for affirmation of our belonging to the sustainable food movement.

Through BUGS and my growing network of Black farmers, I began to see how miseducated I had been regarding sustainable agriculture. I learned that "organic farming" was an African-indigenous system developed over millennia and first revived in the United States by a Black farmer, Dr. George Washington Carver, of Tuskegee University in the early 1900s. Carver conducted extensive research and codified the use of crop rotation in combination with the planting of nitrogen-fixing legumes, and detailed how to regenerate soil biology. His system was known as regenerative agriculture and helped move many southern farmers away from monoculture and toward diversified horticultural operations.[1]

Dr. Booker T. Whatley, another Tuskegee professor, was one of the inventors of community-supported agriculture (CSA),* which he called a Clientele Membership Club. He advocated for diversified pick-your-own operations that produced an assortment of crops year-round. He developed a system that allowed consumer members to access produce at 40 percent of the supermarket pricing.[2]

Further, I learned that community land trusts were first started in 1969 by Black farmers, with the New Communities movement leading the way in Georgia. Land trusts are nonprofit organizations that achieve conservation and affordable housing goals through cooperative ownership of land and restrictive covenants on land use and sale. In addition to catalyzing the community land trusts, Black farmers also demonstrated how cooperatives could provide for the material needs of their members, such as housing, farm equipment, student scholarships and loans, as well as organize for structural change. The 1886 Colored Farmers' National Alliance and Cooperative Union and Fannie Lou Hamer's 1972 Freedom Farm were salient examples of Black leadership in the cooperative farming movement.[3]

Learning about Carver, Hamer, Whatley, and New Communities, I realized that during all those years of seeing images of only white people as the stewards of the land, only white people as organic farmers, only white people in conversations about sustainability, the only consistent story I'd seen or been told about Black people and the land was about slavery and sharecropping, about coercion and brutality and misery and sorrow. And yet here was an entire history, blooming into our present, in which Black people's expertise and love of the land and one another was evident. When we as Black people are bombarded with messages that our only place of belonging on land is as slaves, performing dangerous and backbreaking menial labor, to learn of our true and noble history as farmers and ecological stewards is deeply healing.

Fortified by a more accurate picture of my people's belonging on land, I knew I was ready to create a mission-driven farm centering on the needs of the Black community. At the time, I was living with my Jewish husband, Jonah, and our two young children, Neshima and Emet, in the South End of Albany, New York, a neighborhood classified as a "food desert" by the federal government. On a personal level this meant that despite our deep commitment to feeding our young children fresh food and despite our extensive farming skills, structural barriers to accessing good food stood in our way. The corner store specialized in Doritos and Coke. We would have needed a car or taxi to get to the nearest grocery store, which served up artificially inflated prices and wrinkled vegetables. There were no available lots where we could garden. Desperate, we signed up for a CSA share, and walked 2.2 miles to the pickup point with the newborn in the backpack and the toddler in the stroller. We paid

* Community-supported agriculture (CSA) is a food production and distribution system that directly connects farmers and consumers. In short: People buy "shares" of a farm's harvest in advance and then receive a portion of the crops as they're harvested.

more than we could afford for these vegetables and literally had to pile them on top of the resting toddler for the long walk back to our apartment.

When our South End neighbors learned that Jonah and I both had many years of experience working on farms, from Many Hands Organic Farm, in Barre, Massachusetts, to Live Power Farm, in Covelo, California, they began to ask whether we planned to start a farm to feed this community. At first we hesitated. I was a full-time public school science teacher, Jonah had his natural building business, and we were parenting two young children. But we were firmly rooted in our love for our people and for the land, and this passion for justice won out. We cobbled together our modest savings, loans from friends and family, and 40 percent of my teaching salary every year in order to capitalize the project. The land that chose us was relatively affordable, just over $2,000 an acre, but the necessary investments in electricity, septic, water, and dwelling spaces tripled that cost. With the tireless support of hundreds of volunteers, and after four years of building infrastructure and soil, we opened Soul Fire Farm, a project committed to ending racism and injustice in the food system, providing life-giving food to people living in food deserts, and transferring skills and knowledge to the next generation of farmer-activists.

Our first order of business was feeding our community back in the South End of Albany. While the government labels this neighborhood a food desert, I prefer the term *food apartheid*, because it makes clear that we have a *human-created* system of segregation that relegates certain groups to food opulence and prevents others from accessing life-giving nourishment. About 24 million Americans live under food apartheid, in which it's difficult to impossible to access affordable, healthy food. This trend is not race-neutral. White neighborhoods have an average of four times as many supermarkets as predominantly Black communities. This lack of access to nutritious food has dire consequences for our communities. Incidences of diabetes, obesity, and heart disease are on the rise in all populations, but the greatest increases have occurred among people of color, especially African Americans and Native Americans. These diet-related illnesses are fueled by diets high in unhealthy fats, cholesterol, and refined sugars, and low in fresh fruits, vegetables, and legumes. In our communities, children are being raised on processed foods, and now over one-third of children are overweight or obese, a fourfold increase over the past 30 years. This puts the next generation at risk for lifelong chronic health conditions, including several types of cancer.[4]

At Soul Fire Farm we had to invest in the soil in order for her to yield the food our community needed. Working hard to build up the marginal, rocky, sloped soils using no-till methods, we managed to create about a foot of topsoil. Into this rich, young earth, we were finally ready to plant over 80 varieties of mostly heirloom vegetables and small fruits, centering crops with cultural significance to our peoples. Once per week we harvested the bounty and boxed it up into even shares that contained 8 to 12 vegetables each, plus a dozen eggs, sprouts, and/or poultry for the members of South End Community. The farm share was a subscription program based on the African American Kwanzaa principle of *Ujamaa*, meaning "cooperative economics." As Dr. Maulana Karenga once explained, "In a world where greed, resource seizure, and plunder have been globalized with maximum technological and military power, we must uphold the principle and practice of *Ujamaa* (Cooperative Economics) or shared work and wealth. This principle reaffirms the right to control and benefit from the resources of one's own lands and to an equitable and just share of the goods of the world."[5]

Desiring to move beyond the casual and exploitative relationship between producer and consumer that capitalism celebrates, we developed long-term relationships of mutual commitment with our members. In early spring members signed up for the program and committed to spend whatever they could afford on our farm's bounty. We used a sliding-scale model where people contributed depending on their level of income and wealth. In turn we committed to providing members with a weekly delivery of bountiful, high-quality food throughout the harvest season, which lasts 20 to 22 weeks in our climate. We delivered the boxes directly to the doorsteps of people living under food apartheid and accepted government benefits as payment, such as the federal Supplemental Nutrition Assistance Program (SNAP). This reduced the two most pressing barriers to food access: transportation and cost. Using the farm-share model, we can now feed 80 to 100 families, many of whom would not otherwise have access to life-giving food. One member told us that their family "would be eating only boiled pasta if it were not for this veggie box."

While it was and continues to be essential for us as farmers to maintain a commitment to food access, that work alone is inadequate to address the systemic issues that led to food apartheid in the first place. Racism is built into the DNA of the US food system. Beginning with the genocidal land theft from Indigenous people, continuing with the kidnapping of our ancestors from the shores of West Africa for forced agricultural labor, morphing into convict leasing, expanding to the migrant guestworker program, and maturing into its current state where farm management is among the whitest professions, farm labor is predominantly Brown and exploited, and people of color disproportionately live in food apartheid neighborhoods and suffer from diet-related illness, this system is built on stolen land and stolen labor, and needs a redesign. We were aware that we could not solve the entire problem on our own, but neither could we cast our silent vote for the status quo through complicit non-action. We needed mentorship from our elders.

We invited veteran civil rights activist Baba Curtis Hayes Muhammad to our table to discuss the role of farmers in the movement for racial justice. "Recognize that land and food have been used as a weapon to keep Black people oppressed," he said. "Recognize also that land and food are essential to liberation for Black people."

Muhammad explained the central role that Black farmers played during the civil rights movement, coordinating campaigns for desegregation and voting rights as well as providing food, housing, bail money, and safe haven for other organizers. With his resolute and care-worn eyes, immense white Afro, and hands creased with the wisdom of years, this was a man who inspired us to listen attentively so that we might stand on the shoulders of activists who had gone before.

"Without Black farmers, there would have been no Freedom Summer—in fact, no civil rights movement," he said. He asked us, "How are you contributing to today's movements for racial justice?"

Even as we continued to provide nourishing food to our people living under food apartheid in six Capital District neighborhoods, we knew we needed to do more. So we started to organize. We expanded our work to include youth empowerment and organizing, specifically working with court-adjudicated, institutionalized, and state-targeted youth. Arguably, the seminal civil rights issue of our time is the systemic racism permeating the criminal "justice" system. The Black Lives Matter movement has brought to national attention the fact that people of color are disproportionately targeted by police stops, arrests, and police violence. And once they're in the system, they tend to receive subpar legal representation and longer sentences, and are less likely to receive parole. The 2014 police killings of Eric Garner and Michael Brown were not isolated incidents, but part of a larger story of state violence toward people of color.

Black youth are well aware that the system does not value their lives. "Look, you're going to die from the gun or you are going to die from bad food," one young man said while visiting Soul Fire Farm. "So there is really no point." This fatalism, a form of internalized racism, is common among Black youth. It's a clear sign that this country needs a united social movement to rip out racism at its roots and dismantle the caste system that makes these young people unable to see that their beautiful Black lives do matter.

We started the Youth Food Justice program in our third year, aiming to liberate our young people from the criminal punishment system. Through an agreement with the Albany County courts, young people could choose to complete our on-farm training program in lieu of punitive sentencing. It was imperative that we interrupt the school-to-prison pipeline that demonizes and criminalizes our youth. We felt that young people instead needed mentorship from adults with similar backgrounds, connection to land, and full respect for their humanity.

In the 50-hour training program, young people learned basic farming, cooking, and business skills. More important, they learned that they were necessary, they were valuable, and they belonged. We took our shoes off and placed our bare feet firmly on the warm earth. As we walked past the garlic field, the swallows swooping over the buckwheat flowers, the grandmothers in the ancestor realm whispered their love for us. The most hardened and defended child, who earlier asked, "What's the point?," began to weep. His grandmothers reached for him through the earth under his feet and reminded him that there was a point. He and his peers found meaning in tending the crops that would feed their communities back home, and teaching adults the skills they had garnered. They sat in a circle and analyzed the brokenness of the criminal punishment system, compiling necessary policy changes that would be championed by the New York State Prisoner Justice Network. They made bows and arrows in the forest, threw stones in the pond, and allowed themselves some laughter.

"My original charge was loitering, and then once I was in the system, everything got harder and started getting out of control," shared one young man on the first day of the program. As others spoke, we learned that his story was not unique—in fact, most of the young men's first arrests had been for loitering. I shared with the group that loitering laws were part of the vagrancy statutes included in the Black Codes. These were laws written to control and re-enslave the Black population after Reconstruction, a set of policies that followed the Civil War. Some things have not changed.

Youth program participants at Soul Fire Farm place their bare feet on the soil.

We agreed with the position of Malcolm X in his "Message to the Grass Roots," a speech he delivered in 1963. "Revolution is based on land," he said. "Land is the basis of all independence. Land is the basis of freedom, justice, and equality." We saw the Youth Food Justice Program as an opportunity for these young men to heal relationships with their communities, the land, and themselves, as well as to recognize their potential to be agents of change in society.

However, it was not enough for young people to simply feel connected to land. We knew the land and belonged to the land, but the land did not belong to us. Ralph Paige of the Federation of Southern Cooperatives put it simply: "Land is the only real wealth in this country and if we don't own any then we're out of the picture."

Brutal racism—maiming, lynching, burning, deportation, economic violence, legal violence—ensured that our roots would not spread deeply and securely. In 1910, at the height of Black landownership, 16 million acres of farmland—14 percent of the total—was owned and cultivated by Black families.[6] Now less than 1 percent of farms are Black-owned.

Our Black ancestors were forced, tricked, and scared off land until 6.5 million of them migrated to the urban North in the largest migration in US history. This was no accident. Just as the US government sanctioned the slaughter of buffalo to drive Native Americans off their land, so did the United States Department of Agriculture and the Federal Housing Administration deny access to farm credit and other resources to any Black person who joined the NAACP, registered to vote, or signed any petition pertaining to civil rights. When Carver's methods helped Black farmers be successful enough to pay off their debts, their white landlords responded by beating them almost to death, burning down their houses, and driving them off their land.[7]

Participants in Soul Fire Farm's Black Latinx Farmers Immersion exchange a fist pump while tending to the onions.

According to the think tank Race Forward, even today, Blacks, Latinx, and Indigenous people working in the food system are more likely than whites to earn lower wages, receive fewer benefits, and live without access to healthy food.[8]

Owning our own land, growing our own food, educating our own youth, participating in our own health care and justice systems—this is the source of real power and dignity.

In our fourth year we started the Black Latinx Farmers Immersion at Soul Fire Farm as a humble attempt to rewrite part of this story, to reclaim our ancestral right to both belong to the land and have the land belong to us. We heard time and again from aspiring Black and Brown farmers that agricultural training programs in their communities were, at best, culturally irrelevant, and often outright racist. One young Black farmer was subjected to the white farm owner asking him over bean picking, "Why is it that Black men abandon their families?" Another sister shared that her application had been denied for several apprenticeships and incubator farms, and she could not afford to leave her family to attend a land grant university. We knew that we had to act.

The Black Latinx Farmers Immersion was designed as a rigorous introduction to small-scale sustainable farming that balanced the nerdy explication of concepts like "soil cation exchange capacity" with the cultural and historical teachings necessary for our people to heal our relationships to land. On a typical day we woke at dawn and circled up on the dew-covered grass for morning movement. From there we broke up into teams to learn hands-on farming tasks by doing—bed prep, seeding, transplanting, pest control—and then shared that knowledge with other participants through an "each one teach one" popular education model. We took turns cooking the recipes of our ancestors, substituting locally grown vegetables for their tropical equivalents. We learned the songs and prayers used in the process of slaughtering animals. We learned to take life. Then we engaged in herbal healing baths in the African tradition to cleanse that strong energy and lay down our metaphorical "knives." We used drums and song to encourage the seeds to grow, and we filled the moonlit night sky with the sounds of our dancing to Kendrick Lamar and Nicki Minaj. We bathed ourselves in resiliency.

On the last day of the weeklong program, we sealed our intentions with the words of Assata Shakur: "It is our duty to fight for our freedom. It is our duty to win. We must love each other and support each other. We have nothing to lose but our chains." We chanted it over and over, growing louder each time until the powerful echoes of our shouts mingled with quiet tears—tears for the struggles of oppressed peoples and tears for the hope of liberation.

Alumni of the Black Latinx Farmers Immersion have become a force to be reckoned with in the food justice world. Together we have catalyzed a national reparations initiative to return stolen land and resources to our people. We are working on a northeast regional land trust that will upend private ownership of Mama Earth and increase farmland stewardship by people of color. Our alumni collective of speakers and trainers are waking the public up to the pervasive injustice in the food system and providing concrete actionable next steps to restore justice. We have even expanded our solidarity work beyond the imaginary political boundaries of nation state, to include sibling farms in Haiti, Puerto Rico, Mexico, Ghana, and Brazil, with whom we work, organize, and learn through mutual exchange.

Black Latinx Farmers Immersion was a little peek into what is possible in a mended world. Our people have been traumatized and disoriented. While the land was the "scene of the crime," she was never the criminal. Our people mistakenly strove to divorce ourselves from her in an effort to get free. But without the land we cannot be free.

This book, *Farming While Black*, is a reverently compiled manual for African-heritage people ready to reclaim our rightful place of dignified agency in the food system. To farm while Black is an act of defiance against white supremacy and a means to honor the agricultural ingenuity of our ancestors. As Toni

Participants in Soul Fire Farm's Black Latinx Farmers Immersion look super fly while hanging onions to cure in the barn.

Morrison is reported to have said, "If there's a book you really want to read, but it hasn't been written yet, then you must write it." *Farming While Black* is the book I needed someone to write for me when I was a teen who incorrectly believed that choosing a life on land would be a betrayal of my ancestors and of my Black community.

We have organized and expanded upon the curriculum of the Black Latinx Farmers Immersion to provide readers with a concise how-to for all aspects of small-scale farming, from business planning to preserving the harvest. Throughout the chapters, we include "Uplift" sidebars to elevate the wisdom of the African Diasporic farmers and activists whose work informs the techniques described. For example, in the chapter about accessing funding for your farming venture, you will learn about federal loan programs and also about setting up a *susu*, a traditional Caribbean form of community lending. In the chapter about soil repair, you will read about techniques used by the Haitian community to restore degraded hillsides that can be adapted for use in your climate. At the end of the book, there is a Resources section with contact information for the organizations referenced in the chapters.

Each chapter also reveals an honest and transparent portion of the story of Soul Fire Farm. You will cry with us as we hand-dig our foundation through hard clay, hands bloodied with broken calluses and hearts near defeat. You will celebrate with us as we triple the depth of our topsoil using techniques taught to us by Haitian farmers and defy the naysayers who told us that vegetables cannot be grown in these mountains. This book will also be honest about the limitations of its perspective. As a multiracial, light-skinned, raised-rural, northeastern, college-educated, gender queer, able-bodied, Jewish-Vodun-practicing biological mother who grew up working class, this author will invariably introduce bias into the text and invisibilize crucial experiences of members of our Black farming community. This book is mainly told from my life experience as an activist-farmer, recognizing that each member of the 10-person team that is Soul Fire Farm has their own equally rich story of arriving at this work. For the harm that may come from the limitations in my perspective, I apologize, and pledge to do what I can to uplift the voices of the Black farmers who complete and augment the narrative.

The technical information presented is designed for farmers and gardeners with zero to five years of experience. For those with more experience, we hope that this book provides a fresh lens on practices you may have taken for granted as ahistorical or strictly European. Our Black ancestors and contemporaries have always been leaders in the sustainable agriculture and food justice movements, and continue to lead. It is time for us all to listen.

As Toni Morrison wrote in her 1977 novel *Song of Solomon*:

> *"See? See what you can do? Never mind you can't tell one letter from another, never mind you born a slave, never mind you lose your name, never mind your daddy dead, never mind nothing. Here, this here, is what a man can do if he puts his mind to it and his back in it. Stop sniveling," [the land] said. "Stop picking around the edges of the world. Take advantage, and if you can't take advantage, take disadvantage. We live here. On this planet, in this nation, in this county right here. Nowhere else! We got a home in this rock, don't you see! Nobody starving in my home; nobody crying in my home, and if I got a home you got one too! Grab it. Grab this land! Take it, hold it, my brothers, make it, my brothers, shake it, squeeze it, turn it, twist it, beat it, kick it, kiss it, whip it, stomp it, dig it, plow it, seed it, reap it, rent it, buy it, sell it, own it, build it, multiply it, and pass it on—can you hear me? Pass it on!"*[9]

This book is an honoring of that directive—to pass it on.

CHAPTER ONE

Finding Land and Resources

Land is the only real wealth in this country and if we don't own any then we're out of the picture.

—RALPH PAIGE, Federation of Southern Cooperatives

In our early 20s my partner and I convened a group of land-starved friends to pool resources and find a piece of ground to which we could belong. We failed. Our "100-plus acres with a stream running through it" dream turned out to be a fantasy beyond the combined financial means and persistence of our collective. We had saved a portion of our public school teaching and community garden organizing wages by squeezing our family of four into a low-cost single bedroom in a collective house where we heated water for baths on a woodstove and subsisted on black beans and rice. We had endured interminable consensus-based meetings on Monday evenings and rejected car ownership in favor of four-season bicycling. Still, the market realities negated the pride we felt in our thrift.

A few years later we moved from central Massachusetts to Albany, New York, and learned that the logged, thin-soiled mountainside slopes nearby were within our financial means. So excited were we, the first time we stepped foot on the overgrown, south-facing pastures of what would be Soul Fire Farm, that we blinded ourselves to practical pitfalls of wedding to this piece of land. First, there was no road access. We would have to negotiate an easement with a couple of neighbors who were deeply invested in the Second Amendment and had signs like ANYONE FOUND HERE TONIGHT, WILL BE FOUND HERE IN THE MORNING. Also, the cost of the land would completely deplete our savings, and we had no plan for the hundreds of thousands of dollars required to install a road, electricity, septic system,

The author walks the forested land that will become Soul Fire Farm (2006). Photo by Jonah Vitale-Wolff.

barn, home, and other infrastructure. We didn't think carefully about where our children would attend school. While we were deeply invested in the public school system, an all-white, conservative district was not going to be the right fit for our multiracial Jewish children.

In some ways I am grateful for our eager naïveté. If we had known the road ahead, we might have turned back without trying. Instead we walked reverently across these 72 acres and allowed the land to spool its invisible tendrils of connection around our eager feet. We inhaled possibility and exhaled commitment. We wrestled with the acknowledgment of the Land's original theft from the Mohican people and also the right of the Black and Brown laborers whose blood had mixed with the soil to claim belonging here. With consent of the Land herself we chose to sign the white man's title papers and bound our life to this place.

Aspiring farmers need three essential ingredients to begin: training, land, and material resources. This chapter outlines conventional and unconventional ways to access farmer training, including a specific list of justice-oriented and people-of-color-led educational programs. It explores ways of accessing land, from squatting and informal lease agreements to buying and owning rural land. The chapter explains the state, federal, and private sources of funding for new farmers and discusses ways to tap into our communities and networks for alternative funding. Because our people were often excluded from mainstream farmer programs, we have had to create our own systems of mutual support. We uplift and learn from the examples of New Communities Land Trust, Caribbean susu, Combahee River Colony, and the Rapp Road community.

Accessing Land

We begin this chapter with the most daunting prerequisite for farming: access to land. As a result of decades of discrimination by the USDA, white supremacist violence, and legal exploitation of heir property, Black people have been almost entirely dispossessed of our land. In 1920, 14 percent of all landowning US farmers were Black, and today around 1 percent of farms are controlled by Black people, a loss of over 14 million acres.[1] What our community fundamentally needs is large-scale reparations and land redistribution, and we are organizing for this change through the HEAL Food Alliance and the National Black Food & Justice Alliance. While we advocate for a long-term shift to the structural root causes of unequal land distribution, there are several strategies we can use in the near term to work within and outside of the system to access land.

Find the Right Land

Regardless of whether you decide to buy, lease, work with a land trust, join an incubator, or squat on abandoned land, it is important to make sure that the land you select is a good match for your goals. My partner and I half joke that we have three children, Neshima, Emet, and this Land. Your land becomes a family member and a daily obligation. It is a relationship to enter into after detailed research and with care. Here are some characteristics of suitable land:

- It is geographically accessible to a community where I feel a sense of belonging.
- Zoning and building codes will permit the type of infrastructure I want to create.
- Soil has a favorable "perc test," which means a low-cost septic system can be installed.
- There is direct access to a road and utility lines, without traversing a neighbor's property.
- There are no boundary disputes, back taxes owed, or undesirable easements on the property.
- The deed is clean, meaning that there are no claims on the property and ownership is clear.
- The school district is a good match for my children or the children of my future collaborators.
- Town services such as garbage disposal, plowing, emergency and fire response, and access to cable and Internet meet my needs.
- There are cleared areas, relatively free of rocks, with favorable soil tests for agriculture.

- There is clean drinking water on the property, or I can drill a well at a reasonable cost.
- There is a local or regional market where I can sell my agricultural products.
- Slopes are south-facing for maximum solar gain, benefiting both crops and solar energy projects.
- I have a financial plan for paying for the land, infrastructure, taxes, and insurance.

Land Trusts: A Values-Positive Option for Accessing Land

A land trust is an organization that holds land or conservation easements to protect properties for a public or environmental purpose. The boards that manage land trusts can place certain restrictions on how the land can be used or sold. Land trusts can be urban or rural and have stewardship goals ranging

UPLIFT

New Communities Land Trust and the Pigford Case

Shirley and Charles Sherrod founded New Communities in 1969, a farm collective owned in common by Black farmers and the first community land trust in the United States. According to food justice activist and organizer Dara Cooper, New Communities was "born out of violent attacks against Black people who were working to build power and register Black people to vote during the civil rights movement."[2] As Black activists were dispossessed of their homes, a planning committee of civil rights organizations throughout the South, including SNCC and the Federation of Southern Cooperatives, worked with the Sherrods to create a plan for collective landownership and cooperative farming leases. They fund-raised in their own communities and across the nation to raise capital to purchase land in Albany, Georgia.

At 5,700 acres New Communities became the largest Black-owned property to date in this country. They raised hogs, grapes, sugarcane, melons, peanuts, and other crops on their land in southwest Georgia. The residents of New Communities continued to face white-led violence and discrimination. According to Mrs. Sherrod, "Once white people realized we had the land, they started shooting at our buildings. I mean, we went through so much during our time up there. We even caught a white couple stealing our hogs!" During the severe droughts of 1981–82, New Communities requested emergency assistance from the USDA for irrigation. They were denied while white farmers received relief.

By 1985, after repeated droughts and active discrimination on the part of the USDA Farmers Home Administration, New Communities faced foreclosure and lost their land. They became plaintiffs in the landmark class-action lawsuit against the federal government, *Pigford v. Glickman*. The chief arbitrator of the case, Michael Lewis, opined that the USDA's treatment of New Communities "smacked of nothing more than a feudal baron demanding additional crops from his serfs." The case was settled out of court in 1999 for $1.2 billion, the largest civil rights settlement in US history.

While the average payout was $50,000, New Communities won $13 million and used part of the settlement money to reestablish itself on a 1,600-acre former plantation that they have named Resora.[3]

Southern-facing slopes are conducive to solar technology. Soul Fire Farm uses reclaimed solar panels to heat water.

from the creation of affordable housing to the maintenance of ecosystems. Land trusts are an excellent starting place in your search for affordable farmland.

The Black Family Land Trust (BFLT) in Durham, North Carolina, is the only land trust in the nation dedicated to preserving Black farmland. They facilitate the purchase of perpetual agricultural conservation easements, whereby the farmer sells development rights to their property, thus generating income and ensuring that the land will forever be dedicated to agriculture. Joe Thompson, a tobacco farmer who sold an easement through BFLT, stated, "When my time comes and I die, I can lay my dead bones down in peace knowing that my land will always be used to feed people."[4] BFLT also works with farmers on intergenerational financial management and estate planning, and can help match aging farmers with the next generation.

A growing number of land trusts exist with the explicit goal of preserving farmland. These land trusts seek farmers to lease land, often at low cost, and keep it in agricultural production. The National Young Farmers Coalition has published an excellent guide called *Finding Farmland: A Farmer's Guide to Working with Land Trusts*[5] to help you access land that is under a conservation easement. The American Farmland Trust and the Land Trust Alliance each provide lists of land trusts by geographic area. The national organization Equity Trust has a special project to help CSA farmers and provide loans for growers. Note that leasing land is different from purchasing it, in that you may have limited ability to build equity or make independent decisions about long-term management. Still, a long-term lease with the right provisions can be an affordable way to access land while challenging the very concept of private landownership. Since the European enclosure movement of the 1400s we have been struggling to find ways to reclaim the commons, land trusts are at the forefront of that movement.

For several years our family was part of an urban collective house that sat on property owned by a community land trust in Albany, New York. While the three families in the collective purchased our home together at low cost, we leased the land under the home for a modest fee. Further, we were restricted in resale of the home to make sure that the housing stayed affordable and would not go to absentee landlords or speculators. Urban land trusts are growing and can support community members in finding affordable housing and garden space. The National Community Land Trust Network can connect you with your local urban land trust.

Currently, Soul Fire Farm is working with other members of the Northeast Farmers of Color Network and local Indigenous communities to form a regional land trust dedicated to increasing the amount of farmland stewarded by Black, Indigenous, Latinx, and Asian growers.

Squatting and Temporary Land Tenure

When our family first moved to the South End of Albany, we had no land access. Soil contact was essential to my soul craft, and earth-pulled vegetables were a non-negotiable part of our parenting philosophy, so we had to act. After the requisite amount of asking around for resources and permission, and coming up dry, we decided to squat an abandoned corner lot. Together with neighbors, we cleaned up the trash, planted fruit trees, established garden beds, and built an earthen bread oven for community

UPLIFT
Combahee River Colony

During the Civil War several hundred African American women living in the Gullah/Geechee communities of South Carolina and the Georgia Sea Islands founded the Combahee River Community. These women courageously refused to work for white people while their husbands went off to join the Union army. They found and occupied abandoned farmland, grew crops, and took care of one another. The sale of their cotton and handicrafts sustained them. The Combahee River Colony became known as an example of Black women's independence, perseverance, and collective spirit, and operated free from white oversight.[6]

In 1974 Barbara Smith founded the Combahee River Collective, a Black lesbian feminist organization. The collective highlighted the ways that the white feminist movement was not addressing the particular needs of Black women. The name of the collective was inspired by both the Combahee River military campaign, planned and led by Harriet Tubman on June 2, 1863, which freed 750 enslaved people, and the Combahee River Colony.

potlucks. We kindled fires for the oven using a giant bow drill operated by a dozen people pulling the rope back and forth until friction-induced smoke curled up from the hearth. Then we loaded the oven with the harvest of our modest lot—root vegetables, corn bread, and herbed fish—and shared the bounty with one another over stories and laughter.

At the time we did not know about adverse possession law, which allows "trespassers" to actually become owners of the land after a certain number of years of open, exclusive, and continuous possession including the payment of taxes. The laws vary by state, but in New York we could have filed a claim of adverse possession after 10 years. We encourage you to start documenting your squatting activities now, in the hope of possible future ownership.

Grow Where You Are, an urban agriculture project of Black farmers Eugene Cooke and JoVonna Johnson based in Atlanta, Georgia, creates lease agreements with churches, schools, and private landowners across the city. While they are not squatters, they look for underutilized land and offer to make that land productive at no cost to the owner. In order

Eugene Cooke and JoVonna Johnson, farmers at Grow Where You Are, Atlanta, Georgia, welcome the author for a visit.

to reduce the risk of these temporary arrangements, Grow Where You Are establishes equity partnerships with the landowners, where they hold 10 to 20 percent equity and have right of first refusal in case the owner decides to sell. Landowners have been excited to see land that they had deemed worthless be transformed into beautiful and bountiful community spaces. The organization Equity Trust provides templates for writing solid lease agreements.[7]

Land Link: Keeping Black Land in the Family

The average age of a farmer today is 58, and the average age of a Black farmer is 62.[8] We are an endangered species, and risk losing both land and intergenerational wisdom if we fail to act. Unfortunately, I was infected with the "frontier mentality" when it came to finding land and imagined that the only course of action was to start a new project entirely from scratch. It would have been wiser to connect with a retiring farmer and infuse new life into an existing project.

There are a number of resources to support beginning farmers in finding existing farmland. Every region has several "Land Link" projects that provide matchmaking between sellers and seekers of land; the National Young Farmers Coalition website offers a directory. Some of these programs, like California FarmLink, additionally provide flexible financing with the aim of building a more racially diverse farming community. Farmland Access provides a comprehensive legal tool kit to support growers in making land transfer agreements on its website.[9] You can find contact information for these farmland access organizations in the Resources section on page 321.

Farm Incubators: Supported Start-Up

Farm incubator programs are designed for new farmers with plenty of field experience, but little or no management experience. The farm incubator provides the new farmer with land, access to farm infrastructure, marketing support, and mentoring for a modest fee. The farmer leases these resources for a limited amount of time, from one to several years, and then graduates the business to its own land. The National Incubator Farm Training Initiative maintains a database of incubator programs in North America and tips for applying.[10] While not led by people of color—which, as the next section discusses, can lead to problematic arrangements—the most reputable farm incubator programs in the nation include:

- Agriculture and Land-Based Training Association (ALBA), California
- The Intervale Center, Vermont
- Minnesota Food Association
- New Entry Sustainable Farm Project, Massachusetts
- The Seed Farm, Pennsylvania

Farm Skills Training

An attendee at a recent Black Latinx Farmers Immersion at our farm shared a story of their experience as a volunteer farmer through the WWOOF (World Wide Opportunities on Organic Farms) program. They lived in their own camping tent, worked for no pay without much tutelage, and ate the meager rations provided to them in exchange for their labor. The arrangement was all too reminiscent of the exploitation of their ancestors, and they decided to leave.

Many aspiring Black farmers find that opportunities for agricultural training range from culturally irrelevant to blatantly racist. Small-farm training in the US often demands that learners give up their bodies and labor for free or cheap to a landowner who provides instruction, under the title of an apprenticeship or work exchange. For many Black people, women in particular, the generational trauma of this arrangement is unbearable, and they choose alternative career paths.

While I deeply appreciate the mentorship and kindness of the farmers with whom I trained as a young person, I, too, endured rural isolation, racial microaggressions, unlivable wages, and cultural loneliness during my time on the land. Tragically, there

Black Latinx Farmers Immersion at Soul Fire Farm is led by and for Black and Brown farmers. Participants in the 2014 BLFI harvest cilantro together. Photo by Capers Rumph.

are very few rural farm training programs in the US led by people of color. Those of which we are aware are shared in a sidebar, as are university programs and farm apprenticeships that may be white-led but have an explicit racial justice commitment. Search your region for other people-of-color-led projects on Natasha Bowens's Color of Food Map.

Additionally, screen farm job descriptions that are posted on Good Food Jobs, ATTRA, and the Growing Food and Justice Listserve. Contact information for these organizations can be found in the resources section, page 321.

Self-Study

While the value of mentorship is undeniable, we must also remember that our enslaved ancestors often taught themselves to read with scavenged scraps of text and at great personal risk. There are free and low-cost online courses available through ATTRA and the Cornell Small Farms Program. At Soul Fire Farm we learned how to slaughter and eviscerate chickens by watching

Scholarship Opportunities to Pursue Agricultural Studies

National Black Farmers Association Scholarship Program. NBFA provides scholarships from $2,500 to $5,000 for African American students pursuing agriculture-related studies.

USDA 1890 National Scholars Program. This program offers full tuition, room and board, and employment to US citizens to attend one of the eighteen 1890 Historically Black Land-Grant Institutions and Tuskegee University to study agriculture, food, or natural resource sciences and related majors.

Farm Training Programs Led by People of Color

- Barbara Norman's Blueberry Patch, Covert, Michigan
- Black Dirt Farm Collective, Preston, Maryland
- The Black Oaks Center for Sustainable and Renewable Living, Pembroke, Illinois
- Detroit Black Community Food Security Network, Detroit, Michigan
- D-Town Farm, Detroit, Michigan
- Earthseed Land Cooperative, Durham, North Carolina
- East New York Farms!, Brooklyn, New York
- Family. Agriculture. Resource. Management. Services. (F.A.R.M.S.), Southeast USA
- Farm School NYC
- Farms to Grow, Inc., Oakland, California
- Federation of Southern Cooperatives Rural Training & Research Center
- Foot Print Farms, Jackson, Mississippi
- Grow Where You Are, Atlanta, Georgia
- Mayflor Farms, Stockbridge, Georgia
- MESA, Berkeley, California
- Movement Ground Farm, East Taunton, Massachusetts
- Mudbone Farm, Portland, Oregon
- National Black Farmers Association's Let's Get Growing Program, Baskerville, Virginia
- National Hmong American Farmers, Fresno County, California
- RID-ALL Green Partnership, Cleveland, Ohio
- Rocky Acres Community Farm, Freeville, New York
- Small Farmer Leadership Training Institute, Baton Rouge, Louisiana
- Soilful City, Washington, DC
- Soul Fire Farm, Grafton, New York
- Soul Flower Farm, El Sobrante, California
- Southeastern African American Farmers' Organic Network (SAAFON), Decatur, Georgia
- Three Part Harmony Farm, Washington, DC
- Truly Living Well, Atlanta, Georgia
- Tuzini Farms, Lusaka, Zambia
- Urban Farm Institute, Boston, Massachusetts
- Urban Growers Collective, Chicago, Illinois
- The Yisrael Family Urban Farm, Sacramento, California

Soul Fire Farmers originally learned chicken processing through YouTube videos and, after years of practice, passed these skills on to learners in our programs. Photo by Neshima-Vitale Penniman.

Farm Training Programs with Stated Racial Justice Commitment, Not Necessarily Led by People of Color

- ALBA (Agriculture and Land-Based Training Association) Farm Incubator, Salinas, California
- Flats Mentor Farm, Lancaster, Massachusetts
- The Food Project, Boston, Massachusetts
- Groundswell Center for Local Food & Farming—Incubator Program, Ithaca, New York
- Hudson Valley Farm Hub ProFarmer Training Program, Hurley, New York
- Land Stewardship Project, Minneapolis, Minnesota
- Minnesota Food Association Big River Farms Farmer Education Program, St. Paul, Minnesota
- New Entry Sustainable Farming Project, Lowell, Massachusetts
- Southside Community Land Trust, Providence, Rhode Island

Agricultural Universities with Racial Diversity or Equity Commitment

Alabama A&M University College of Agricultural, Life and Natural Sciences, Normal, Alabama. Booker T. Whatley, pioneer of community-supported agriculture, is an alumnus of this institution. The current dean is Dr. Lloyd T. Walker, an accomplished Black food scientist.

The Southern University Agricultural Research and Extension Center (SUAREC), Baton Rouge, Louisiana. Dr. Owusu Bandele, co-founder of SAAFON, author, and community organizer, is professor emeritus at SUAREC.

The Sustainable Agriculture Consortium for Historically Disadvantaged Farmers Program (SACH). SACH is a cooperative marketing initiative led by five 1890 land-grant universities: Alcorn State University (Mississippi), Fort Valley State University (Georgia), Prairie View A&M University (Texas), Tuskegee University (Alabama), and the University of Arkansas at Pine Bluff.[11]

Tuskegee University, Alabama. This historically Black college was founded by Booker T. Washington and continues to be the intellectual home of notable Black scholars including Tasha Hargrove, research assistant professor of agricultural and resource economics, and Walter Hill, vice provost of land grant university affairs, dean of the College of Agriculture, Environment and Nutrition Sciences, and the 1890 research director of the George Washington Carver Agricultural Experiment Station.

how-to videos and practicing on our own. We need not underestimate our capacity to self-teach using the abundant informational resources in our networks.

Gathering Material Resources

With our meager savings exhausted by the land purchase, we turned to the local bank to apply for a mortgage to build our home and educational center. I am not sure if it was the mention of the word *farm* or perhaps the description of the straw-bale construction that most offended the bank employee, but she summarily hung up the phone on me.

Rejected by the conventional credit infrastructure, we turned to our community for alternative financing. We asked around about alternative credit, and it turned out that a multiracial group of our neighbors had organized a lending society and kept it operating since the 1960s. Similar to a Caribbean susu, everyone contributed a set amount on a monthly basis and could apply to take out loans from the group for homes, vehicles, or tuition. We joined the lending society and were granted a mortgage to cover the material cost of constructing our timber-frame, straw-bale, passive solar home and education center. The interest we paid on the loan was recycled back to our friends and neighbors.

Many of us do not have access to conventional credit because of a legacy of structural racism. In this section we consider alternative means for accessing the financial resources necessary to build our land-based projects.

Government Funding

At the conclusion of the *Pigford v. Glickman* case in 1999, the United States Department of Agriculture (USDA) was forced to hand over $1.2 billion to Black

UPLIFT

Caribbean and African Susu

The susu is a microfinance strategy thought to have originated in Nigeria and spread to Ghana in the early 20th century.[12] Susus are now common across the Diaspora, including in Jamaica, Mexico, Brazil, and the United States. Susu groups pool money on a daily, weekly, or monthly basis by collecting set amounts from each susu member, putting all the money together, and distributing the lump sum to one person at a time. Once all group members have received their money, the cycle starts anew. For market women, susu groups are vital because they allow access to much-needed capital to sustain small businesses. Upon receiving susu money, the women purchase goods to garnish their stalls. Market women's susu groups typically have a leader called the susu mother, or *susu ma*, and 15 to 20 members on average. Older women who have been known in the community for a long time and have established a reputation of trustworthiness are often preferred as susu ma. The susu ma is responsible for running a susu group; she makes sure that all members pay in a timely manner and that the money is redistributed to the member who is set to receive it at a given moment. Susu groups perform complex financial operations by enabling members to double, triple, quadruple, or quintuple their dues, allowing them to receive a proportional return on investment.[13] The electronic platforms eMoneyPool, Monk App, Puddle, and Partnerhand are based on the susu model.

farmers in the largest civil rights settlement in US history. Since then the USDA has made efforts to correct its institutional bias and has set aside specific funds to support "historically disadvantaged farmers and ranchers." Specifically, the Farm Service Agency (FSA) is responsible for distributing loans and the Natural Resources Conservation Service (NRCS) is responsible for distributing environmental preservation grants to farmers. Both agencies of the USDA have several offices in each state that can be found online at https://offices.sc.egov.usda.gov/locator/app.

The first step in accessing funding is to visit the service center and fill out required paperwork. It is very helpful if you have already established your farm as an official entity by reporting agricultural income on your tax return. Even if you are just selling cut flowers on the side, if you can report $1,000 of farm income or production, you are considered a "farm" by the USDA and become eligible for many programs. Soul Fire Farm has had a positive experience applying for and receiving USDA funding for two high tunnels.

Since USDA resources are distributed proportionately depending on the number of farmers in a given region, it is essential that all Black farmers, urban and rural, register for the USDA Census of Agriculture.

The following USDA programs have special provisions for farmers of color. Please note that as of this writing, the Farm Bill is up for reauthorization, and some of these programs will be adjusted by the time of publication.[14]

Direct and Guaranteed Farm Ownership Loan. Provides funding for the purchase of farmland and buildings. The maximum direct loan amount is $300,000; the maximum amount for a guaranteed loan is $1,399,000 (adjusted annually for inflation).

Community members meet to discuss the formation of a lending society. Photo by Neshima Vitale-Penniman.

Down Payment Loan Program. Provides funding for land purchase. The maximum purchase price for the loan cannot exceed $667,000.

Land Contract Guarantee Program. Provides support for retiring farmers to transfer their land to beginning farmers. The purchase price or appraised value of the farm or ranch that is the subject of the contract sale cannot be greater than $500,000.

Direct and Guaranteed Farm Operating Loans. Provides funding for annual operating expenses and minor improvements. The maximum direct loan amount is $300,000; the maximum amount for a guaranteed loan is $1,399,000 (adjusted annually for inflation).

Microloans. Provides funding for land and operating expenses for both urban and rural farms. The process is streamlined. The maximum loan amount for microloans is capped at $50,000.

Noninsured Crop Disaster Assistance Program (NAP). Provides low-cost insurance to protect farms in the case of crop loss. Payments are capped at $125,000, and there is a gross income limit of $900,000.

Environmental Quality Incentives Program (EQIP). Provides funding for conservation improvements on farms, such as high tunnels, wildlife habitat, cover crops, fencing, irrigation, and transitioning to organic. Funding amounts vary by project. A match of 10 to 25 percent of the project cost is required.

Conservation Stewardship Program (CSP). Provides payment for advanced conservation efforts, such as cover cropping, buffer strips, and rotational grazing. Funding amounts are per acre and vary by project.

Value-Added Producer Grants (VAPG). Provides funding for marketing, collaborative product distribution, and value-added products like hot sauce and jam. Planning grants are capped at $75,000; working capital grants are capped at $250,000. Matching funds are required.[15]

Soul Fire Farm's high tunnel was purchased with a grant from EQIP, a USDA program. Photo by Neshima Vitale-Penniman.

Many states also provide support for new and beginning farmers. For example, in New York State, Empire State Development's New Farmers Grant Fund provides grants of up to $50,000 for farmers to improve the profitability of their farms. A detailed business plan and 50 percent match are required.[16]

Reparations

Forty acres and a mule would be at least $6.4 trillion in the hands of Black Americans today according to *YES! Magazine*.[17] The economic offenses committed by this nation against Black people are numerous. They include hundreds of years of unpaid wages under slavery, discriminatory fees and lending rates imposed upon African American business owners under the Black Codes, and the exclusion of Black people from the social safety net and government housing programs. While we cannot hold our breath for wholesale reparations from the federal government, we can encourage conscious individuals in our community to take reparations into their own hands. To that end, Soul Fire Farm alumni have created a reparations map to catalyze voluntary transfers of land and resources from people with inherited privilege to Black and Brown farmers.

Wildseed Community Farm & Healing Village in Millerton, New York; Harriet Tubman Freedom Farm in Durham, North Carolina; and Harmony Homestead & Wholeness Center in Hillsdale, New York, are people-of-color-led projects that were born out of explicit acts of reparations. In each case, a European-descendant person with means recognized that their wealth was not rightfully their own and transferred a portion to the descendants of those who created that wealth. Increasingly, there are woke white folks who want to consciously right the wrongs of history. Many are organized in a collective called Resource Generation, a group of young people with wealth who have stated their aim to address

The land for Wildseed Community Farm, New York, was donated as an act of reparations. Wildseed members celebrate their oyster mushroom harvest. Photo courtesy of Wildseed Community Farm & Healing Village.

racism through redistribution of resources.[18] Others may be your neighbors.

While it may seem like a long shot, I encourage you to write up your request for reparations outlining your specific needs for money, land, and equipment. Share this written request with everyone in your networks, framing the share as a request for advice on refining the language, not a direct ask for money. A mentor of mine said, "If you want advice, ask for money. If you want money, ask for advice." Additionally, send your request to facilitators leading anti-oppression trainings in your area. There is often a "what you can do" portion of that training, and the right person might be listening.

Crowdfunding

Many of us are rich in social capital, while limited in financial capital. Online crowdfunding platforms like YouCaring, Kickstarter, and Indiegogo can help us harness the energy and resources of our networks. Soul Fire Farm has raised over $70,000 for infrastructure using crowdfunding. We are certainly not experts in online marketing, but found a few strategies that motivated our community to give. First, we built an insider team that was invested in the campaign and asked them to support in a few crucial ways. Everyone on the team made a donation and shared the campaign on day one. That way we could show early momentum and buy-in, motivating others to join the effort. Each person on our team was also assigned a particular day during the campaign to post a 30- to 60-second video explaining why they love Soul Fire Farm. These were low-tech, unedited videos, but full of heart. We also wrote personal testimonials to everyone in or networks via email, sharing vulnerably about the challenges of our work and our need for community solidarity to fortify our resolve

To build interest in our crowdfunding campaign, we snapped this photo of our whole team sleeping in one bed, emphasizing the need for funds to build additional staff lodging. Photo by Jonah Vitale-Wolff.

to continue the project. We put the money we raised to good use immediately, and posted playful updates on the progress of our buildings. For example, we posed with our children climbing dramatically on scaffolding to show how the barn construction was progressing even as the fund-raiser was under way.

Lending Society

Black people have a deep history of supporting one another through pooling resources. We started our first bank, the True Reformers Bank, in 1888. One of our first credit unions was established in 1939 by 25 neighbors of the Tyrrell County Training School to save their farms from foreclosure and purchase farm equipment.[19] Before these formal institutions, we had informal credit associations such as susus, where we could pool our meager funds and take turns receiving a lump sum.

Creating your own susu, or lending society, is simpler than you might think. Invite 5 to 12 people to join your susu and work together to establish criteria for the income, employment, and credit status of members. Set a regular deposit amount and deposit interval. For example, you may decide that each of 10 members needs to pay $100 per month. Deposit these funds in an accredited financial institution. Each month, one of the members receives the full "hand" of $1,000. Traditionally, susus do not collect interest, and the hand is given on rotation. However, our variation of the susu gives out loans on an application basis and does charge interest, which is recycled back to the collective. The currency of the susu is trust, and this is maintained by having face-to-face meetings and being open and transparent about all dealings of the susu. We personally used a lending society to provide our mortgage, and had to present detailed financial information to all members of the group in order to be approved for the loan. We paid back our loan with interest, all the while contributing the regular membership dues of our collective, which were $150 per year. Should you desire to take your susu beyond the informal, and create a recognized credit union, the National Credit Union Administration and Caribbean American Chamber of Commerce and Industry can support you in that process. You may also choose to use an online platform like Kiva to facilitate your loan process.

Frugality, Patience, and Faith

Our ancestors purchased 16 million acres of land with the meager wages they saved through sharecropping, tenant farming, and hiring themselves out on Sundays as laborers. Through thrift and restraint, they gathered the resources necessary to get off the plantation and farm with dignity. While I am extremely privileged relative to the hardships endured by my ancestors, it is also true that I was raised in relative poverty with no inherited wealth.

UPLIFT

Rapp Road Community in Albany, New York

The Rapp Road community was a model of thrift, patience, and fortitude. Black sharecroppers from Shubuta, Mississippi, moved to Albany, New York, in 1930 with their pastor, Louis Parson, in the hope of building a life free of debt and the threat of lynching. They purchased a marshy 14-acre parcel for $400, started to farm, and used their earnings to build homes. "People would come with whatever little bit of money they had and little by little they built their houses and created this community," explained Sara DeWitt, a fourth-generation community member. With no access to credit or family wealth, this community of 28 families worked together, investing blood, sweat, and tears to build a new life on land.[20]

Table 1.1. Costs for Land, Infrastructure, and Equipment at Soul Fire Farm

Permanent Infrastructure	Cost
Land—72 acres	$185,000
Driveway, septic, well, electricity	$55,000
Family home/education center	$310,000
Shop/storage	$30,000
Barn/Apartment	$120,000
Semi-Permanent Infrastructure	**Cost**
Hoop house	$3,000
High tunnel	$9,800
Chicken housing	$2,400
Walk-in cooler	$2,200
Irrigation system	$1,000
Electric fencing	$3,000
Equipment/Supplies	**Cost**
Tractor	$16,000
Tractor implements	$10,000
Hand tools	$750
Power tools	$4,000
Benches/tents/tables	$750
Delivery van	$8,000
Total	**$760,900**

At Soul Fire Farm we are building our infrastructure over time and with the help of friends. Photo by Capers Rumph.

By living collectively, limiting expenses, and setting aside a large portion of my wages, I was able to save over $20,000 each year for 10 years. This was enough to purchase land and get started on building a home. In retrospect, our expenses would have been much lower if we had connected with a retiring farmer to revive an existing project rather than starting with undeveloped land. However, we knew no better at the time and charged ahead with building a farm "from scratch." In table 1.1, we share the approximate costs of the farmland and infrastructure, which were paid over time from 2006 until the present.

According to African American philosopher and civil rights leader Howard Thurman, your task is not to "ask what the world needs. Ask what makes you come alive and go do it, because what the world needs is people who have come alive." I believe that the thing that makes you come alive is integral to your destiny and will manifest if you put your prayers up and your hands to work. Do not be intimidated by the entirety of the journey; just take one step in the direction of your dream and let your ancestors help you with the rest. The Land is calling you home, and will help you get back to her.

CHAPTER TWO

Planning Your Farm Business

Men anpil chay pa lou.
Many hands make the load lighter.

—Kreyol proverb

Between my partner and me, we had two college degrees and 15 years of farming experience, and yet were unable to provide our newborn and toddler access to nourishing food. There were no farmers markets, co-ops, or grocery stores on Grand Street in Albany's South End, just a liquor store, bodega, and McDonald's. The community garden plots were all taken for the season, and we did not find neighbors willing to share their paltry growing spaces with our family. Temporarily unemployed, I was accepted into the Women Infants and Children (WIC) nutrition assistance program, which rationed conventional cereal, milk, and eggs when I was willing to endure the glares and, once, the spit of impatient customers in line behind me.

As I child, I experienced hunger and food scarcity, growing up in a trailer in rural Massachusetts with a single parent holding down multiple jobs in order to provide for three children. However, I imagined that those days of exiguity were past and expected that my own children would enjoy abundant, life-giving food.

Eventually we found a community-supported agriculture (CSA) program that offered a pickup point 2 miles uptown. The price of the share became our second largest expense after the rent. Since we were carless, collecting our share involved a weekly walk with one child strapped to my back and the other in the stroller. On the return trip, the stroller-bound child protested mildly as she balanced the heavy vegetables in her lap.

As we built relationships with our Albany neighbors, we learned that our struggles to access food were echoed throughout the community, with the most significant barriers being cost and transportation.

Soul Fire Farm was born in response to the South End community's yearning for life-giving food and our family's passion for connecting land stewardship and social justice. We believed that to free ourselves collectively, we must feed ourselves. Rooted in the principle of Ujamaa, cooperative economics, we decided on a subscription-based farm-share model that would reduce the access barriers of transportation and cost, while providing the farm with guaranteed income.

In 2011, the first year of our Ujamaa Farm Share CSA, fewer than 20 families signed up, almost all of them our former neighbors from the block. I did most of the growing in the evenings and delivered vegetables and eggs on Sundays, working around my full-time public school science teaching job. We accepted SNAP/EBT (formerly food stamps) and charged a sliding scale based on self-reported income. To address the transportation barrier, we brought bushel boxes stuffed full with a rainbow of mineral-rich bounty right to people's doorsteps. Each year we grew the program, expanding beyond the South End to other neighborhoods impacted by food apartheid, including West Hill, Arbor Hill, and South and North Troy. My partner, Jonah, paused his natural building business to work on the farm full-time, and eventually we welcomed apprentices, then staff. We currently have 80 to 100 families in the program who affectionately term it "Netflix for vegetables."

Just as it takes a village to raise a child, so does it take a village to steward a farm. Soul Fire Farm was, in many ways, the third biological child of Jonah and me. This beloved farm child needs at least as much resource, time, tenderness, 24/7 care, and deep commitment as our two human children. However, it would be deeply dangerous and counter mission to keep this child trapped in our nuclear family. We grew our village slowly, first with another family of close friends who built a *bohío* (Puerto Rican roundhouse) and lived with us part-time. Then we welcomed apprentices and interns to live in our home from anywhere from a few months to a few years. We incorporated and built a board of 15 directors for shared accountability and ownership. We have a committed and brilliant staff of 3 to 10 people depending on the season. The farm and the CSA are now managed by the land-loving power trio of Larisa, Damaris, and Lytisha; while Amani and I facilitate the programs; and Jonah keeps the human habitat in order. Over time the Soul Fire Farm village has grown into concentric circles of hundreds of committed individuals to whom the project is accountable.

In the first year of Soul Fire Farm's Ujamaa Farm Share CSA, the packing operation was small enough to fit in our living room. Photo by Jonah Vitale-Wolff.

In this chapter we explore models for business planning that provide financial sustainability while questioning capitalism. We delve into models for worker-owned cooperative businesses, explore community-based marketing, uplift communal labor strategies, and end with templates for creating your own mission-driven business plan.

Worker-Owned Cooperative Business Model

Nana Ofori, chief of the Akim Abuakwa people of the Ashanti kingdom in Ghana (1912–43), said, "I conceive that land belongs to a vast family of whom many are dead, a few are living and a countless host are still unborn."[1] Our people have stewarded land in common for much longer than we have acquiesced to the Western notions of enclosure and private property.

At the time of writing, we are working to transition Soul Fire Farm from private landownership to communal landownership and stewardship. Jonah and I have experience living and working in two cooperatives previously. We spent three years in a collective house on 5 acres at the edge of Worcester, Massachusetts, that was owned in common through a trust. The 8 to 13 residents made decisions through consensus and split the costs of property maintenance. We also lived cooperatively with two other families in a three-unit property that was owned by an urban community land trust in Albany. In both cases we were affirmed that ownership in common is not only ethically desirable, but profoundly practical.

UPLIFT
Fannie Lou Hamer's Freedom Farm Cooperative

Fannie Lou Hamer was the granddaughter of enslaved Africans and the daughter of sharecroppers in the Mississippi Delta. Hamer was tricked into picking cotton when she was only six years old and then forced by the plantation owner to continue picking, once he had seen that she was capable. She later worked as the plantation's timekeeper, keeping official records of the number of bales sharecroppers had picked. Realizing that the plantation owner had adjusted the scales to maximize his profit, Hamer surreptitiously set the scales right. In that act she became a rebel, later reflecting that "I didn't know what to do and all I could do is rebel in the only way I could rebel."[2]

While Hamer is best known for her organizing with the Mississippi Freedom Democratic Party, her rebellion extended to food and agriculture. In 1969 Hamer founded the Freedom Farm Cooperative on 40 acres of prime delta land. Her goal was to empower poor Black farmers and sharecroppers, who had suffered at the mercy of white landowners. She said, "The time has come now when we are going to have to get what we need ourselves. We may get a little help, here and there, but in the main we're going to have to do it ourselves."[3] The co-op consisted of 1,500 families who planted cash crops, like soybeans and cotton, as well as mixed vegetables. They purchased another 640 acres and started a "pig bank" that distributed livestock to Black farmers. The farm grew into a multifaceted self-help organization, providing scholarships, home-building assistance, a commercial kitchen, a garment factory, a tool bank, agricultural training, and burial fees to its members.[4]

Owning a farm business cooperatively has many potential advantages. Multiple individuals can join their purchasing power and share the upfront investment. The administrative burden of marketing, record keeping, and tax filing is shared. Members can delegate responsibilities for different aspects of the business, both distributing risk and allowing for specialization. In an era where loneliness is on the rise, cooperatives can also provide the solidarity, fellowship, and deep human connection that we require as human beings.

Choosing and Creating a Legal Entity

The first step in creating your farm is to access and honor legacy knowledge. At the 2017 Black Farmers Conference, farmer Matthew Raiford of Gilliard Farm explained, "I planted 100 trees in my first year on the land and they all died, because I didn't bother asking my grandma what grows well here." We honor our family and ancestors by building upon their wisdom, rather than imagining we can invent new truth all on our own. Pour your elders a cup of tea, sit at their feet, and ask them to give you advice.

Under the guidance of your elders, the next step in establishing your cooperative business is to choose a legal entity. Creating a legal entity to own your business cooperatively allows for greater access to bank accounts, loans, and investor funds and also allows a vehicle for smooth transition among members. It is crucial to determine your business goals and select a legal entity that will serve those goals, rather than letting the legal entity define your operation. In the case of Soul Fire Farm, we started as a sole proprietorship for its simplicity and then transitioned to a worker-run nonprofit organization to be able to access resources

(continued on page 33)

Global Village Farms is a cooperatively owned project led by Indigenous people in Massachusetts. Youth members of the Community Environmental College program of the Environmental Justice League of Rhode Island help with the Global Village harvest. Photo by Matt Feinstein, Future Focus Media Co-operative.

Legal Entities for Cooperative Farm Businesses

Contractual agreement. A contractual agreement may be used for trial or temporary ventures with limited risk. While not legal entities, contractual agreements outline the conditions for a particular project. For example, two farmers who want to share land for a season could write a contract outlining how costs and labor will be shared. A contract can be a precursor to joint ownership and the creation of an enduring legal entity.

Sole proprietorship. A sole proprietorship is a business owned by one person that provides no distinction between the finances of the owner and the business. To create a sole proprietorship, you need to register the name of your business (called a DBA—doing business as) at the county clerk's office. Farm income is reported on the owner's personal taxes, on Schedule F. Sole proprietorships do not provide liability protection, and your personal assets could be seized if the business incurs unpayable debt.

Partnership. Partnerships are not recommended as a legal entity, because they do not provide liability protection and expire at the death or departure of one of the members. However, they are simple to create and ensue automatically whenever two or more people go into business together for a profit. To formally create a partnership, you need to write a partnership agreement that describes each partner's contribution, the allocation of profits and losses, rules for decision making, and the process for partners joining or leaving. Extension personnel, land-grant university staff, farm credit institutions, and other agricultural service agencies can also help you set up a partnership agreement. The profits from a partnership are taxed on your individual income tax return.

Limited liability corporation (LLC and L3C). An LLC is the simplest and most versatile business entity available for cooperative businesses. The members of the LLC determine how it is owned and managed, the allocation of profits and losses, and the decision-making process. To create an LLC you need to choose a business name, file Articles of Organization with the secretary of state followed by a Statement of Information, and create an Operating Agreement with basic operational rules. Filing fees for an LLC are modest, generally under $100. LLCs provide limited liability, meaning that your personal assets are protected when the business incurs an unpayable debt. The profits of an LLC are taxed on the individual tax returns of the members. This means that members of an LLC do not receive a salary, rather a share of profits based on percent ownership or relative contribution to the project. Note that LLCs can be organized with internal structure adhering to cooperative principles, namely "one member—one vote" and the distribution of profits back to members according to use, allowing them to be taxed as cooperatives by the IRS. Some states allow a low-profit limited liability company (L3C) to facilitate investments in socially beneficial, for-profit ventures. A significant advantage of an LLC or L3C is that undocumented people can be owners of a business in the USA, while they are not allowed to be employees. Consider creating a business entity that uplifts immigrant justice.

Cooperative. A cooperative is a business that is member-owned, member-controlled, and generates member benefit. Cooperatives are democratically controlled by their members, with each member having an equal vote. All profits of the cooperative are distributed back to members in proportion to how much they have participated in the cooperative. To create a cooperative, you first choose a business name that includes the word *cooperative*. Next, identify initial members and a board of directors. Prepare bylaws and membership agreements, which define the internal rules of operation. Prepare and file Articles of Incorporation and a Statement of Information with the secretary of state. As a type of corporation, cooperatives have required management practices, including processes for electing a board of directors, holding board and member meetings, keeping written records of meetings, and filing annual reports. Cooperatives are required to distribute dividends to their members, but can retain a certain percentage of their profits in the business bank account. These retained assets are taxed as corporate profits. The USDA Office of Rural Development provides services to rural co-ops and will help a group of farmers form a cooperative. They have sample bylaws, information on how to run a cooperative, loans, and grants.

Nonprofit organization 501(c)3. While farming is not often profitable, it is not in itself a charitable purpose that qualifies for nonprofit designation by the IRS. If you are organized for a charitable purpose—the alleviation of poverty, education, research, or religion—you can consider forming a 501(c)3 nonprofit organization. Having a tax-exempt nonprofit qualifies you to apply for grant funding and means that individuals can make tax-deductible donations to your organization. Nonprofits are only permitted to generate income from activities related to the mission of the organization. Nonprofits are owned by no one and do not distribute income to their members. Members of the nonprofit are employees who can receive a reasonable salary but cannot retain the assets of the organization, even if it disbands. Any personal funds put toward starting the organization would be considered donations, as members cannot build equity.

There are significant requirements and legal restrictions for creating a nonprofit. You must elect a board of directors to manage the organization, and there are restrictions on decision making for those who have personal relationships or conflicts of interest. To create a nonprofit, you first file Articles of Incorporation with the state, create and file bylaws, and then file for tax-exempt status with both the state and federal government. The filing process is complex and can take over one year and approximately $1,000 before approval is granted. We found it essential to engage the assistance of a lawyer in the process of applying for 501(c)3 status. Many law schools have free or low-cost legal clinics to help community members with the paperwork.

A less common form of nonprofit is the 501(d), a structure used by monasteries and other entities that support themselves with a cottage industry and divide income equally. Another type of nonprofit is the worker-run nonprofit. This can be a 501(c)3 or 501(d). In a worker-run nonprofit, the board of directors can grant the employees power to influence the realms and programs in which they work, the conditions of their workplace, their own career paths, and the direction of the organization as a whole.

Source: Adapted from Faith Gilbert, *Cooperative Farming: Frameworks for Farming Together*, *Northeast SARE*, accessed April 14, 2017, https://www.cooperativefund.org/sites/default/files/Cooperative%20Farming%20Greenhorns%20Guidebook%202014.pdf.

to support our educational programs. We are putting our land into an LLC that will be run as a cooperative, to maintain maximum flexibility and the ability for members, including children, to build equity.

Agreements for Sharing Resources and Labor

While it may seem that informal agreements are more in keeping with our culture as African-heritage people, the reality is that contractual agreements have a deep and enduring history among our people. When I was living in Ghana in 2002, I remember market women outlining detailed procedures for how they would locate their stalls, share market upkeep, and collect fees for their collective business. These agreements were as thoroughly understood and strictly enforced as the written contracts of the West. Further, tacit agreements are inherently biased toward founders and those closest to them, and can perpetuate bias. When agreements are explicit, written down, and accessible to current and prospective cooperative members, greater transparency and agency ensue.

Membership

One of the first things to decide as a collective is the process for becoming a member and the responsibilities of membership. Here are some questions to consider:

- What are the criteria for membership?
- What is the process for applying to be a member?
- Is there a trial period for new members?
- What financial, work, or other obligations do members have?
- What decision-making authority do members have? Do all members have an equal vote or are there tiers of membership?
- What authority do members have to act on behalf of the group? Can they sign contracts, make purchases, take on debt, give public lectures?
- What is the process for dismissing a member?
- When a member leaves voluntarily, what happens to contributions that the member has made to the organization?
- If several members leave at the same time, what structures are in place to ensure that the assets of the collective are not excessively drained?

Roles and Responsibilities

Last winter our family was forced to cancel a trip to visit my mother when the solar hot-water panels malfunctioned. Just as we were about to pull out of the ice-crusted driveway, we realized that the pumps were not circulating properly. We needed to cover the solar panels with reflective tarps immediately or risk the panels exploding. This was just one example of the countless times that our collective business project offered an unexpected and obligatory task for which no one was clearly responsible.

Rather than being caught off guard by the variety of roles and responsibilities present in a collective business, it is best to plan ahead and decide who will

Wildseed Community Farm

Wildseed Community Farm & Healing Village, a new people-of-color-led, queer-centered land and justice project in Millerton, New York, developed a detailed application process for joining their collective based on the qualities they most desire in collaborators.

Desirable Qualities for Collective Members

- Emotional maturity
- Strong resonance and alignment with vision and purpose
- Meaningful skills and experience to contribute to, such as:
    - Farming and gardening skills
    - Construction skills
    - Plumbing and electric skills
    - Group process and conflict resolution skills
- Hardworking, with physical and emotional stamina
- Collective-minded, collective responsibility and accountability
- Commitment to self-awareness and self-reflection
- Nourishes successful long-term relationships
- Experience in group living situations
- Financially responsible with ability to honor commitments
- Long-term vision
- Eagerness to work on ongoing deep personal growth and relationship to others
- Anti-oppressive framework and willingness to challenge power dynamics, internalized oppression, and privilege
- Skilled at handling conflict
- Clear communication
- Attitude of gratitude

Delegation of tasks supports efficiency and clarity of expectations. For example, Emet is responsible for weighing and packing beans for the farm share, while other members of the team take care of marketing and delivery.

specialize in each task. For a cooperative farm business, here are some likely roles:

- Bookkeeping and tax reporting
- Business and preseason crop planning
- Certifications and filings
- Communications with customers, maintaining contact information database
- Day-to-day operations, manual labor
- Managing visitors and volunteers
- Outreach, marketing, emails, web maintenance
- Procurement and budgeting
- Record keeping of production, sales, and maintenance schedules
- Repairs, maintenance, and upkeep of infrastructure

While specialization is now part of our efficiency strategy at Soul Fire Farm, we have found that it builds empathy and ownership in our team when we each participate in some aspect of all of the roles. Specifically, we have a commitment that everyone in the organization spend at least some time each week with their hands in the soil. Everyone also takes responsibility for one aspect of the administration, whether marketing, program logistics, or fund-raising. Mistrust can quickly build between team members if there is a perception that others are not pulling their weight. Each person having deep knowledge of and participation in all aspects of the collective work is an antidote to suspicion.

Finances

Having a desire for material security is not inherently capitalist. The traditional petition in the religion of Nigerian Yoruba Ifa includes the phrase, "Efun mi no owo . . . Ko si ailowo," which means "Please give me money . . . keep away poverty."[5] As someone who grew up in relative poverty, I find that my attention to material security is greater than that of my friends who were raised upper- or owning-class. Those of us who know lack do not assume that the world owes us anything. Those of us who know suffering and exploitation understand that we need to pay careful attention to our agreements, lest we perpetuate our difficulties. This is why determining how to share money in your collectively owned business is perhaps the most difficult negotiation.

Here are some questions to consider as your draft financial agreements for your cooperative business:

Who will provide the upfront costs for starting the business? How will they be reimbursed, if at all? If you take out a loan, who is responsible for paying it back?

When income is generated, how much of it will stay in the business bank account to cover costs and reinvestments in the business? How much will be distributed to the members?

Does everyone make an equal hourly wage or are wages different based on longevity, experience, or need? Are additional profits distributed equally or based on how much total investment each member put into the collective?

Can members build equity in the business? In other words, when they leave or the business is sold, do they receive some return on their investment?

What checks and balances will be instituted to prevent cheating, stealing, or financial dishonesty? How will you maintain transparency?[6]

Organizations like The Working World provide capital, loans, and technical support to help worker-owned cooperatives get established.

Decision Making and Meetings

The 15 board members of Soul Fire Farm stood with their palms raised facing the center of the circle. The three new staff of Soul Fire Farm were invited into the middle of the circle to receive the blessings, welcome, and gratitude of the current team. These new farmers would be doing the sacred work of tending the land, feeding our beloved community, and building the "we" of the project. The new team members then joined with the wider circle, where we offered a song: "We are an old family, we are a new family, we are the same family, stronger than before. We

This meeting of the Freedom Food Alliance opens with a reflective moment to thank the ancestors and the land for the bounty we enjoy.

honor you, inspire you to be who you are." This symbolic ritual of welcome was part of our quarterly board meeting. During the meeting we also ate delicious food, danced, approved budgets, and updated bylaws. We believe in meetings as opportunities for authentic and productive connection, not a time for tedious and rushed recitations of disembodied data. We strive for a balance of heart and head.

There is a minimum of three types of meetings that your business cooperative should consider: logistics, strategic discussion, and reflection. Logistics meetings can be as short as 15 minutes and deal with the "what" of the project. Strategic discussion meetings last a minimum of 60 minutes and consider the "why" and "how" of the project. Reflection meetings are used for evaluation, gratitude, or conflict resolution and take a minimum of 90 minutes. Offering clarity as to which type of meeting is occurring helps to avoid frustration and impatience. For example, if during a logistics meeting conversation on who will feed the chickens over the weekend, someone asks, "Do we really want to be using this conventionally grown feed?," the group can suggest postponing the feed sourcing discussion to the next discussion meeting. The following are sample agendas for each type of meeting:

Logistics Meeting Sample Agenda

1. How is everyone feeling today? Do you need any support? (5 minutes)
2. Task delegation for today. (10 minutes)
3. Reminder to update chalkboards with the weight of harvested items. (2 minutes)
4. Upcoming calendar events for the week. (5 minutes)

Strategic Discussion Meeting Sample Agenda

1. What are you feeling grateful for in this moment? (5 minutes)
2. Crop planning discussion: What varieties worked well this past season? Do we want to adjust the

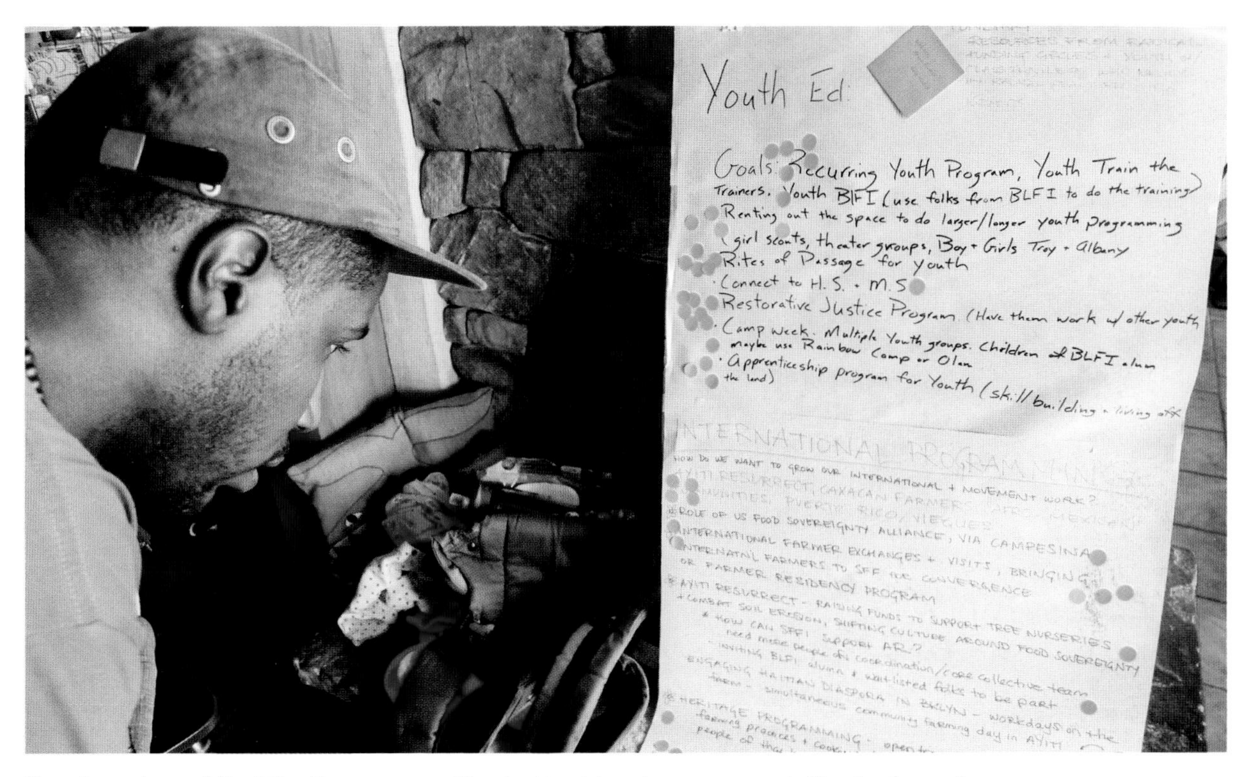

Board members of Soul Fire Farm use a ranking tool to determine program priorities for the next season.

timing of plantings to reduce the labor burden of bean picking? What did our CSA member survey reveal about preferences? Vote to approve crop plan. (25 minutes)
3. Budget discussion: What new infrastructure and equipment do we need to purchase? Are we willing to incur debt for those purchases? Vote to approve budget. (25 minutes)
4. Closing circle. (5 minutes)

Reflection Meeting Sample Agenda

1. What is a rose (something going well), thorn (a challenge), and bud (a goal or opportunity) for you in your role in this project? (10 minutes)
2. "Real talk" feedback session: Each person receives feedback from each member of the team on how they are performing in their role. Three types of feedback are offered: positives, deltas, and requests. The recommended phrasing is, "Something I appreciate about working with you is . . . A challenge that I observe in our working relationship is . . . To continue to develop our working relationship, I request that . . ." (60 minutes)
3. Evaluation review: We read evaluations written by our members or customers and have an open discussion about what we are doing well and how we can improve. (30 minutes)
4. Closing story: Each person shares a brief story of a moment when they felt fully alive in their role. (10 minutes)

Regardless of the type of meeting, it is important to be clear how decisions are going to be made. Many cooperatives use different types of decision making depending on the circumstances. At Soul Fire Farm, for example, a person can make a unilateral decision about the tactical way to implement a task for which they are responsible, like choosing paint colors or planning a menu. We use consultative decision making when we redesign programs. This means

Types of Decision Making

Consultative. The leader or manager has the ability to make decisions, but gathers input from all those who are affected by a decision.

Majority and supermajority vote. Members vote democratically on proposals after discussion of the pros and cons. The group decides what constitutes enough of a majority to move forward with a decision, from a narrow majority (51 percent) to a near-total supermajority.

Consensus. Members raise proposals that are then discussed, modified, and passed by the whole group. Members seek agreement among all decision makers to move forward with a proposal. Members are encouraged to listen to others and adapt their ideas, which can result in a much stronger proposal and one with strong community backing. To limit the power of one individual to block a decision, modified consensus procedures exist. These include consensus-minus-one, when all but one member need to agree in order to approve a proposal; principled objection, when blocking is only permitted if the proposal goes against the core value of the group; and sunset clause, which allows a set trial period for a proposal that does not yet have full consensus. Seeds for Change UK and the Olympia Food Co-op have both produced excellent guides for consensus decision making.

Delegation. Committees or working groups are granted authority to make certain decisions within parameters set by the entire cooperative. For example, the cooperative may decide that they want to grow more crops that are culturally relevant to African American people and set a budget of $500 for seed. The group may then empower a committee to research crops, select varieties, purchase the seed, and integrate the new crops into the farm plan.

that everyone gives input and the person in charge of curriculum integrates that input according to their best judgment. For decisions that impact the whole organization, like hiring or budgeting, we operate by consensus.

Farm-Share, Community-Supported Agriculture

I am quite introverted and usually eager to get home, so my farm-share delivery strategy looks a lot like "ding dong ditch." I pull the big white van with the cute Soul Fire Farm magnet up to the curb, double-check the spreadsheet for instructions as to where the veggie box should be left, and then place it as quickly as possible, ringing the bell on my way down the stairs. Larisa, Damaris, and others on our team are warmer and more extroverted, taking time to chat with our members, happy to spend an extra hour or two on deliveries nurturing the relationships we cherish. Whatever the delivery strategy, the end result is the same: Over 95 families, or 350 individuals, living under food apartheid receive a box full of fresh veggies at their doorstep weekly from June until November. For those who live outside our delivery area, we offer group pickup locations on back porches and in church garages.

Our Ujamaa Farm Share CSA is founded on the principle of food as a basic human right and the belief that community food access is compatible with financial solvency for the farmer. The program has

UPLIFT

Booker T. Whatley's Clientele Membership Club

Booker T. Whatley, professor at Alabama's Tuskegee Institute and student of Dr. George Washington Carver, was one of the original thinkers behind community-supported agriculture (CSA). From 1974 until 1981 he refined his formula for a successful small farm. He advocated for a Clientele Membership Club with a pick-your-own operation. Whatley explained, "That's the twist that makes this whole project innovative. The farmer has to seek out people—city folks, mostly—to be members of the club. The annual membership fee, $25 per household, gives each of those families the privilege of coming to the farm and harvesting produce at approximately 60 percent of the supermarket price . . . One of these 25-acre farms should be able to support 1,000 member families, or around 5,000 people. The clientele membership club is the lifeblood of the whole setup. It enables the farmer to plan production, anticipate demand, and, of course, have a guaranteed market. However, that means the grower had better work just as diligently at establishing and maintaining the club as at producing the crops. Put it this way: If you fail to promote your club, something terrible happens—nothing!"[7] Further, he planned crop successions such that there was year-round income and never too much to harvest at one time. His members enjoyed grapes, sweet potatoes, black-eyed peas, blueberries, strawberries, and mustard, collard, and turnip greens. He prioritized crops with an annual income of at least $3,000 per acre, which eliminated lettuce, onions, and white potatoes. He tried to make subscribers feel as if it was their own farm by sending out a newsletter that reported on picking dates and other farming activities.[8]

four central components: community-based marketing, wealth redistribution, flexible payment, and doorstep delivery.

Community-Based Marketing

The initial 20 families who joined the inaugural Farm Share were the same people who encouraged us to start Soul Fire Farm. We did not need to market to our "founding angels," as we affectionately termed them. Since then, we have decided to build our customer base using a grassroots organizing model. Certainly, if we wanted to sell all of our food to suburbanites connected to the so-called good food movement, we would be sold out with a waiting list. Our vision is different. Seeing food as a basic human right and not a privilege compels us to do the hard work of getting our harvest to those marginalized by food apartheid.

Every winter we offer free lectures and workshops in the community on food justice topics, including self-determined health, structural racism in the food system, and reconnecting to land. Through these workshops we connect with public school children, neighborhood association members, Little League parents, church deacons, and other powerful community leaders. We listen as well as speak, asking what barriers stand between people and food and how a farm like ours can deepen our solidarity work with community. Through these conversations, our youth restorative justice program was imagined, which is explained in detail in chapter 13. Many of

Soul Fire Farm provides free workshops on nutrition, farming, and food justice as one way to engage community members in the Ujamaa Farm Share program.

our Farm Share members find out about the program through these workshops in their neighborhoods.

Ujamaa, cooperative economics, imagines a nonextractive relationship between producer and consumer, rooted in authentic relationships, not just the casual exchange of the market. This is what we believe we cultivate in our farm share. We make a long-term commitment to nourish each member with the bounty of the land and they, in turn, make a commitment that Soul Fire Farm will be their source of produce for the season. We foster and honor these relationships by welcoming feedback, sharing thoughtful newsletters with recipes and updates, hosting monthly community days on the land, and saying thank you to those who volunteered with a bonus box of food in mid-November.

Wealth Redistribution

In the United States, we spend less on food, as a percentage of our income, than any other country in the world.[9] This is because the true costs of industrial food on the environment and human health are externalized. While we recognize the need to realign the economy to value food at its true worth, we are also heartbreakingly aware that members of our community do not have the means to access even this cheap food. We address this at a neighbor-to-neighbor

Table 2.1. Soul Fire Farm Ujamaa Farm Share Sliding Scale

Membership Category	Weekly Cost	20-Week Season Cost
Solidarity Share for immigrants, refugees, and those impacted by state violence	$0	$0
EBT/SNAP users	$23	$460
Low income/wealth	$25	$500
Middle income/wealth	Up-front payment expected	$560
Upper income/wealth	Up-front payment expected	$660
"Soul on Fire!" Contributor Share	Up-front payment expected	$800–1,000

level, asking that people with more wealth pay more for their farm-share subscription to subsidize shares for those with less wealth. See table 2.1.

In addition to working with individuals to redistribute wealth and make food access possible, we are starting to explore institutional partnerships to the same end. Our family doctor is a member of a board of directors. She noticed that patients who had recently immigrated to the area through a refugee resettlement program were quickly relinquishing their healthy, vegetable-rich cultural diet when relegated to food apartheid conditions. If we could provide a reliable, affordable supply of vegetables to these neighbors, they could continue to cook their traditional foods and maintain their health. This doctor leveraged institutional money from the community health clinic to help subsidize the farm shares and to provide translation services.

New Entry Food Hub in Massachusetts and Rock Steady Farm in New York also use institutional partnerships. For example, Rock Steady partners with the North East Community Center (NECC) to distribute over 70 CSA shares at three emergency food shelters throughout Dutchess County. NECC raised the money through a grant from the Community Foundations of the Hudson Valley and pays the farmers the full price of the CSA share. This supports the financial viability of the farm while ensuring food access for the most vulnerable members of the community. Nonprofit organizations, schools, daycares, medical facilities, elder care facilities, and residential facilities are potential institutional partners that may be able to leverage funding sources to purchase farm-fresh food for their clientele.

Flexible Payment

The classical CSA model requires members to pay the full share price at the beginning of the season, providing farmers with necessary income at the time when most expenses are made. Spring is the time when farmers purchase seed, equipment, supplies, and compost, often taking on large debt in the hope that a bountiful harvest will ensue. The CSA model mitigates that risk by welcoming community members to invest up front and accept whatever dividends are reaped, whether abundant or scarce. In times of hurricane, flood, or pest outbreak, the customer bears part of the burden of the crop loss, democratizing the food economy.

We implement the pay-up-front requirement for our middle- and high-income farm-share members. For those who cannot pay in advance, we offer payment plans with automatic billing through QuickBooks Payment. We also accept Supplemental Nutritional Assistance Program (SNAP) Electronic Benefits Transfer (EBT), previously known as food stamps. Run by the United States Department of Agriculture (USDA), SNAP is an entitlement program that offers eligible low-income people a set budget to purchase food. The process of applying to accept EBT/SNAP is cumbersome but surmountable. You can complete an application online (https://www.fns.usda.gov/snap/retailer-apply), though I

A typical summer farm share box is full of squash, greens, tomatoes, melons, and other colorful veggies. Photo by Larisa Jacobson.

Emet poses with a farm-share box in front of the delivery van. Photo by Jonah Vitale-Wolff.

recommend that you call your local SNAP Retailer Service Center (1-877-823-4369) to walk you through the process. A CSA farm is neither a retail store nor a farmers market, so your work may not fit neatly into the categories outlined on the application. Since we do not require our members to come to the farm for food, we cannot rely on them swiping their EBT card in our reader. Rather, we ask members to fill out paper vouchers in advance of the season, which we redeem on a monthly basis.

Doorstep Delivery

Lack of reliable transportation is the second most important barrier to good food access, after poverty.[10] Like many farmers in our area, we first considered joining a farmers market as a way to distribute our produce. We did not realize at the time that transportation and cultural barriers made farmers markets all but inaccessible to the communities that we loved. Our South End neighbors told us that it was nearly impossible to walk a full mile with small children in tow to the nearest city bus stop, journey to the market where people look at them sideways for their dark skin, and carry all those expensive vegetables and tired children back home. It was more feasible to survive on pasta and sauce from the corner store.

We decided to give doorstep delivery a try. Our first delivery vehicle was a beat-up diesel station wagon that fit 20 bags of food only if piled on top of the children, who were also delivery assistants. We now use an unrefrigerated 2005 Chevy Express half-ton all-wheel-drive cargo van that can hold up to 90 bushel boxes. We only offer doorstep delivery to families living in neighborhoods under food apartheid. People living in more privileged environments can join the program, but need to pick up their food from the back porch of someone in a targeted neighborhood. Through detailed spreadsheets and efficient route design, we are able to travel 40 minutes into Albany, make about 60 individual stops, and get back to the farm in under five hours. All told, this takes less time than working a farmers market stand, and there is no surplus crop to manage at the end. Our members tell us that doorstep delivery is the non-negotiable element of the Ujamaa Farm Share that makes it possible for them to participate.

Food Hubs

The Food Hub Collaboration defines a *regional food hub* as "a business or organization that actively manages the aggregation, distribution, and marketing of source-identified food products primarily from local

UPLIFT
Food Hubs of the Black American South

Black farmers have been organizing food hubs since 1973, beginning in Mississippi. The farmers used churches as hubs to combine, store, distribute, and market their produce. They would combine their harvest and load a truck with as much as 30,000 pounds (13,600 kg) of aggregated produce and ship it to churches in Chicago, Illinois. In 1981 the Indian Springs Farmers Cooperative Association of the Mississippi Association of Cooperatives was established to formalize these operations. They aggregated, processed, and distributed produce from more than 30 farmers in the area. They provided the cooler space, washing tubs, sorting tables, and other equipment to process produce for the farmers. Additionally, they offered the farmers training and education in cooperative management. At the time the project was called a packing shed, not a food hub, but the concept was the same.

Black North Carolina dairy farmers Phillip and Dorathy Barker started their food hub in the early 1980s. They worked with 10 mostly Black farmers within a 50-mile (80 km) radius. Their enterprise, named Operation Spring Plant, provided aggregation, processing, packaging, grading, and marketing support to the farmers. It was part of the Federation of Southern Cooperatives. Operation Spring Plant sold the goods to schools, grocery stores, hotels, and community members. Additionally, the Barkers supported Black farmers in transitioning from conventional tobacco growing to organic food production.

No analysis of food hubs and food systems can exist without acknowledging the pioneering work of southern Black farmers in general, and the Federation of Southern Cooperatives in particular.[11]

and regional producers to strengthen their ability to satisfy wholesale, retail, and institutional demands." While food hubs can be dizzyingly complex, at their root the concept is lean. Ben Burkett of the Mississippi Association of Cooperatives explains, "I like the idea of a food hub. Get us farmers growing, and we know we have a place that's going to buy it. That idea, that ain't nothing new though."

One potential starting place for your food hub is to identify a market that would be difficult for you to satisfy on your own, but possible with the aggregated growing power of several farmers. For example, Soul Fire Farm was approached by a major cosmetics company with a request to grow herbs for a natural product line. We were also approached by a prison requesting hundreds of pounds of dried spices, a request that was more ethically complex. In both cases the quantity required by the potential wholesale market exceeded our growing capacity. As part of a food hub, instead of declining a customer, we could coordinate with other farmers to meet the demand. Consider that there are several potential wholesale markets in your community, including schools, universities, grocery stores, elder care facilities, and food processing companies. It is important to note that all of the participating farmers need to decide upon a uniform variety, growing method, and harvesting technique so that the finished product has the uniformity demanded by the wholesale customer.

Another potential starting place for your food hub is to gather your farming community and work together to find a market for your existing crops.

In 2009 Black social entrepreneur Dennis Derryck founded Corbin Hill Food Project, a 501(c)3 social enterprise with a mission to provide food to those who need it most. He, together with 10 social investors, raised $700,000 in equity with 72 percent coming from Black and Latinx people, and more than half coming from women, and purchased a 95-acre farm in Schoharie County, New York. Derryck observed that Schoharie County farmers had limited markets while the majority of NYC residents were ignored as customers by the so-called good food movement. Over its first four years, CHFP's food hub increased aggregation of fresh produce from 4 farmers to 42. In the same period CHFP's success was further exemplified by its ability to scale fivefold from 200 shareholders to 1,000, with 66 percent of produce going to people of color with household incomes less than 200 percent of the poverty line. Further, he is working to create micro food hubs to aggregate the products of Bronx community gardeners and turn them into value-added products like hot sauce.[12]

Excepting its erasure of Black farmers' contributions to food hubs, the United States Department of Agriculture published a useful and comprehensive primer for planning your food hub, called *Regional Food Hub Resource Guide*, which offers logistical advice for establishing your aggregation project.[13]

Communal Labor Practices

It's 7:55 AM and the first eager volunteers roll tentatively into the driveway, where the farm collie,

UPLIFT
Haitian Konbit Communal Labor Practice

The Haitian communal labor practice of *konbit* originated in West Africa with the Dahomian practice of *dokpwe*. Pre-revolution, each enslaved adult was allotted a small plot of land for personal subsistence farming. All cultivation tasks had to be completed on Sundays, and a system of mutual cooperation developed to make efficient use of that limited time. Following independence, konbit continued within the structure of the extended family and was centered on the *lakou* (family compound). Konbit are semi-permanent organizations comprising men and women, typically 3 to 15 individuals who take turns hosting work events on their respective farms. Members of the konbit bring food and water to share, which is often supplemented by the host as a courtesy. Traditional foods are preferred: coffee and root vegetables for breakfast, and cornmeal with bean sauce and vegetables for lunch. As one member of Mouvman Inite Ti Peyizan Latibonit (The United Movement of Small Peasants in the Artibonite) explained, "Before in konbit we boiled yam, sweet potato, and had cornmeal or pitimi (millet). Spaghetti was never here! But now everyone leaves their food culture for spaghetti! [We still maintain our traditions]."[14]

For particularly challenging tasks, like sugarcane harvest, the host may hire musicians. Singing is also a common means of passing the time and establishing a cadence by which to coordinate work. The members of a mutual aid konbit often plan far in advance to avoid conflicts in scheduling and ensure that tasks are timed efficiently. For example, one member seeds in week one, the next in week two, and so on. Then, when harvesting, the crops are ready at one-week intervals.[15]

Rowe, greets them with eager wagging and sniffing. The agenda for the community *konbit*, or collective work party, is displayed in English and Spanish on a salvaged wooden easel. Each member of our farm team has prepared a project suitable for guests with various skill levels. I will be planting four beds of garlic and mulching them with straw. Larisa has 100-foot silage tarps ready to install over tired crop beds as a means to exhaust weeds and encourage biological tillage. Amani has benches arranged in a circle around a tall stack of crates, laden with garlic and maize that need to be processed. Jonah is ready to remove tomato and cucumber trellises, put away irrigation line, and generally organize the infrastructure before winter. Our youngest child, Emet, greets the visitors and invites them to sign a waiver and choose their preferred work team. People arrive throughout the morning, some having traveled four or more hours to be present with us on this sacred land. We speak, laugh, and labor together until midday when we gather in a wide gratitude circle, sometimes 90 people thick. Each person calls into the center one word that encompasses their thanks and everyone echoes it back to them—for instance, "Love," then "*LOVE!*" We share an abundant and scrumptious potluck meal before heading on a tour of the farm. Lingerers help clean or make music or catch grasshoppers. All is complete by 4 PM.

Inspired by the communal labor practices of my Haitian family, we harness the collective love and power of our community through monthly konbit work parties. Even before there was Soul Fire Farm,

Volunteers at Soul Fire Farm's konbit process garlic while sharing stories.

Novice and Expert Tasks for Konbit

Novice Tasks: Great for Konbit

Crop processing. Cleaning garlic, shucking maize, shelling beans.

Simple weeding. Removing weeds from crops that are much bigger than weeds or clearly distinct.

Cleanup. Removing stones from pathways, taking down fencing, uninstalling irrigation line.

Mulching. Laying down tarps or organic mulch to suppress weeds.

Planting big seeds. Direct seeding of big-seeded crops like potatoes, garlic, or beans.

Amendment spreading. Spreading limestone or rock dust. (Note that the host needs to create a grid of the area and measure out quantities in advance.)

General harvesting. Harvesting "easy" crops like flowers, brussels sprouts, tomatoes, or greens.

Expert Tasks: Workable for Konbit with Extra Support

Detail harvesting. Harvesting crops that need to be counted, bunched, discerned, or forked out of the ground.

Detail weeding. Weeding crops that are a similar size or appearance to the weeds, or weeding a polyculture.

Planting small seeds. Planting seeds that are tiny and hard to see.

Transplanting. Putting seedlings into the ground.

Food preservation. Canning, fermenting, freezing, and drying the harvest while maintaining a sanitary work environment.

Cutting firewood. Splitting firewood with a wood splitter, then stacking the wood in the woodshed.

Fence repair. Clearing brush from the fence row, tightening wires, and baiting the fence.

the hands of our extended community laid the foundation for what would emerge. We invited people to help raise the timber frame of the house, install the straw-bale walls, and apply the natural plasters. Together we cleared the field of brush and celebrated around a canopy-high bonfire. As our project matures, we find that a monthly konbit is a beautiful container for welcoming new friends into our space, reconnecting with alumni and friends, and getting necessary tasks done with joy. It took some time to settle on a model for konbit that was nurturing. Along the way, we made several mistakes that resulted in such outcomes as the accidental weeding of an entire crop of maize and rushing a guest to the hospital who was in anaphylactic shock. The following are some strategies that we now employ in operating our konbit.

Work Task Selection

It is important to select work tasks that are manageable for guests who may have limited farm experience. This ensures that the tasks are done well and also reduces stress and disappointment for the host and the volunteers. We have found that thinking through each task in advance and laying out all the necessary tools and supplies is essential for flow. It is better not to have the host continually leaving the group to go fetch additional supplies. Assign a knowledgeable member of your team to facilitate and supervise each work project. For safety reasons, it is best to avoid asking guests to use tractors, power tools, chain saws, or other machinery. Potential konbit tasks are listed in the "Novice and Expert Tasks for Konbit" sidebar.

Leah and Amani tend the beans during konbit at Soul Fire Farm. Photo by Neshima Vitale-Penniman.

Food and Accommodations

When we started hosting monthly konbit work parties, attendance would average 20 and consist mostly of close friends and family. Currently, we host over 80 people at each konbit, many of whom are first-time visitors to the farm. We share lunch together potluck-style, asking each person to bring food to share. We supplement these generous offerings with a giant pot of chili and rice. One of the work teams heads inside about an hour before the meal to heat up and set out the lunch. We generally do not host people overnight before or after the konbit, as it would mean no family time or downtime for the farm team. We have, however, compiled a list of affordable local accommodations for people traveling from out of town.

Safety

As the host of the konbit, you are responsible for the health and safety of your volunteers. People accustomed to city life may not know how to navigate the particular risks of farmwork. With each close call we have added to the safety information that we share when people arrive. Currently, we advise people on ergonomic use of tools to avoid injury, keeping track of children near the pond, avoiding wasp nests, avoiding the electric fences, checking for ticks, staying hydrated, preventing allergic reactions, and wearing sunscreen. We also emphasize the importance of wearing protective clothing that is suitable for mud and rain, such as work gloves, rain boots, a waterproof jacket, and a clean change of pants. We make sure that there is a fueled vehicle ready to transport someone to emergency care if needed. If you decide to welcome strangers to your konbit, you may want to have them sign a liability waiver indicating that they are aware of the inherent risks in farm labor and will not take legal action against you if they are injured.

Reciprocity and Appreciation

In Haiti the farmers say, "*Sonje lapli ki leve mayi ou*," which means, "Remember the rain that made your corn grow." The sacred law of reciprocity binds us in a circle with those who contribute to our lives and our projects. At Soul Fire Farm we honor this reciprocity first by saying words of thanks to those who gather at konbit. When other farmers in our network put out a call for collective work, we respond in turn. We also host a volunteer appreciation lunch each winter for those who have made an enduring commitment to our project.

Writing Your Farm Business Plan

Nikki Giovanni reminds us that "Style has a profound meaning to Black Americans. If we can't drive, we will invent walks and the world will envy the dexterity of our feet. If we can't have ham, we will boil chitterlings; if we are given rotten peaches, we will make cobblers; if given scraps, we will make quilts; take away our drums, and we will clap our hands. We prove the human spirit will prevail. We will take what we have to make what we need. We need confidence in our knowledge of who we are."[16] We always begin our business planning sessions with a conversation on this prophetic quotation.

As survivors of racial trauma, we often limit our imagination of what is possible. Babalawo Enroue Halfkenny explains that we carry a great fear of confronting the *no* that might destroy us. So we create a safe box in which to exist, pre-settling for a life that is much smaller than our capacity. Our internalized oppression is so powerful that the oppressor no longer needs to act. The prison of our own skewed sense of inadequacy, of not deserving, of not being enough, is all that is required to keep us in bondage.

The first step in writing a business plan, which is really a plan for the tangible love you want to manifest in the world, is to imagine widely what is possible. We invite you to light a candle, call upon your ancestors, close your eyes, and dream into your biggest visions. The following are specific considerations that we found useful in creating a mission-driven business plan.[17]

Values and Mission

Values are the core beliefs and philosophies that guide your life and change very little over time. Your project needs to be firmly rooted in your central values, whether those be family, environmental stewardship, thrift, or hard work. To get in touch with your values, a helpful exercise is to draft your own obituary. While this may seem morbid, the reality is that all of us will transition from this life, and taking time to reflect on how you want to be remembered can put the big picture in clear focus. From your list of values, you

Business Planning Resources

Growing Farms: Successful Whole Farm Management Planning Book, published by Oregon State University Extension Service
SCORE Business Plan Resources
AgPlan
Whole Farm Planning: Ecological Imperatives, Personal Values, and Economics by Elizabeth Henderson and Karl North, published by Chelsea Green
Exploring the Small Farm Dream: A Decision-Making Workbook, published by the New Entry Sustainable Farming Project

can draft a mission statement. A mission statement describes the specific purpose of your business and its guiding principles. See the "Soul Fire Farm Mission" and "Soul Fire Farm Goals" sidebars for ours.

Way of Life and Personal Needs

Consider aspects of your quality and style of life that are important to you. To what extent do you want your personal life and work life to be separate? Do you prefer to work alone or have a lot of interaction? How much time do you need for rest and for being with family and friends? Do you enjoy marketing and interacting with customers? What are your favorite tasks on the farm? Are you comfortable with uncertainty and risk? What does financial security mean to you? What other time commitments are you juggling and how does this farm fit into those responsibilities?

Existing Resources

What assets do you currently have that can be leveraged toward the manifestation of your project? Consider not only your financial resources, but your existing knowledge and prior experiences. Who in your social networks might contribute skills or labor to your vision? Do you currently have access to land or infrastructure? Is there a ready market for your harvest?

Strategic Goals

What would you like to accomplish within the first three years of your farm project? Strategic goals should be SMART: Specific, Measurable, Accountable to Community, Realistic, and Time-bound. Consider writing goals related to what you will produce, your impact on the ecology of the land, your financial sustainability, and your quality of life.

Income and Expenses

Investigate markets in your area to determine how much income you are likely to generate from your products. At Soul Fire Farm we convened a group of our neighbors to gauge interest in vegetable delivery and determine a reasonable price point. Next,

Farmers work together on their whole-farm business plans during the Black Latinx Farmers Immersion. Photo by Neshima Vitale-Penniman.

Soul Fire Farm Mission

Soul Fire Farm is a people-of-color-led community farm committed to ending racism and injustice in the food system. We raise and distribute life-giving food as a means to end food apartheid. With deep reverence for the land and wisdom of our ancestors, we work to reclaim our collective right to belong to the Earth and to have agency in the food system. We bring diverse communities together on this healing land to share skills on sustainable agriculture, natural building, spiritual activism, health, and environmental justice. We are training the next generation of activist-farmers and strengthening the movements for food sovereignty and community self-determination.

Table 2.2. Sample Annual Expense Budget, Soul Fire Farm

Category	Item	Amount
Income		
Vegetable CSA	90 shares, average $500 each	$45,000
Pasture-raised chicken	250 birds, average $20 each	$5,000
Farm training	4 events at $1,000 each ($25 pp, 40 at each)	$4,000
	Income Total	**$54,000**
Expenses		
Personnel	Farmer (1 person, 40 weeks, $15/hr)	$24,000
	Self-employment tax (15 percent)	$3,600
Administration	Marketing	$200
	Utilities	$1,200
	Insurance	$2,272
	Phone and Internet	$1,000
Farm supplies	Vehicle maintenance	$3,152
	Vehicle insurance and registration	$725
	Chicken feed and expense	$3,500
	Soil amendments	$2,500
	Tools	$1,000
	Fuel	$590
	Seeds	$2,100
	Farm supplies	$5,000
	Soil tests	$150
	Certification and dues	$200
	Tractor maintenance	$1,500
	Expense Total	**$52,689**

Soul Fire Farm Goals

1. **Train and empower aspiring Black, Latinx, and Indigenous growers** so as to reverse the dangerously low percentage of farms being owned and operated by people of color and to increase the amount of good food being grown by and for marginalized people.
2. **Advance healing justice for individuals and communities impacted by racism and other oppressions,** by engaging land-based and ancestral healing methodologies, so as to uplift hope, resiliency, agency, and efficacy in the movement.
3. **Train and empower young people,** especially those targeted by state violence, to build relationships with the land, shift to healthier diets and self-determination around their bodies, and learn organizing skills to correct injustice in their own communities.
4. **Offer popular education workshops, lectures, and publications to activists and community members** to increase awareness and skills on environmental justice, food sovereignty, ending racism, transformative justice, and other concrete tools to increase the impact of movement work.
5. **Provide affordable weekly doorstep deliveries of in-season, farm-fresh, naturally grown food** to families living in food apartheid neighborhoods. Focus especially on the needs of people criminalized by the injustice system—incarcerated people, those impacted by police violence, immigrants, and refugees—to uplift the right of all to access life-giving food regardless of social or economic status.
6. **Refine and share our model of a just and sustainable farm** that stewards biodiversity, captures carbon, pays its workers a living wage, uplifts community wholeness, dismantles racism, inhabits sustainable structures, and achieves financial solvency.
7. **Collaborate with regional, national, and international networks** for Black land justice and community food sovereignty to advance structural changes necessary for a more just food system.
8. In line with our work to advance healing justice and liberation in the wider community, **commit to an organizational culture that cares for the well-being of its workers** through ample rest, compassionate communication, distributed leadership, and investment in personal and professional development.

calculate your start-up and operating expenses. Start-up expenses include land, equipment, tools, and infrastructure. Ongoing operating expenses include seeds, compost, mulch, utilities, supplies, transportation, veterinary expenses, feed, insurance, certifications, labor, and maintenance. A sample operations budget for Soul Fire Farm is provided in table 2.2, which demonstrates how a farmer can make a reasonable living intensively cultivating less than 2 acres (0.8 ha) of land.

Production

Will you produce annuals, perennials, livestock, and/or value-added products, and what varieties of each?

Annual crops include vegetables, grains, herbs, and flowers and require intensive labor during the prime growing season. Perennial crops include nuts, fruits, hops, and grasses and take several years to earn a return on investment. Livestock systems include raising animals for meat, dairy, eggs, or honey, and involve daily chores for the entire year. Consider your personal needs and existing resources in deciding what to produce. Think about whether you will use organic, conventional, or other production techniques and what the certification requirements are for your selected method. Consider what culinary historian Leni Sorensen terms "home provisioning," the act of growing and preserving food for the nourishment of your family as a priority over market sales.

CHAPTER THREE

Honoring the Spirits of the Land

Blefono ke; ena ni hluu, se yomoyo ngo tovi kepa le.*
The white man says he is rich, but the old lady with her black bead surpasses him.

—KROBO PROVERB, Ghana

Rue, hyssop, yarrow, and other fresh herbs wilted in the steaming pot of scalding river water. Eighteen vulnerable hearts gathered on rough wooden benches around the nearby fire. There was quiet and reverent trepidation in their muted conversations. This was the midpoint of the Black Latinx Farmers Immersion (BLFI) program at Soul Fire Farm, marked by a spiritual bath rooted in Vodou tradition. Earlier in the day, many participants had taken life for the first time, transitioning 50 chickens from field to freezer. As is the custom of our people, we marked the boundary between times of aggression and times of peace with a ceremonial washing. In this optional, all-gender space, people brought with them an intention about what they wanted to release and what they wanted to invite into their lives.

We began by thanking the spirits of this land and our ancestors with song and offerings of cornmeal and water. Believing that the universe is governed by laws of reciprocity, we sacrificed just as we hoped to receive. Then, two by two, people came to the rock at the edge of the pond to be gently and reverently washed with the herbal mixture. My sister Naima and I, matching in white clothing, poured the warm bath over the heads of our people, calling to Legba to open the gate, calling to Ogou to fortify our agency, calling to Ezili to fill our hearts with love. We brushed the bath downward, head-to-toe, willing burdens and negativity to be released into the earth and composted there. Others witnessed tenderly as glistening droplets accented the black and brown hues of their friends' skin. They held towels ready to envelop their peers as they finished and resisted the urge to pull bits of clinging fern and flowers from one another's hair. Two by two, until all were bathed, and then we sang and reflected by the fire.

In African cosmology we believe that there is no separation between the sacred and the everyday. Just as we punctuate BLFI with a spiritual herbal bath, so do we mark our days and seasons at Soul Fire Farm with ritual. Each spring, before breaking ground with the hoe or planting the first seed, we ask permission from the Spirit of the Land and make offerings of gratitude. Each fall we celebrate the yam festival to give thanks for the harvest and spiritually

* Tonal accents and other pronunciation guides for words in Krobo, Yoruba, and other African languages are not provided in this text. For Yoruba language, please consult *Fama's Ede Awo (Orisha Yoruba Dictionary)* by Chief Fama.

Ewe (herbs) and *omi* (water) are infused with prayer and used for spiritual cleansing.

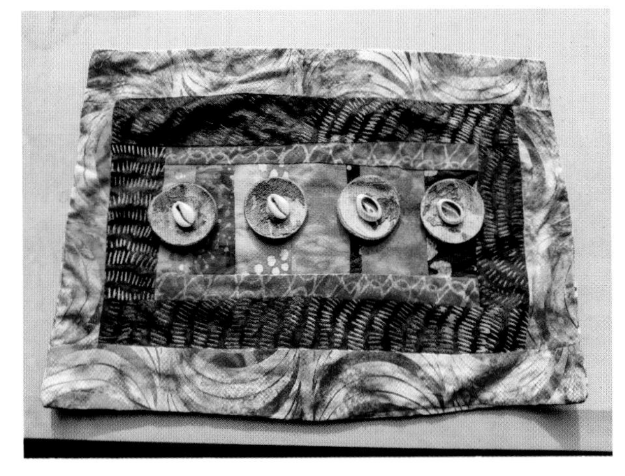

New World obi obata are one of the divination tools used to communicate with the ancestors and orisa (deities, forces of nature).

refortify ourselves with the strength of our ancestors. In between these bookends of the season, we maintain an intimate connection with the Divine through singing and dancing in the fields, making food and drink offerings, and asking permission before new enterprises. We are guests and stewards on this Earth, not owners, and need to behave as such.

To acknowledge that we as humans are not the most powerful force in nature and to act accordingly is not always roses and sunshine. For example, when we first came to this land, we were very excited about the potential for renovating an existing overgrown swamp into a pond for swimming and irrigation. Intuitively, we could tell that the swamp held a concentration of spiritual energy, so we felt that we needed to ask permission before moving earth and disturbing the ecosystem. We used a simple divination tool taught to us by elders in the Yoruba Ifa tradition to determine whether the spirits of the land wished for the pond to be dug. The response was a firm no. We honored this boundary and waited another 10 years until the spirits gave us a conditional yes. In order to have their blessing for the pond digging, they required us to put certain safety features in place to protect children and also to make regular offerings to Nana Buruku, the grandmother of the universe whose energy dwells in forest wetlands. We now have a beautiful pond as well as the good graces of Spirit. We half-jokingly say that the relative scarcity of ticks, poison ivy, and biting insects is a sign that we have a harmonious relationship with the land.

Our spiritual practices at Soul Fire Farm are informed by our specific lineages while making space for everyone to bring their own belief system to share with the community. My personal connection to nature-based religion began when I was a small child inventing praise songs with my sister in the forest. This connection was formalized in my young adulthood during a five-month visit to Ghana. There I was initiated and enstooled by the Manye of Odumase-Krobo in Ghana in the Vodun/Vodou religion. *Manye* (Queen Mothers) are spiritual leaders in the community, arbitrators of conflict, and keepers of the genealogical knowledge of our people. We are responsible for upholding ethical standards of behavior in our circles, especially in terms of the protection of women, children, and the vulnerable. Our role as clergy is both secular and spiritual. As custodians of cultural-religious traditions, we take leadership in ceremonial rites of passage marking birth, puberty, marriage, death, and honoring of ancestors. As spiritual activists, we work for economic justice, social welfare, and environmental protection. As Manye, we carry the *ase* (spiritual force) of the Orisa-Lwa

Nana Buruku, the primordial grandmother of the Universe. She is envisioned as an ancient dark-skinned woman, energy healer, defender of justice, protector of children, guiding light of the moon, dweller of forest wetlands, and source of all waters.

While my practice is primarily rooted in Vodou, my spirituality is as diverse and complex as my family's mixed ethnic heritage: Haitian, Black American, Hebrew, French, English, Cherokee, Taino. I am Manye, yes. I am also a Jewish, Unitarian Universalist, Haitian Vodouisant, initiate of Oya, and worshiper of Ifa. I make spiritual herbal baths, chant Torah in synagogue, dance to orisa, divine with shells and bones, and receive dream messages from the ancestors. For me the Truth (capitalization intentional) is at the intersection of our multiplicitous ways of knowing the Sacred. Haitian Vodou instructs us in this, as our ancestors made space in the religion for the practices of the Ewe, Fon, Kongo, Ibo, Yoruba, Taino, Catholics, and dozens of others in an elegant and integrated whole, without diluting the integrity of each lineage.

In this chapter we explore how to honor the spirits of the land, primarily through the lens of Vodou (Haiti, Ghana, Benin) and Yoruba Ifa (Nigeria, Benin, Togo), since those are most practiced at Soul Fire Farm. We will explore how verses of Ifa guide us as farmers, learn how to honor the Yoruba agricultural spirit of Orisa Oko and the Haitian Vodou farmer spirit of Azaka, discuss how to prepare a cleansing herbal bath, explore how to celebrate planting and harvest rituals, and learn how to enliven our days on the land with ancient and contemporary work songs.

Vodou and Ifa, however, are not the only indigenous African religions that persist. Even as we focus this chapter here, we acknowledge and shout out the myriad spiritual paths curated by our ancestors and relatives. Mojuba, we pay homage to: Bushongo, Lugbara, Baluba, and Mbuti (Congo), Dinka (Southern Sudan), Hausa (Chad, Gabon, Niger, Nigeria), Kotuka (South Sudan), Akamba (Kenya), Maasai (Kenya, Tanzania, Ouebian), Kalenjin (Kenya, Uganda, Tanzania), Dini Ya Msambwa (Bungoma, Trans Nzoia, Kenua), Lozi (Sambia), Tumbuka (Malawi), Zulu (South Africa), San (South Africa), Manjongo (Zimbabwe), Akan (Ghana, Ivory Coast), Dahomey (Benin, Togo), Efik (Nigeria, Cameroon), Edo (Nigeria), Odinani (Igbo people, Nigeria), Serer (Senegal, Gambia, Mauritania), Vodun (Ghana, Benin, Togo, Nigeria), Dogon (Mali), Kemet (Egypt, Sudan), Berber (Burkina Faso, Mali, Chad, Niger), Waaq (Ethiopia, Somalia, Eritrea), Candomble, Umbanda, and Quimbanda (Brazil), Santeria, Palo, Vodu, and Abakua (Cuba,Puerto Rico, Dominican Republic, Venezuela, Panama, Colombia), Obeah, Kumina (Jamaica), Winti (Suriname), Spiritual Baptist (Trinidad and Tobago), Louisiana Voodoo and Roots (southern US), and all other spiritual practices and religions created by African people.[1]

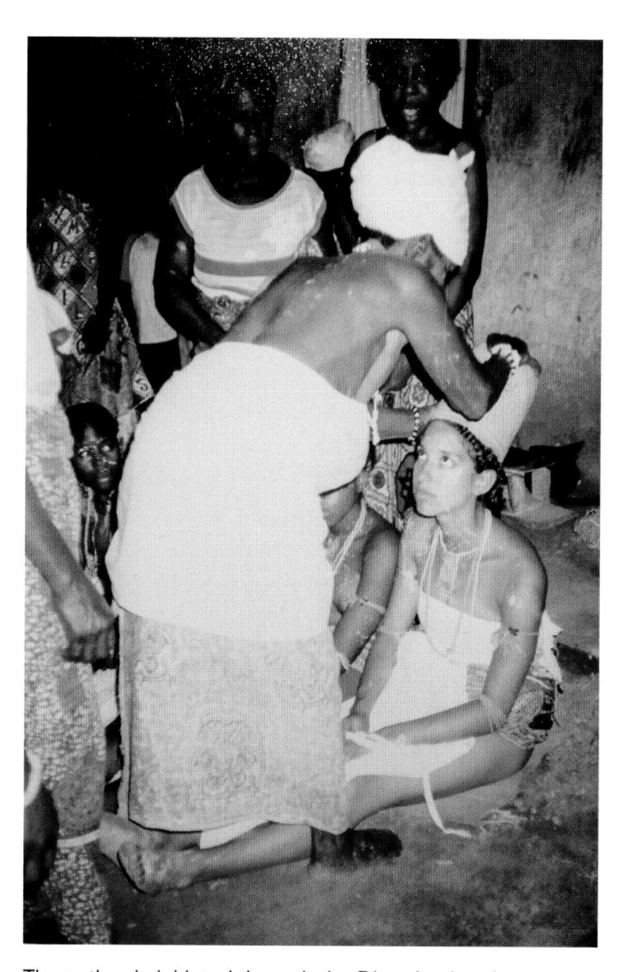

The author is initiated through the Dipo ritual and enstooled as a Queen Mother in Odumase-Krobo, Ghana.

Sacred Literature

I grew up with the myopic misunderstanding that only the Abrahamic faiths—Christianity, Judaism, and Islam—offered sacred texts to guide successive generations in terms of ethics and spiritual practice. While I study and cherish the Torah, I felt a great and expansive joy to learn that my Indigenous African ancestors had an oral sacred literature as long and deep as several Torahs combined. Each of our lineages carry sacred texts rich with the stories of our ancestors and guideposts for living. In this section we will look at the Odu Ifa, originated with the Yoruba people, and listen to its teachings for us as farmers.

Broadly, we learn from Odu Ifa that nature is sacred. "The basis of Yoruba religion can be described as a worship of nature," explains Professor Wande Abimbola, the *Awise Awo ni Agbaye* (Spokesperson of Ifa in the Whole World).[2] In the Yoruba religion

Babalawo Onigbonna Sangofemi (Enroue Halfkenny) consults Ifa through divination. Photo courtesy of Enroue Halfkenny.

UPLIFT
Odu Ifa

The Ifa divination system, practiced among Yoruba communities and throughout the African Diaspora, was inscribed in the Intangible Cultural Heritage of Humanity list by the United Nations Educational, Scientific, and Cultural Organization (UNESCO) in 2008. The word Ifa refers, among other things, to Orunmila, the *orisa* (deity, force of nature) of wisdom, destiny, and intellectual advancement. Ifa divination is rooted in a literary corpus, the Odu Ifa, which contains the Yoruba history, cosmology, beliefs, language, and social analysis. There are 16 books, each containing 16 chapters, for a total of 256 *odu* (chapters) to the Odu Ifa, each chapter containing around 600 to 800 *ese ifa* (verses). The Ifa priest(ess) (Awo or Babalawo or Iyanifa) uses sacred *ikin* (palm nuts) and an *opele* (divining chain) to generate a system of signs that correspond to specific odu of the literary corpus. The Awo chants the ese ifa corresponding to the Odu revealed during the divination. The wisdom in the ese guides the client in making decisions in line with their destiny and in harmony with the orisa. The Awo also uses the divinatory tools to determine what offerings the orisa require to secure the blessings or remove the challenges revealed during the divination. The knowledge of Odu requires extensive training, memorization, and spiritual discipline. As with most Indigenous practices, Ifa divination is under attack by Christianity, Islam, and Western colonization.[3]

it is believed that divinities, known as orisa, changed themselves into forces of nature, such as thunder, lightning, rain, rivers, oceans, and trees, after they completed their work on Earth. Oya became the Niger River. Sango became thunder, lightning, and rain. Olokun changed herself to become oceans, while Osun and Yemoja became the Osun and Oogun Rivers, respectively. There are at least 64 trees whom the Yoruba people worship as divinities. Every hill, mountain, or river of Yorubaland is a divinity worshiped by some people. Numerous birds and animals are sacred to Yoruba people who worship or venerate them. The Earth herself is a divinity. Human beings are ourselves divine through our *Ori* (personal divine nature) and *Emi* (divine breath encased in our hearts), which are directly bestowed on humans from Olodumare, the Supreme God.

Central to the recognition of the sanctity of the Forces of Nature is the practice of reciprocity through *ebo* (sacrifice). Numerous verses in the Odu Ifa (see the Uplift sidebar) speak to the importance of prayers and sacrifice by farmers. An excerpt from an ese in Ogbe Iwori reads, "Cast divination for the Farmers, the one that was going to choose a new land for annual farming. They asked him to take care of the ground [by performing sacrifice to the land before his departure, and pledging an ongoing relationship of stewardship with the Earth]. The Farmer heard about the sacrifice and performed it . . . Offering of prescribed sacrifices and free gifts to Esu. Life so please us aplenty."

Ifa also explains the consequences of taking without reciprocity, of failing to make sacrifices. Ofun Oturupon teaches, "Incidentally, once the farmers refuse to offer sacrifices to Sango, there would be the onset of drought. The sun would be very intense . . . As soon as they offer sacrifice to Sango, rain must fall . . . Their farm products would be very fine. They would sell and buy." Further, Ika Irosun warns, "He was going to choose a virgin land to plow . . . he was asked to perform a sacrifice . . . yet did not perform it. Something suddenly occurred and it took [him] away. They said the snake uses its mouth as a vise. The scorpion uses its tail as a ring. The farmer . . . would not witness its harvest . . . He died and used his hoe as a pillow on the farm." It is scientifically justified to see rampant wildfires in California and devastating hurricanes of the South as consequences of human disregard for the laws of nature, specifically the balance of the carbon cycle. This may also be true in the spiritual realm. As humans increasingly worship ourselves and the material products of our own making, raping the Earth in the process, and taking without giving back, metaphorically (and perhaps literally) we invite a kind of death. Of course, consequences for failing to make ebo certainly exist, but are not always death. By violating the natural laws of reciprocity, we may invite spiritual poverty, impairments to our physical and emotional well-being, or a sense of disconnection from our purpose.

Ifa further describes the characteristics of a successful farmer: efficiency, punctuality, and honoring of agricultural tradition. Odi Ogunda relates, "Cast divination for the man who used to leave early from the farm. The one that would not subscribe to time wasting. He would not use his seasons to loaf around. The Punctual Farmer is a zealous worker. We are grateful to God for the Punctual Farmer. We rejoice with him. Fortune and wealth overwhelm us." Ofun Irete warns us not to be greedy in the harvest, disregarding the requirements of the plants upon which we depend. In this ese the only surviving farmer is the one who does not climb and overharvest the fruits of the Jua tree. "The Jua tree, however, is a cash crop. No one climbs it . . . I have only picked the dropped ones. I did not climb the Jua tree on my own. I picked the ones that dropped." On sharing our harvest with the community, Ogbe Iwori teaches, "Let us combine hunting with farming for our world to be better. The farmer was returning from the farm. He gave a small piece of yam to the hunter. The Hunter was coming back from the deep forest. He brought a small piece of meat and gave it to the farmer. Let us combine hunting with farming for life to be better for us." These teachings are particularly poignant to me, as they dispel the myth that efficiency in work and specificity in technique are Western notions. In fact, our farming traditions gift us with strategies for working hard, fast, and with high attention to detail. This ensures an abundant harvest that we can share and exchange in community.

Ose Otura affirms the importance of prayer: "The perennial farmer is the one that prays fervently to God. A torrential rainfall would clear and soften up the soil." While there are traditional prayers for each Force of Nature, those simply spoken from the heart can be just as powerful. My son recently affirmed, "Mom, I like how you go outside in the storm and say, 'Thank you for watering our crops, and please be gentle on our house!'" The essence of prayer and sacrifice is the acknowledgment of the being-ness of the nonhuman energies in our world.

According to Ifa, the farm is also a place of refuge from violence. Owonrin Obara teaches, "The warfare in the city does not get to the farm. When everyone heard about Cricket's house in the farm, they paid him homage. He shrieked, 'The warfare in the city did not get to the farm.'" I believe that in our exile from the red clays of the South, to the paved streets of the West and North, we left behind a little piece of our souls. Forced by structural racism into overcrowded and under-resourced urban neighborhoods, many of us have grown up with profoundly traumatizing exposure to violence. Ifa invites us to reclaim rural land as a haven of peace, even as we do the essential work of uprooting violence from our urban communities.[4]

Farming While Black contains exactly 16 chapters to honor the 16 major books in Odu Ifa literary corpus. From just a few verses of the sacred literature of Odu Ifa we receive essential lessons for approaching our work as farmers.

- Offer prayers and sacrifices to the Land, honoring the laws of reciprocity.
- Rise early and work hard.
- Cultivate and harvest crops according to their prescribed needs; never overharvest.
- Share the bounty of the land with others in the community.
- Use the farm as a refuge from violence and strife.

Participants in Black Latinx Youth Immersion rites of passage program offer rum and songs to the forest. Photo by Neshima Vitale-Penniman.

Offerings to Azaka and Orisa Oko

We ask a lot of this land. Each year we coax around 80,000 pounds (36,500 kg) of mineral-laden vegetables from the hard clay soils to nourish our community, plus several cords of wood to keep our spaces comfortable and showers hot. We also make steep spiritual requests of the land, welcoming thousands of strangers each year, along with their traumas and burdens, to come and heal on this ground. The Land generously and relentlessly composts their pain into hope. While most of our programs center Black, Latinx, and Indigenous learners, we recently held an Uprooting Racism Immersion for people with white privilege who were interested in dismantling oppression in the food system. It was perhaps the first time in decades that so many European-heritage people had slept, worked, and taken meals on the Land. Partway into the program, I had a dream that the Land was restless and so took my divination bones to the forest to inquire about the trouble. The divination revealed that Azaka, Haitian Spirit of Agriculture, Guardian of Farmland, Friend to Peasants, was confused and upset that there were so many "police" on his farm. Further, Azaka expressed discontent about the arrogance of the visitors, as evidenced by their use of forks during meals. With some hesitancy, I shared

UPLIFT
Bwa Kayiman, Haitian Revolution

Haiti was once the wealthiest colony in the Western Hemisphere and also the most brutal. The average life expectancy for an enslaved African in Haiti was 21 years. Abuse was dreadful, and routine: "Have they not hung up men with heads downward, drowned them in sacks, crucified them on planks, buried them alive, crushed them in mortars?" wrote one former slave some time later. "Have they not forced them to eat excrement? Have they not thrown them into boiling cauldrons of cane syrup? Have they not put men and women inside barrels studded with spikes and rolled them down mountainsides into the abyss?"[5]

On August 14, 1791, the Haitian Revolution began with a secret Vodou ceremony held in the forest near Le Cap. Dutty Boukman, a Vodou *houngan* (priest) and a respected community leader, convened the gathering as a planning meeting for revolt as well as a religious ceremony. Boukman prayed, "God who has made the sun that shines upon us, that rises from the sea, that makes the storm to roar, and governs the thunders . . . you have seen what the whites have done . . . give strength to our arms and courage to our hearts. Sustain us . . . Harken until Liberty!" During the ceremony, the priest (*Mambo*) Cecile Fatiman sacrificed a Creole black pig to Ezili Danto, the spiritual mother of Haiti. It is of note that Ezili Danto was a spirit born in Haiti from the marriage of the traditional lineages of the indigenous Taino and the African Dahomey. Those present at the ceremony swore to take up arms against their enslavers unto death.

They went on to lead a successful insurrection from French colonial rule that resulted in independence in 1804. It was the only uprising of enslaved Africans to date that led to the founding of a state. The self-liberated Haitians outlawed slavery and inspired a cascade of subsequent revolts throughout colonized lands across the globe.

the revelations of the divination with participants, explaining that even if they were not literally officers of the law and did not intend to come with arrogance, that energy was being perceived by the Land. I let people know that we would need to make a cornmeal offering to Azaka with humility and gratitude in our hearts. We would also need to begin eating with our hands. To my relief, the community understood, complied, and all was well with the Land again.

In Haitian Vodou we believe in a supreme, unknowable, and singular God, Bondye. Aspects of this Divinity are called *lwa* (parallel to Yoruba orisa) and manifest as forces of nature and spirits. Azaka, affectionately called *Kouzen* (m.) or *Kouzin (f.)* (cousin) by Haitians, is the multiple-gendered lwa of agriculture and the head of the family of earth spirits. Azaka came into existence after the Haitian Revolution when enslaved Africans were finally allowed to own land. Azaka is personified as a hardworking farmer who wears blue pants and shirt, a red neckerchief, and a woven sack, and carries a machete or sickle. Azaka walks with a limp due to their heavy workload, and has an insatiable appetite for food. Supplications to Azaka fortify the land for a bountiful harvest.[6]

To connect with Azaka, you need only to come to the farmland with humility and make offerings from your heart. Because the reality of working hard to earn a fruitful harvest is so onerous, Azaka requires abundant offerings to believe in the sincerity of the supplicant. Offerings to Azaka include corn in various forms: cassava bread, sugarcane, rice and beans, tobacco, herbs, cereal, and rum. You can place these offerings directly on the ground or dig a hole and bury them. For those initiated to Vodou, there are more elaborate communal rituals that can result in the union of the supplicant with the lwa, a process called mounting or possession. The act of mounting underscores the ultimate unity and nonduality of the universe, and the democracy of the Vodou faith; any person can unite with the Divine and receive revelation directly and personally.[7]

Azaka's counterpart in Yoruba Ifa is Orisa Oko. Orisa Oko is the orisa of agriculture, prosperity, and rural land. Orisa Oko started life in mortal form in the city of Ife-Ooye, where he was a hunter and a fisher. He rescued Yemaya Agana, the drowning daughter of Obatala and Yemoja, and subsequently married her and moved into the home of his in-laws. The people of Ife began to ridicule him for living off the fortunes of his parents-in-law, saying:

Orisa-Oko plants no melon; yet he eats its egusi [seeds].
He lives on the fortunes of his parents-in-law.
He gulps down the delectable egusi soup with no thought of family obligation.

—OGBE-ODI

To regain his good reputation, Orisa Oko consulted Ifa through divination, who told him to leave town and settle on a farm in present-day Irawo, Nigeria. Orisa Oko did as advised and became wealthy and well regarded in his new home. Then he "entered into the ground" (an expression for the attainment of immortality) and became the orisa of farming and prosperity. His shrines are often depicted with a phallic sculpture representing fertility, and his ritual objects are painted red and white, representing the fecund liquids of blood and semen, respectively. In devotees' shrines a 5- to 6-foot-tall (1.5 to 1.75 m) iron or metal staff may also be leaning against the wall.[8] Orisa Oko is one of the orisa *Funfun* (orisa of the white cloth), whose devotees wear white garments.

As one might imagine, Orisa Oko takes egusi soup as an offering. In the Diaspora, Orisa Oko also favors sweet potatoes, nyame, taro roots, corn, and dishes seasoned with palm oil and smoked fish. Orisa Oko, however, should only be given new yams at the annual yam festival, discussed below. You can leave offerings for Orisa Oko in a fertile field or a hole dug in the earth.

Due to the association of agriculture with forced labor, the veneration of Orisa Oko has dwindled in the New World. There is a danger in confusing the oppression that our people experienced on land with the Spirit of the Land itself. The truth is that we require a spiritual union with the living earth to maintain our well-being. As Baba Malik Yakini

Prayer for Azaka

Azaka Mede! Kouzin Zaka! We come to you.
You support all the believers. We call on you.
We need your help.
The doors to success are closing on us.

Azaka Mede! Hear our prayer.
Show us the path to peace and comfort.
Show us the path to success,
The path to truth, the path to dignity
The path to prosperity.

Azaka Mede! Kouzin Zaka!
Show us the means to earn a decent living.
Make us strong and disciplined.
Lead our way to what we need.
Help us become prosperous.

Azaka Mede! Kouzin Zaka!
Help us so we can help others.
Whatever we possess, may it also be of service
To those in need and the deserving.

Azaka Mede! Kouzin Zaka!
Free us from our fears.
Bring order in our lives. Give us faith.
Make us better servants of the lwa now and forever.
Ayibobo! Ayibobo! Ayibobo![9]

Prayer for Orisa Oko

Orisa Oko you are the one who cultivates the land so it yields hearty harvests. Cultivate within us a desire to live our purpose and achieve our destinies.

Orisa Oko you are the one that brings fertility to the Earth so that it may give sustenance to all the living. Allow our spirits, minds, and bodies to become fertile with the creativity needed to manifest our hearts' desires.

Orisa Oko you are the one that heals aulments that plague so many with misery and dis-ease. Allow us to become healers of our families, our communities, and ourselves and help spread light and love throughout the world.

—from *Orisha Oko—Orisha of Fertility, Progress and Evolution* [10]

Right, Soul Fire Farmers prepare an offering for Azaka each spring in supplication for a bountiful growing season.

explains, "The Earth is not just the third rock from the sun. The Earth is alive. To be whole, we need to be connected with that spiritual energy."

Planting and Harvest Rituals

A pile of newly harvested sweet potatoes, the yams of the Diaspora, rested on a white sheet to the east of our gathering space. We had harvested these yams from the cool soil to the rhythm of a drum, as is the custom at the start of Manje Yam, the Haitian festival for the eating of new yam. Now it was time to offer praise songs and adorn the yams with offerings of oil, rum, and candlelight. Offerings and prayers complete, we then covered the floor completely with banana leaves, representing the surface of the water that the magic boat of the lwas crosses to reach the holy city of Ife. Banana leaves are selected because they perpetually self-renew from the roots, representing the eternal nature of the Divine.[11]

Most of our friends were new to the ritual of Manje Yam, so they giggled a bit before taking the mystic journey across the sea to visit the land of their ancestors. One at a time we saluted the four directions and kissed the ground to the west. Then we lay down on the banana leaves and rolled ourselves toward the yams in the east, the land of Ginen, our ancestral home where the lwa would fortify us spiritually for the year ahead. After the ritual return back across the Middle Passage to receive the blessings of our ancestors, we could prepare and eat the new yam. We cooked the yam with salted fish and passed the pots of cooked *kwi* (yam) over the heads of those present.[12] The fish represented the bounty of the sea, and the yam represented the bounty of the land. Conveniently, Manje Yam takes places on or near November 25, so family

UPLIFT

Ngmayem Festival for the Millet Harvest, Ghana

Each October, the people of Manya Krobo, Eastern Region, Ghana, hold a weeklong festival to celebrate the abundant harvest of *ngma* (millet) and other crops. They offer gratitude and thanksgiving to Mau (the Creator) and supplicate for continuous bounty and protection in the coming years. The festival marks the beginning of a period of rest from the intense agricultural season and the resuming of cultural and social activities that have been on hold during the arduous months of rain, such as marriages, funerals, and reunions. During Ngmayem, the Krobo people journey to their ancestral home on Krobo Mountain (Klo yo) for spiritual renewal and then share in a community-wide harvest feast. The Koda festival, a time of planting and focus on spiritual development, precedes Ngmayem by 20 weeks.[13]

The millet harvest festival is also celebrated in the Diaspora, namely on the Caribbean island of Curaçao. The *seu* is a communal labor society built around the harvesting of the millet. The seu workers harvest the millet to the rhythm of the drum, cow horns, and *agan* (a piece of a plow fashioned into a musical instrument). They then process to the *manganzina* where the millet is stored, while singing songs. Once the work is finished, the people gather at the home of the farmer to sing and dance to the rhythm of the drum, a celebration called *seu sera*. Their dance called *wapa* mimics the movements used in planting and harvesting. Today seu is also celebrated with a national Harvest Parade, a colorful event with music and folkloric groups.[14]

members who had release time from work due to Thanksgiving were able to gather.

The Haitian harvest festival of Manje Yam is rooted in Igbo tradition. The Igbo believe that the yam is the sovereign of all crops and has a spirit called *njoku*. The veneration of the new yam marks the beginning of the harvest season. It is a time to give thanks to the Spirit of the Land for a good yield and successful harvest. During the festival, people who share the same ancestors travel to their homeland to eat the new yam together. Elders teach the children about family history. Relatives take time to settle quarrels and make peace, as the new yam cannot be consumed with bitterness. The Igbo believe that quarreling on a yam farm, throwing a yam in anger, eating yams with others whom you despise, or defecating near the yam desecrates the yam spirit.[15]

We have yet to find an exception to the harvest festival tradition among our African ancestors. To give thanks for the harvest in community with family and friends is an integral part of farming while Black. Some of the common elements that you can draw upon in convening your own harvest festival include: gathering extended family, waging peace, making prayers and offerings to the Land, renewing yourself spiritually, and enjoying the fruits of your labor with joy.

Of course, there is no reaping without sowing. It is equally important to ritually mark the beginning of the farming season with a festival. In our mixed Black-Jewish family, we have evolved the Passover seder into this ritual, affectionately termed AfroSeder. We gather to celebrate and tell stories of freedom, both ancestral and personal, and to honor our ancestors who were lost to slavery in all of its forms. We sing the freedom songs of the Black American past—"We who believe in freedom, cannot rest!"—interspersed with a reenactment of Harriet Tubman's leadership on the Underground Railroad. As the sun sets, we head outside and plant the first seeds of the season into the eager ground, infused with our prayers for freedom and abundance. We remind ourselves of the true words of Malcolm X, "Revolution is based on land. Land is the basis of all independence. Land is the basis of freedom, justice, and equality."

There are also many models for African spring planting festivals that are not syncretized with Jewish ritual. Farmers along the Niger River make

During the Manje Yam festival, we make offerings to the spirit of the yam in gratitude for an abundant harvest.

We celebrate stories of freedom and ritually plant the first seeds of spring during our community AfroSeder. Photo by Neshima Vitale-Penniman.

abundant offerings of wine and wild game to the land as part of a yearly yam planting festival.[16] In Burkina Faso farmers emerge from the *Bwa* (sacred forest) wearing masks that represent forest creatures including ox, serpent, warthog, antelope, and owl. These animal spirits purify the village and protect the community from harm, in anticipation of the farming season.[17] The International Coalition to Commemorate African Ancestors of the Middle Passage (ICCAAMP) encourages us to plant a tree each spring in honor of our ancestors and the living earth.

Herbal Baths

The use of spiritual herbal baths for healing, protection, and divine revelation is a cross-cutting African indigenous practice. We use herbal baths at transition points in the season: first planting, welcoming of new team members, first harvest, and so forth. Baths are a restorative and grounding way to fortify our connection to the land, ancestors, and our own divine light.

In the Vodou tradition the guardian of spiritual baths and all plant medicine is *Gran Bwa* (Great

Before harvesting wild plants to make a spiritual bath, we present offerings to *Gran Bwa*, guardian of the forest. Photo by Neshima Vitale-Penniman.

Forest). Gran Bwa is the lwa of the forest and owner of medicinal leaves. Gran Bwa has great healing powers and can alleviate suffering and redress harms. As the tree is the element connecting the three realms of ancestors, living beings, and spirits, Gran Bwa is one of the lwa responsible for the ancestors. Gran Bwa is only mentioned with the greatest respect. This lwa is linked to the Bwa Kayiman ceremony that catalyzed the Haitian Revolution.

The first step in making a spiritual bath is to give offerings to Gran Bwa. The offerings for Gran Bwa are leaves and plants in a *makout* (straw bag), as well as honey and *kleren* (liquor).[18] We tie a beautiful cloth around a tree and leave gifts at the base.

From there we can harvest the plants that are required for the bath. Depending on the purpose of the spiritual cleansing—to fortify power, gain love, secure protection—there are different recipes. A simple bath for purification includes just two ingredients, basil and rue.

It is also possible to directly ask and listen to the plants of the forest to determine which are appropriate to use in your bath. This direct relationship with herbs was used by our ancestors to determine the bath recipes we inherit today. We can approach a native plant, leave a small offering of cornmeal or flowers, and listen energetically to whether it is calling to be harvested. With some practice you will be able to "hear" the intentions of the plants. We uphold the law of the honorable harvest as explained by Robin Wall Kimmerer; to never harvest the first or second individual found, only the third and beyond. Never harvest more than a third of the plants.

We then activate the leaves by rubbing and crushing them in water while singing certain songs, a process called *pilèy fey*. You may sing or recite the songs to Gran Bwa below, or share the prayers of your heart in whatever form. Stay focused on your intention and your gratitude for the support provided by these Forces.

Once thoroughly crushed, the bath should be applied to the entire, unclothed body. Clean your skin first with soap and water. Pray over the medicine, give thanks, and ask for what is needed. Then apply the bath head-to-toe, from top to bottom, using about a quart of it. Rub it in vigorously and allow the little bits of leaf to stay on your body. It's best to air-dry and then rest, wearing white clothing for the remainder of the day. The color white is sacred to spirits of Dahomey origin and also reflects negative energy away from the devotee. The following are two praise songs sung for Gran Bwa.

Fèy Nan Bwa Rele Mwen!

Fèy nan bwa rele mwen!	Leaves in the woods call me!
O fèy nan bwa rele mwen!	Oh leaves in the woods call me!
O depi m piti, m ap danse la!	Oh since I was small, I have danced here!
Fèy nan bwa e, rele mwen e!	Leaves in the woods hey, call me hey!
O fèy nan bwa e rele mwen o!	Oh leaves in the woods hey, oh call me!
Ago-Tchi! Ago-Tchi! Loko-Tchi, Loko-Tchi!	Ago-Tchi! Ago-Tchi! Loko-Tchi, Loko-Tchi!
O fèy nan bwa rele mwen o!	Oh leaves in the woods call me!
O depi m piti, m ap rele!	Oh since I was small, I have cried out!
Fèy nan bwa! Rele mwen!	Leaves in the woods! Call me!
Fèy nan bwa e!	Leaves in the woods hey![19]

Nou Menm Rasin Gran Bwa

Nou menm rasin Gran Bwa . . .	We ourselves are Gran Bwa's roots . . .
Nou se papa tour bwa ki gen o	Oh we are the fathers of all the trees
Nan Gran Bwa.	In Gran Bwa.
Le Bwa Kayiman . . .	At the time of Bwa Kayiman . . .
Rasin Gran Bwa te la.	Gran Bwa's roots were there.
Boukmann o . . .	Oh Boukmann . . .
Rasin Gran Bwa nan men ou.	Gran Bwa's roots are in your hands.[20]

Songs and Chants

We are a singing people. Whether working the fields of our ancestral homelands as free bodies or toiling under enslavement, we have maintained our souls through our voices. We have used our song traditions to remind us of home, to keep our spirits high, to express our discontent, and to plan resistance and rebellion. Many of our songs are in a call-and-response format, where the caller sings a verse and then the others respond with a chorus. This format promotes dialogue, inclusion, and improvisation. While many of our songs are rooted in a specific religious tradition, others are decidedly secular. Thus, work songs are an inclusive starting point for elevating the soul energy on your farm. Below are some of our favorite work songs at Soul Fire Farm. You can listen to audio recordings of these songs at www.farmingwhileblack.org.

Hoe Emma Hoe

Caller: Hoe Emma Hoe, you turn around dig a hole in the ground, Hoe Emma Hoe
Chorus: Hoe Emma Hoe, you turn around dig a hole in the ground, Hoe Emma Hoe
Caller: Emma, you from the country
Chorus: Hoe Emma Hoe, you turn around dig a hole in the ground, Hoe Emma Hoe
Caller: Emma help me to pull these weeds
Chorus: Hoe Emma Hoe, you turn around dig a hole in the ground, Hoe Emma Hoe
Caller: Emma work harder than two grown men
Chorus: Hoe Emma Hoe, you turn around dig a hole in the ground, Hoe Emma Hoe
Caller: Master he be a hard hard man
Chorus: Hoe Emma Hoe, Hoe Emma Hoe
Caller: Sell my people away from me
Chorus: Hoe Emma Hoe, Hoe Emma Hoe
Caller: Lord send my people into Egypt land
Chorus: Hoe Emma Hoe, Hoe Emma Hoe
Caller: Lord strike down Pharaoh and set them free
Chorus: Hoe Emma Hoe, Hoe Emma Hoe, Hoe Emma Hoe[21]

Thank You for This Land

Caller: Thank you for this LAND
Chorus: Thank you for this Land (2x)
Chorus: This healing, this healing, this healing land (2x)
Caller: Thank you for this FOOD [or choose any word]

We Are an Old Family

Sung as a round

Group 1: We are an old family, we are a new family, we are the same family, stronger than before
Group 2: We honor you, inspire you to be who you are

Celebrating Me

Caller: When I think about myself, I want to celebrate my BEAUTY [or choose any word]
Chorus: I want to celebrate my BEAUTY

How Could Anyone

by Libby Roderick

How could anyone ever tell you
You are anything less than beautiful
How could anyone ever tell you
You are less than whole
How could anyone fail to notice
That your loving is a miracle
How deeply you're connected to my soul

When You Were Born

Sung as a round

When you were born you cried
And the world rejoiced
Live your life so that when you die
The world cries and you rejoice

Hold My Hand

Popularized by Dr. C. J. Johnson

Hold my hand while I run this race (3x)
Cause I don't want to run this race alone

UPLIFT
Zora Neale Hurston

In the years before the 1937 publication of her acclaimed novel *Their Eyes Were Watching God,* anthropologist Zora Neale Hurston traveled extensively in the Black American South and Caribbean during the 1930s to document folk songs, stories, and culture. She gathered these songs primarily by learning them herself. As Hurston explained, "I just get in the crowd with people as they are singing, and I listen as best as I can then I start to join in with a phrase or two. Finally, I get so I can sing a verse and keep on until I can get all the verses. Then I sing them back to the people until they tell me that I can sing them just like them. Then I try it out on different people who already know the songs until they are quite satisfied that I have it. Then I carry it in my memory. I learned the song myself, and then I can take it with me wherever I go."[22] Her extensive anthropological work, including blues, work songs, ballads, tales, dance music, circle games, prayers, and sermons, is housed in the Library of Congress's "Florida Folklife" collection.[23]

Integrating song and dance into the work of farming elevates the spirit, builds connection, and honors the living earth. Photo by Neshima Vitale-Penniman.

Chineke I Dinma (My God You Are So Good)

Igbo song from Nigeria

Chineke I dinma o
I dinma
I dinma e
Idinma o
Idinma, Idinma o

Set on Freedom

Popularized by Reverend Osby

Woke up this morning with my mind
Set on freedom (3x)
Hallelu, Hallelu, Hallelujah
I'm walking and talking with my mind
Stayed on freedom . . .
I'm singing and praying with my mind
Stayed on freedom

Omiwa

Traditional Haitian Kreyol song

Omiwa ye
Hounsi yo mande houmble
Omiwa ye
Omiwa chante

Omiwa ye
Sèvitè yo mande houmble

English translation

The devotees ask for blessings.

A Sower Went Out to Sow Her Seed

by Toshi Reagon

A sower went out to sow her seed (2x)
And as she sowed
Some fell by the wayside
And it was trodden down
And as she sowed
Some fell by the wayside
And of it the birds did eat
A sower went out to sow her seed (2x)
And as she sowed
Some fell upon the rock
And as soon as it was sprung up
It withered away
Because it lacked water
It withered away
A sower went out to sow her seed (2x)
And as she sowed
Some fell among the thorns
And as soon as it was sprung up
It withered away
There was no air to breathe
It withered away
There was no room to grow
It withered away
A sower went out to sow her seed (2x)
And as she sowed
Some fell on GOOD GROUND (2x)
From it the plants did grow
From it the flowers bloomed
And in due time
Came forth bearing fruit
A hundredfold (4x)[24]

Heart Song

Sung as a round

Listen, listen, listen to my heart song (2x)
I will always love you, I will always serve you
Listen, listen, listen to my heart song (2x)
I will never forget you, I will never forsake you

Ibo Le Le

Traditional Haitian Vodou Song

Ibo Lele, Latibonit, granmoun pa jwe o
Ibo Lele, Ibo Lele o
Ibo Lele Latibonit, granmoun pa jwe o
Si'm te la le grann mwen te la
Si'm te la le grann mwen te la
Si'm te la le grann mwen te la
Li ta montre'm danse Ibo
Ibo, Ibo Lele[25]

English translation

Ibo Lele (lwa/spirit of the Igbo people in Haiti), in the Artibonite (a river and region in Haiti), the elders don't mess around. If I were there when my great-grandmother was there, she would have taught me to dance Igbo.

Evolutionary-Revolutionary

by Amani Olugbala

Caller: I feel evolutionary
Chorus: Revolutionary!
Caller: I will [state an action commitment]
Chorus: Yes, you will!

It Is Our Duty

by Assata Shakur

Caller: It is our duty to fight for our freedom
Chorus: It is our duty to fight for our freedom
Caller: It is our duty to win
Chorus: It is our duty to win
Caller: We must love each other and support each other
Chorus: We must love each other and support each other
Caller: We have nothing to lose but our chains
Chorus: We have nothing to lose but our chains (repeat 3x, softer to louder)

A Note on Appropriation and Appropriate Use

Cultural appropriation is the taking of intellectual property, traditional knowledge, or cultural expressions from someone else's culture without permission and compensation, and in a way that reinforces historically exploitative relationships, or deprives others of the opportunity to control or benefit from their own cultural material. While cultural exchange and sharing bring wholeness, cultural appropriation is destructive. It is best to look to one's own lineage for our spiritual inheritance before adopting the practices of others.

The practices and information shared in this chapter are part of Haitian Vodou, West African Vodun, and Yoruba Ifa religious heritages. While there are different customs in each lakou, I was taught that the uninitiated may make offerings directly in nature and give themselves spiritual baths. However, only a trained, initiated leader (Mambo, Houngan, Manye, Santera, Awo, and so on) should offer spiritual services to others. This honors the tradition and our people, and ensures that practices are done in a way that is helpful and not harmful to giver and receiver. When in doubt about appropriate use, we ask the lwa directly through divination or *iluminasyon* (dream work).

The Queen Mothers of Manya Krobo Ghana are the spiritual activists of their society, entrusted with moral education for the youth, environmental protection, conflict mediation, ceremonial leadership, and griot work. In 2002 the author was enstooled as a Queen Mother and asked to carry out this work in the Diaspora.

Acknowledgments

It is our tradition to honor and pay homage to our teachers whenever we discuss spiritual matters. I give thanks to my spiritual teachers, mentors, and friends from the diverse sacred paths that inform my spiritual Truth. I cherish each of you and the precious lessons you have imparted.

- Revered Adele Smith-Penniman, my mother, Unitarian Universalist minister, and spiritual activist, who has guided me throughout my life
- Keith Penniman, my father, lay minister in the Unitarian Universalist church and spiritual activist, who has guided me throughout my life
- Nene Zogli, Manye Nartike, Manye Maku, Manye Esther, and the Manya-Krobo Queen Mothers Association, Odumase-Krobo, Ghana, for overseeing my initiation, enstoolment, and training (2002)
- Rabbi Jordan Millstein, who welcomed me as a Jew through training, mikveh, and beit din (2003)
- Rabbi Deborah Gordon, who taught me to interpret Torah and to lead Shabbat services
- Cantor Terry Horowit, who taught me trope and the chanting of Torah
- Iyanifa Patrisse Cullors, who taught me how to hear my ancestors and how to divine with obi obata
- Awo Fabayo and Iya Adekoya, for welcoming me into their spiritual house and guiding me through study of Odu Ifa and initiation to Oya
- Oluwo Ifakolade Obafemi, for training in Egungun worship, Hand of Ifa, and initiation to Oya (2016)
- Babalawo Onigbonna Sangofemi and Apetebi Okekunmi for their example of good character, Ifa knowledge, and heart-centered living, as well as for teaching me to pray in my own words
- Houngan Onelieu Wilsage, Houngan "Papa Loko," and the community of Komye, Haiti, for Milocan and ancestor ceremony, and instruction in spiritual medicine making, prayer, and divination in Haitian Vodou tradition (2012–17)
- Jun Yasuda, Japanese Buddhist nun of the Nipponzan Myohoji, for her mentorship in mindfulness and spiritual activism

CHAPTER FOUR

Restoring Degraded Land

If the yam does not grow well, do not blame the yam. It is because of the soil.

—GHANAIAN PROVERB

The first time I plunged a shovel into the mountainside soil that would become Soul Fire Farm, I was able to dig precisely 7 inches (18 cm) before hitting hardpan gray clay, virtually impenetrable to crop roots and resistant to even the metal blade of the tool. In contrast, the prime river bottom soils of the Midwest have a typical topsoil depth over 40 inches (1 m).[1] The open, south-facing portion of our land was on a steep slope, dangerous to navigate by tractor, and highly susceptible to erosion. In fact, we discovered that much of the topsoil had washed down the slope over decades and was trapped on the uphill side of the stone wall at the property's edge. The New York State Department of Agriculture and Markets ranks agricultural soils by their potential to yield nourishing crops, and our soils were listed near the bottom as "marginal, unsuitable."[2]

Farmers in Leogane, Haiti, head back to the nursery to gather more tree seedlings for their reforestation project.

UPLIFT
Haitian Farmers Remediate Eroded Hillsides

Haiti's soil was stolen by the Spanish, who cut down the trees for sugarcane plantations, and then by the French, who continued the deforestation to make way for coffee, indigo, and tobacco. Without tree roots to hold the soil in place, the earth washed into the ocean. Farmers say, "*Tè a fatige*," which means, "The earth is tired."

Peasant farmers across Haiti are leading grassroots solutions to soil erosion, with success. Haiti's forests have made a comeback and now cover 30 percent of the land area. At Soul Fire Farm we have been working with farmers in Komye, Leogane, Haiti, since the 2010 earthquake and have learned about their soil conservation strategies. They plant a densely rooted perennial grass called vetiver in contour strips on the mountainside. Once the vetiver is established, it holds the soil in place and the farmers can plant mango, moringa, and cherry trees in the stable soil behind it. Approximately every 20 feet (6 m), they dig 3-foot-deep (1 m) trenches along the contour to catch runoff and soil that inevitably washes down the hillside during the heavy spring rains. Periodically, these trenches are dug out and the soil returned to the hillside. The combination of vegetated strips and soil conservation trenches increases water infiltration and maintains soil health.

Of course, our people are no strangers to marginal lands. I worked for years in Black and Latinx communities with Worcester Roots Project in Massachusetts, remediating urban soils that had been contaminated with lead. I regularly discovered gardens, yards, and playgrounds with lead levels as high as 11,000 parts per million (ppm), which is above the threshold for a highly toxic Superfund site, as designated by the US Environmental Protection Agency.[3] Our own daughter, at just one year of age, was poisoned by soil lead. I had also seen firsthand how my family in Haiti was forced off fertile land by colonizers and made to retreat to degraded mountainsides.[4] I held in my heart the countless stories of Black, Latinx, and Indigenous people driven off prime lands by colonizers and onto the margins, from the Gullah-Geechee to the Standing Rock Sioux.[5] While our fight for the literal and metaphorical "river bottom soils" must continue, it also behooves us to know how to remediate marginal land and bring back life where it has been banished.

Because almost all wealth is inherited, and because our ancestors were among the dispossessed, many of us are not independently wealthy. This means that the land we are able to access may be contaminated with lead, eroded, or otherwise degraded. In this chapter we explore specific strategies for restoring degraded land to health, including lead remediation, no-till raised beds, terracing, and agroforestry. This chapter highlights the soil restoration techniques of the Haitian Peasant Movement, the terracing practices of Kenya, and the raised beds of the Ovambo people.

Remediating Soil Contaminated with Lead

Lead is a pretty magical element. Before science realized that it causes brain damage in children, the West was wowed by its capacity to enhance the color quality of paint and prevent "knocking" in engines. Now that it's been released into the environment, it's ubiquitous. As an element, it cannot be broken down, only moved from one place to another. In many urban areas, lead paint continues to flake off of old houses and contaminate nearby soils.

According to the EPA, soil lead levels over 400 ppm are hazardous to humans.[6] The best way to determine if your potential garden or farm site contains lead is to get a professional soil test. Your state cooperative extension or land-grant university will offer soil tests at a nominal cost; in our area it's about $12 to $20 per sample. The testing agency will give you specific instructions for how to conduct the test, but the basic process is the same. You create a "composite" sample by taking soil from a number of locations throughout your potential growing area. Dig down and gather your sample to a depth of 6 to 8 inches (15 to 20 cm) below the surface in each location; remove the rocks, roots, clumps, and debris; thoroughly mix the subsamples; spread about one cup on a clean paper to air dry; and then pack up the soil in a plastic bag and mail it to the testing agency.

If your test comes back and shows less than 100 ppm of lead, you can safely use the soil in your growing area for crops. If the test reveals that you have over 1,200 ppm of lead, it's best to call in professionals to remove the contaminated soil or find another piece of land. However, if the test reveals that your soil is in the sweet spot between 100 and 1,200 ppm of lead, you can try phytoremediation or immobilization.

Phytoremediation is the process of using plants to remove contaminants from the soil. It takes at least one year and very careful monitoring, so it's not for everyone. If you decide to try phytoremediation, you need to first make sure you have at least five months of above-freezing weather ahead of you so that the hardworking plants can live out their full life cycles. The next step is to fence off the area and put up signs indicating that it's unsafe to touch the soil during the phytoremediation process. You need to protect yourself and your loved ones from interacting directly with contaminated soil. Identify designated shoes, clothing, and gloves for this project. This contaminated work attire should be bagged up at the end of each use and stored away from children.

Soil from 12 points around the field is collected and mixed in a bucket to achieve a composite sample.

Always change into clean clothes before going back into your home.

Once you are ready to work with the soil, you need to chelate the lead, which means to make it more mobile by acidifying the soil to a pH of 6.0 (see chapter 5 for more on soil pH). The fastest way to acidify soil is to add elemental sulfur in the quantities described in table 4.1. Sulfur is also sold as ammonium sulfate, iron sulfate, and aluminum sulfate, which are slower-acting compound forms of the element. It's also possible to acidify soil by using pine bark mulch, urine, or peat moss, but these methods require a lot more time and material.[7] Apply your amendments and gently fork in to a depth of 6 to 8 inches (15 to 20 cm). Ideally the sulfur is added months before planting, but you can get away with planting your phytoremediation crops four weeks after applying the acidifying agent.

Table 4.1. Acidifying the Soil with Elemental Sulfur

Current Soil pH	Pounds of Elemental Sulfur Needed to Reduce pH to 6.0, Per 100-Square-Feet (9-square meters) Garden Area
7.5	3.5 (1.5 kg)
7.0	2.0 (1.0 kg)
6.5	1.0 (0.5 kg)

Source: Sandra Mason, "How to Lower Soil pH," in *The Homeowners Column*, University of Illinois Extension, July 5, 2008, http://web.extension.illinois.edu/cfiv/homeowners/080818.html.

The best plants for lead phytoremediation in our trials at the Worcester Roots Project included scented geranium (*Pelargonium* spp.), sunflower (*Helianthus annuus*), Alpine pennygrass (*Thlaspi caerulescens*), and mustard greens (*Brassica* spp.).[8] Pelargonium is an African indigenous crop with a rich, enduring scent. During our years working on urban soil remediation in Worcester, our bedroom was completely taken over with cuttings of this sacred plant. The growing requirements for pelargonium and the other bioaccumulators are listed in table 4.2. Sow these accumulating plants densely throughout the growing area, give them 1 to 2 inches (2.5 to 5 cm) of water per week, and wait until they are fully mature or your growing season has ended. Remember that you have now mobilized soil lead through chelation, so your growing site is temporarily more hazardous than it was before you acted. Make sure that neighbors, especially children, stay out. Continue to wear designated attire whenever working onsite.

At the end of the season, harvest the accumulator plants and put them into double trash bags. Check with your local waste disposal facility to determine the proper procedures for disposing of lead-contaminated plants. Most municipalities have a hazardous waste collection area. Do not compost these plants! Again, they are full of lead, and lead cannot be broken down; it's an element.

Now the moment of truth. It's time to send your soil in for another test. Be sure to use the same procedure for your composite sample so that you are accurately comparing your pre- and post-remediation contamination. If your soil lead levels are now below 400 ppm, you can go ahead and use the area for growing crops. If lead levels are above 100ppm, it's best to use at least 1 inch of phosphate-rich compost in all of your growing areas. Phosphorus binds with the lead and forms a compound that is less water-soluble.[9] If someone were to accidentally ingest lead from phosphorus-rich soil, it's more likely that the toxins would just pass through the body rather than penetrate the blood–brain barrier and cause neurological problems. Poultry manure, both fresh and composted, is by far the richest organic source of phosphorus, with cow manure a close second.[10] Apply composted manure at 1 to 3 inches (2.5 to 7.5 cm) per year. If you use fresh manure, allow 120 days to pass before planting food crops. In lieu of manure, you can also use rock phosphate, an

Table 4.2. Growing Requirements for Bioaccumulators

Name of Plant	Direct Seed or Transplant?	Spacing—On Center	Growing Considerations
Scented geranium	Transplant	12 inches (30 cm)	Full sun, well-drained soil, not frost-tolerant
Sunflower	Direct seed	Seed at 6 inches (15 cm), thin to 18 inches (45 cm)	Tolerant of light frost, prefers full sun, tolerant of poor soil
Alpine pennygrass	Direct seed	Seed at ½ inch (1.25 cm), thin to 3 inches (7.5 cm)	Tolerant of poor soil, tolerant of light frost
Mustard greens	Direct seed	Seed at ½ inch (1.25 cm), thin to 3 inches (7.5 cm)	Prefers cool weather or partial shade, well-drained soil

The sunflower is a hyperaccumulator and can trap lead in its tissues. Photo by Neshima Vitale-Penniman.

organically approved mineral dust, at 10 pounds per 100 square feet (4.5 kg per 9 square meters).

If your soil still has lead levels higher than 400 ppm, you can try another round of phytoremediation, or you can take measures to isolate the contaminated soil from contact with plant roots or human feet. This latter process is called immobilization. If you choose immobilization, all of the areas that will become walking paths or crop beds need to first be covered with a layer of landscaping fabric, which is impenetrable to roots but porous to water. This fabric will hold the contaminated soil in place where it can do no harm, while you bring in layers of new soil for growing.

Just as our Ovambo relatives in Namibia and Angola practice, you will grow your crops in raised beds to avoid contact with the contaminated soil. There are many right ways to build raised beds, and the simplest uses 2-by-12-inch (5-by-30 cm) wooden planks, framing angles, and fasteners. If you can source them, oak or cedar planks are best, because they are rot-resistant. We often settle for hemlock,

UPLIFT

Raised Beds of the Ovambo

The Ovambo people of Northern Namibia and Southern Angola are clear that soil fertility is not an inherent quality, but rather something that is nurtured through mounding, ridging, and the application of organic matter. When the colonial government attempted to force the Ovambo farmers off their land, offering them equivalent plots with "better quality" soil, the farmers countered that they had invested substantially into building their soils and doubted that the new areas would ever equal their existing farms in fertility. The practice of the Ovambo is to demarcate the field, clear brush, then build raised rectangular mounds about 10 feet (3 m) long, 5 feet (1.5 m) wide, and 1 foot (0.3 m) tall. The pathways double as irrigation ditches. The Ovambo farmers add ample manure, ashes, termite earth, cattle urine, muck from wetlands, and other organic matter to increase the fertility of their mounds. This system concentrates fertile topsoil, aerates soil, and prevents waterlogging. The Ovambo also integrate a rotating fallow after a few seasons' harvest of millet. During the fallow, cattle and goats graze the brush of the resting cropland and deposit additional manure and urine to replenish the soil.[11] The Susu-Jalonke speakers of Guinea also planted their crops in mounds to decrease soil erosion. The practice of mounding the earth to grow tubers and other crops is common throughout West and West Central Africa.[12]

because it is affordable and more durable than pine. Avoid using railroad ties or pressure-treated wood, because they contain arsenic and creosote, which can leach into your soil. You can make your raised bed any length, but keep your width to a maximum of 5 feet (1.5 m) so that you can easily reach to the middle for weeding and harvesting. An 8-by-4-foot (2.5-by-1.25 m) bed is a convenient size, since lumber is often sold in 8-foot lengths. Arrange your wooden planks in a rectangular design and use the framing angles and fasteners to hold then together. You can also make your raised bed out of cement blocks, stones, or logs from the forest. Once your structure is built, fill it almost completely with a blend of 50 percent topsoil and 50 percent compost. A raised bed that is 8-by-4-by-1-foot (2.5-by-1.25-by-0.3 m) holds exactly

Rosa and Rae construct a raised bed using hemlock boards and framing angles as part of the Soul Fire in the City initiative.

Youth at the Sherman Park Community Garden construct new raised beds. Photo by Camille D. Mays.

one cubic yard (0.75 cubic meter) of soil. Many cities have a municipal composting program where you may be able to get free or low-cost compost. Call your city parks department or community gardens coordinator and ask for support. In the case where you do not have ready access to compost, you can fill the raised bed halfway with raked leaves and then top it off with only ½ cubic yard (0.4 cubic meter) of the compost-soil blend. This reduces the expense and hassle of acquiring so much soil.

The pathways between the raised beds need to be thickly mulched with wood chips or other organic material to prevent foot traffic from stirring up contaminated soil. There is a common misconception that most exposure to soil lead happens by eating plants that have absorbed this lead and stored it in their edible parts. In fact, most soil lead contamination comes from ingesting the soil itself. People get soil trapped in the treads of their shoes and under their fingernails and then bring that lead back into their homes where children become exposed. To minimize this risk, the pathways should be covered with landscaping fabric and at least 2 inches of wood chips, pea stone, or other mulch material.

On the garden edges and in between places, it's acceptable to grow fruiting bushes and trees directly into the contaminated soil. The fruiting parts of the plants accumulate the least lead and pose little human health risk. Again, direct exposure to soil is the highest risk, so the soil around the bushes and trees should be covered with landscaping fabric and a minimum of 2 inches (5 cm) of mulch material.[13] When working

This community garden in Frederick, Maryland, immobilizes contaminants using landscaping fabric and wood-chip mulch. Photo by Natasha Bowens.

in soils with over 400 ppm of lead, remember to use designated shoes, clothing, and gloves that are stored separately from your everyday clothes.

Healing Erosion with Terraces

In the rural landscape, lands that we are likely to be able to afford may be sloped, leached, eroded, rocky, and otherwise marginal. When George Washington Carver arrived at Tuskegee in 1897, he observed that "much of the topsoil had been carried away by a stream that cut across the property." He and the students filled in the gullies and ditches with pine tops, hay, bark, old cotton stocks, leaves, and anything that would decay and eventually become soil. By 1905 Carver reported that "the injurious washing has been almost completely overcome."[14]

UPLIFT
Terraces of Kenya

The practice of terracing sloped farmland is an agricultural technique perfected by Indigenous farmers in several regions of the world. In East Africa this practice is called *fanya-juu*, meaning "throw it upward." In practice, farmers revive degraded land by recovering soil from the bottom of the slope and throwing it upward to form terraces, or steps of roughly level land. The practice of fanya-juu results in soil organic matter levels 35 percent higher than found on conventional farms, comparable to natural ecosystems. This practice also results in 25 percent higher crop yields than from conventional farms. Atmospheric carbon dioxide is sequestered in the terraces, mitigating climate change.[15]

The soils at Soul Fire Farm were also severely eroded when we arrived, and we had to dedicate the first couple of growing seasons just to getting the soil back in place. Local farmers in our community confidently informed us that it was "impossible" to grow decent crops given the condition of our soil. Fortunately, we are stubborn. After years of employing African ancestral land restoration processes, namely terracing and no-till mounding, the quality and nutritional value of our crops is excellent.

Knowing that we had an upcoming community volunteer day and many of our experienced farmer friends from Oaxaca, Mexico, would be in attendance, we decided to start our agroforestry terracing project. The highly sloped areas on the farm with a grade over 15 percent were not safe to navigate with a tractor and susceptible to erosion. To bring these areas into food production, we needed to essentially create an exaggerated staircase that shapes the soil into a series of flat planting areas suited for trees or annual crops. We decided to lay out terraces 25 feet (7.5 m) apart from one another, allowing plenty of space for fruit trees to spread their branches.

Terraces need to be built in line with the contour of the hill. Of course, you can use a transit level to determine elevations and stake out the terraces. However, our ancestors made use of clues from the environment to accomplish the same goals. Without any fancy tools, it is possible to accurately lay out the contours. Trees generally grow straight up and can be used as a reference point; the terraces should be perpendicular to the trees. You can also walk along the edge of your future terrace, tracing a path where you neither climb nor descend, but rather maintain your elevation. For a more accurate determination of slope, you can invite two friends over and use a string and level as described in the sidebar. Once the contour lines are determined, use wooden stakes, approximately 4 feet (1.25 m) apart, to mark and identify where each new terrace will be built.

The secret to making durable terraces on marginal land is organic matter. In our case, we observed that soil had washed down the hillside and accumulated against a stone wall at the bottom of the field. We decided to reclaim this precious resource and dedicated several

Finding Slope

To find the slope of your hill, you need a line level and string, and usually three people.

1. Start at the high point. Person 1 stands at the high slope and holds one end of the string at the ground.
2. Person 2 walks to the low point of the slope with the other end of the string, holding it tight.
3. When the line level and string are level, person 2 measures the distance from where they are holding the string to the ground.
4. Person 3 measures the length of the string from person 1 to person 2.
5. Now you have the measurements you need to calculate the slope. Divide the vertical distance (person 2's measurements) by the horizontal distance (person 3's measurement), then multiply by 100 to get percent.
6. To create a contour line, work your way across the land, keeping both the vertical and horizontal distances equal between terraces. In our case, on a 16 percent slope, terraces were built 25 feet (7.5 m) apart (horizontal) and 4 feet (1.25 m) apart (vertical).

A line level and a string can be used to determine slope. Photo by Jonah Vitale-Wolff.

days to schlepping the soppy earth uphill. In Kenya this practice is called *fanya-juu*, meaning "throw it upward" (see the Uplift sidebar). Other sources of soil include the muck from the bottom of ponds, topsoil or compost from municipalities, or topsoil suppliers who sell by the yard. Regardless of its source, pile the reclaimed soil behind the stakes that mark your contour lines, until there is enough earth for a level terrace. In the absence of an abundance of soil, you can use undecomposed organic matter. Gather leaves, sticks, grass, wood chips, hay, or other organic material and pile it up behind your contour line. Get your friends to help you stomp it down and compact it, then cover with several inches of topsoil or compost until the organic debris is no longer visible. It will take several months to a year for the debris to break down, but shallow-rooted crops can be grown in the interim.

Once the stakes and soil are in place, the next step is to weave your terrace basket. Gather ample branches and sticks from the forest and bring them up to the terrace. Using the stakes as uprights, weave the branches in and out to form the terrace wall that will hold the soil in place. The terrace wall should extend to the height of the bottom of the next terrace. The tighter the weaving, the more powerful soil conservation potential it will have.

With the terrace wall secure, use shovels and hoes to shape the soil into a flat bed. Pull soil toward the terrace wall and backfill completely. Pull soil up to the bottom of the next wall and use your skills in contour determination to get the soil as level as possible. To hold the soil in place, it's best to plant a fast-growing, strong-rooted cover crop right away, like buckwheat, rye, or oats. If you used pure soil rather than a mix of soil and forest debris, you may be able to plant fruit tree saplings immediately.

Agroforestry for Soil Restoration

The forest is a superorganism comprised of trees that "talk" to one another using the internet of fungal mycelium, which quite literally send warning

Volunteers at our *konbit* (community work day) helped build terraces on the sloped areas of the field.

UPLIFT

Sahel Reforestation

In the 1980s Yacouba Sawadogo of Burkina Faso revived a traditional tree-based approach to farming called *zai* or *tassa*. This technique uses pits dug in the preseason to catch water and concentrate organic matter. Sawadogo innovated on the genius of his ancestors by filling the pits with manure and compost to attract termites, whose tunnels further decompose the organic matter.

At the time there was a mass exodus from his community after severe droughts slashed food production and turned the savanna into a desert. Millions of people starved. However, Sawadogo's father was buried on his farm. Leaving was not an option.

Sawadogo dug zai pits to capture rainfall and nutrients for his millet and sorghum crops. He found that his grain yield increased; additionally, native trees started to grow out of the zai. These trees anchored the soil, buffered the wind, and helped retain soil moisture. They also provided mulch for the crops and fodder for the livestock. As others adopted Sawadogo's technique, water tables across the Sahel began to rise for the first time in decades. Sawadogo explained, "My conviction, based on personal experience, is that trees are like lungs. If we do not protect them, and increase their numbers, it will be the end of the world."[16] His leadership is transforming the Sahel desert into a landscape of green.

Fruit trees are planted in early spring as soon as the ground can be worked.

messages, share carbon and minerals, and take turns helping one another out when environmental conditions are rough. Not only is the cooperation of the forest a profound guide for how we need to exist in human community, it's also a practical survival strategy. We want our cultivated lands to be part of that network with the native forests. One of the best ways to do that is to establish an agroforest between the wild forest and your annual crops.[17] Our sibling farmers in Haiti understand the value of agroforestry systems and plant *jaden lakou* (courtyard gardens) in every compound. Jaden lakou are planted with vegetables, herbs, and fruit trees in a closely intertwined system. Livestock provide manure for fertilization, and kitchen wastes are composted to enrich the soil.

We have made a few missteps in our attempts to establish a jaden lakou agroforest, and are finally homing in on the best methods. We plant fruit trees on the well-drained terraces, including apples, peaches, plums, Asian pear, persimmons, cherry, and apricot. In the marshy wet areas at the bottom of the hill, we have a row of happy elderberries. Closer to the house we grow the berries that require more frequent picking and care, including blueberries, raspberries, strawberries, jostaberry, juneberries, and honey berries. We also have a hedge of hazelnuts that just started producing fruit, and trellises laden with hardy kiwis and seedless grapes. Interspersed with these woody perennials we have medicinal and pest-preventing herbal plants as well as a thick layer of wood-chip mulch to encourage the fungal mycelium to connect with the tree roots.

Table 4.3. Spacings for Fruit Trees and Vines

Fruit Tree	Spacing (feet)
Apple, semi-dwarf	15–20 (4.5–6 m)
Apple, standard	20–25 (4.5–7.5 m)
Apricot	15–20 (4.5–6 m)
Elderberry	4–6 (1.25–1.75 m)
Gooseberry	3 (1 m)
Grape	8 (2.5 m)
Hardy kiwi	10 (3 m)
Hazelnuts	4–6 (1.25–1.75 m)
Honey berry	3-6 (1–1.75 m)
Juneberry, large	10–15 (3–4.5 m)
Juneberry, medium	6-8 (1.75–2.5 m)
Nut trees	35 (10.5 m)
Pawpaws	15–25 (4.5–7.5 m)
Pear	20 (6 m)
Pear, Asian	20 (6 m)
Persimmon, American	35–50 (10.5–15 m)
Plum, hybrid	10–15 (3–4.5 m)

If you choose to plant trees, it's best to do so early in the spring, as soon as a hole can be dug. If you wait until too late in the season, the plants will be unduly stressed. Prepare the ground in the fall, months before planting. Pile a full wheelbarrow load of well-aged compost over the area where the tree will go. Additionally, add limestone (if needed) to bring the pH to 6.5. In the spring dig a hole right into the compost, twice the width and about as deep as the tree roots. The tree should be planted at the same depth that it was in the nursery; look for a dirt mark or change in bark color to indicate that point. "Puddle in" the young tree by adding 2 to 3 gallons (8 to 12 L) of water to the hole. Use the soil around the tree to form a berm that will direct water toward the tree roots. Add a thick layer of wood chips to mulch the tree, allowing a 6-inch (15 cm) mulch-free circle around the trunk itself to discourage rodents. In the interim years before the tree roots establish and hold the terraces in place, it's crucial to maintain your mulch layer and fast-growing cover crops around the trees.

Around and between the trees on our terraces, we plant concentric circles of perennial herbs that have a beneficial relationship with the trees. Some of our favorite companion plants for the orchard are:

Chives. Plants in the allium family attract pollinating insects, discourage burrowing animals, and repel insects.

Sage. Sage grows horizontally to provide a satisfying ground cover, and its aroma wards off pests.

Mint and lemon balm. Members of the mint family are powerful insect repellents and contain organic insecticides. The citronella in lemon balm even deters mosquitoes.

Comfrey. A dynamic nutrient accumulator, comfrey has a knack for mining the subsoil for minerals and bringing them to the surface to improve the topsoil. Comfrey can be cut back several times in a season and its large leaves used to mulch the trees.

We also grow chamomile, echinacea, bee balm, arnica, skullcap, and other medicinal herbs in the understory of the orchard. The care and use of herbs will be discussed in more detail in chapter 10.

The first time we attempted to plant fruit trees, we naively provided them with no pest protection and lost almost all of our orchard to deer browsing. Young fruit trees have a tender, tasty inner bark that is irresistible to rodents and deer, especially in the long, sparse winters of the North. It is important to put a small cage of mesh hardware cloth around the base of the tree, buried 1 inch into the ground. This prevents burrowing rodents from girdling and killing your tree. Further, a 6-foot-tall (1.75 m) 2-by-4-foot (0.6-by-1.25 m) welded wire pen encircling the entire tree is recommended in areas with deer.

There is a Chinese proverb that says, "The best time to plant a tree was 20 years ago. The second best time is now." I would like to qualify this wisdom by adding that the planting of small fruits now to complement the larger, slow-growing trees gives the farmer something to eat in the interim. Raspberries, blueberries, and strawberries are comparatively easy to grow and fruit within just a year or two.

Strawberries. Strawberries love to be planted in areas that have already been gardens for a couple of years and have lots of organic matter. They can be planted in early spring or in the fall, provided they are given a thick mulch layer to get them through the winter. Space strawberry plants 10 to 12 inches (25 to 30 cm) apart on center and mulch thickly with straw or leaves. The mulch helps the soil retain moisture, smothers weeds, retains soil heat, and keeps the berries clean. In the first year remove the blossoms so that the plant puts

Medicinal and culinary herbs are planted in the understory of the *jaden lakou* agroforest. Photo by Neshima Vitale-Penniman.

A wire-mesh tree cage protects the sapling from girdling by rodents.

energy into root and foliage growth. The heaviest berry production will be in year two. After three years the berry production drops and it's time to rotate the strawberries to a new area. Strawberries reproduce by sending out runners. You can harvest these and plant them each year to ensure an ongoing supply. At all times, we have one-year, two-year, and three-year strawberries rotating through different beds.

Blueberries. Blueberries love naturally acidic soil with a pH of 3.5 to 4.5 and a heavy mulch of pine needles, wood chips, or sawdust to encourage fungal growth. Plant blueberries early in the spring. Depending on the variety, the plants are spaced 4 to 6 feet (1.25 to 1.75 m) apart. Their roots grow shallow and require oxygen, so they are very sensitive to disturbance of the soil at their base. Take care when weeding not to expose these tender roots. With proper pruning and care, blueberry plants live to be 40 to 60 years old.

Raspberries. Raspberries thrive in a wide variety of soils but prefer a rich, well-drained loam with a pH of 5.5 to 6.5. Plant them in the spring in rows 2 to 3 feet (0.6 to 1 m) apart. Use 6 to 8 inches (15 to 20 cm) of mulch in the pathways, and paper mulch or no mulch in the row. The root crowns must have access to light, so take care not to plant the raspberries too deeply. Raspberries produce fruits on a two-year cycle, so canes that have borne in their second year can be selectively pruned in the fall or winter. Alternatively, half of your raspberry patch can be mowed each fall in an alternating pattern. After seven years, it is best to restart your raspberry patch in a new area.

We grow other less common fruits at Soul Fire Farm, from pawpaws to hardy kiwi. While the details of the cultivation of these crops is beyond the scope of this volume, we recommend Michael Phillips's book *The Holistic Orchard* and Lee Reich's *Uncommon Fruits for Every Garden* for further details. We have also found that the tree and shrub planting guides of Fedco Trees and St. Lawrence Nurseries have ample information for the novice orchardist.

No-Till and Biological Tillage

Digging, stirring, overturning, and other methods of mechanical agitation are known collectively as "tillage." While tillage is an effective way to manage weeds in the short term, we have completely fallen in love with alternative strategies, such as tarping and sheet composting. How could any farmer say no to the opportunity to turn our soil into a carbon sink, reduce labor, and have a perfectly weed-free area in which to sink our crop seed? The major downside to our commitment to minimal tillage is the challenge of wind wrestling. Even on a day when the air feels calm on the skin, laying out 100-by-50-foot (30-by-15-meter) silage tarps looks more like parasailing than farming. We probably move several tons of rocks from the stone walls to hold our tarps in place, and we are happy to do so because the end result stirs our hearts.

In the case of marginal soils, digging down is not prudent; rather, you build up. This building-up process is also known as no-till and has the incredible side benefit of slowing climate change. In fact, if all farmers adopted the practices described in this section, we could capture 106.25 gigatons of carbon into the soil in less than five years, which would halt climate change.[18]

Sheet Composting

For small areas of the land with a slope less than 15 percent, you can grow annual vegetable crops in raised beds using this no-till process. To get started, the first step is to cut the grass or shrubs as short as possible using a push mower or machete. Next, establish layers of organic matter on top of the cut grass for the dual purpose of smother-killing the sod and building up the soil. Add a thick layer of paper (newsprint, craft paper, cardboard, or feed bags all work) over the entire planting area to prevent any light from reaching the ground. As you spread the paper, place shovels of soil at regular intervals to prevent the paper from blowing away. Then add 2 to 7 inches (5 to 18 cm) of compost on top of the

An opaque tarp helps smother weeds without tillage. Photo by Neshima Vitale-Penniman.

paper, taking care to distribute it evenly. Finally, add a 12-plus-inch (30-plus-centimeter) mulch layer, placing grass clippings, hay, straw, leaves, and/or shredded paper on top. Allow the organic matter sandwich to decompose for three to six months, over winter if you have that season in your region. In the spring, or once the paper has decomposed and the grass is killed, it's time to dig the beds. We use shovels to dig 4-foot (1.25 m) beds on contour with 1-foot (30 cm) pathways. Essentially, the pathway is a shallow trench, 6 to 8 inches (15 to 20 cm) deep. Shovel out the pathway and place that rich, fresh soil onto the beds on either side of the pathway.

Tarping

We found that this no-till method works well for growing areas up to ½ acre (0.2 ha) with two strong, skilled laborers committed to the task. For larger areas, consider tarping or cover cropping.[19] We like to thoroughly prepare our beds before tarping. This means that we dig out the pathways and turn them up onto the beds to give them a nice shape. We then add 1 to 2 inches (2.5 to 5 cm) of compost over the top of the soil. Finally, we cover the crop area with large black silage tarps, weighing them down with ample large stones along the edges and in each pathway. The maximum size tarp that we recommend is 50 feet by 100 feet (15 meters by 30 meters) because the larger tarps are nearly impossible to move by hand. Water pools on the larger tarps, as well, and can cause a "parking lot" effect where the weight of the water compacts the soil beneath. If you are unable to purchase silage tarps, you can use old billboards, black garbage bags, camping tarps, landscaping fabric, or any other opaque cover that will not disintegrate under UV light.

We leave the tarps on for a 6-plus-week period in the cooler weather, or 3-plus weeks in the hot seasons to thoroughly destroy weeds and weed seeds. In northern climates, we recommend against tarping over the winter if you need to remove the tarps for early spring planting, as this serves only to promote perennial weed growth. Assuming that you leave the

Sheet mulching with paper, compost, and straw is a no-till method for preparing your growing area. Photo by Jonah Vitale-Wolff.

tarps on for long enough in warm weather, the weeds germinate and then encounter light deprivation and die back. The weed seeds in the top few inches of soil cook slowly in the heat trapped by the tarp. The amount of time recommended to leave the tarp in place varies depending on the season and the climatic region. You are encouraged to experiment to determine the minimum amount of time necessary to achieve a growing area free of weed seeds. After you remove the tarp, the bed should be ready for planting immediately. While a light raking is fine, take care not to turn over the soil after tarping as you do not want to bring up weed seeds from the subsurface.

Cover Cropping

Cover crops are planted to feed the soil, which in turn feeds our people. The best cover crops for no-till systems naturally die back in the winter and are high in biomass so that they can serve as mulch for the vegetable crops. Our preferred cover crops for this purpose are buckwheat, oats, field peas, soybeans, and millet.[20] We plant these tender soil feeders in mid- to late summer, so they have time to grow to maturity before winter. The crops die back over the winter, and we can plant spring vegetable transplants directly into the rich mulch layer they leave behind. If weeds are properly managed throughout the season, you may not need to repeat the tarping with black plastic. More details on cover cropping can be found in the next chapter.

The commonality in these three no-till methods is that they never leave the soil bare and eliminate turning over of the soil. As our ancestors taught us, we learn about the Earth by observing and imitating her. The soil in the forest is never exposed; rather it is covered with humus, leaves, and growing plants. The forest does not stir up its earth; rather it enriches from above. In restoring degraded soils, it is essential that we heed these lessons.

CHAPTER FIVE

Feeding the Soil

Something slow moves through him, watched by hills.
Something low within each rock receives
His noonday wish, then crumbles rich; so fills
Each furrow that the prairie year upheaves.
His arm has lain with boulders. His copper hand
Has mused on roots, uncaring of barbed wire.
His fist has closed on thistle, and dug the land
For corn October snows have whelmed entire.
Something flows within him in stubborn streams,
And in the parted foliage something lives
In upright green, stirred by the rhythmic gleams
Of his hoe and spade. From worn-out arms he gives;
The earth receives, turns all his pain to soil,
Where he believes, and testifies through toil.

—JAMES A. EMANUEL, "For a Farmer"

Three distinct mounds of earth rested on empty feedbags in the middle of our circle of benches. We were gathered for a soils class during the Black Latinx Farmers Immersion, and the participants held notebooks in ready hands. "Put your notebooks down and go interact with the soil, find out its secrets. Use all of your senses. Get close," the facilitator encouraged.

We looked, touched, and the braver among us smelled and tasted. After only five minutes we produced a detailed description of each soil. The first was gray, compacted, dense, waterlogged, sour tasting, and brittle. The second soil was friable, dark brown, slightly sweet in taste and smell, moderate in density, and sticky. The third soil was undecomposed, structureless, black, and smelled richly of humates. The participants, none of whom had studied soil, were able to accurately predict which would be best for growing annual crops and which needed amending.

We revealed that the first sample was the soil we found on this land when we arrived in 2006—heavy clay, rocky, impenetrable to tools. The third soil was what we found in the top layers of the forest—high in humus, rich, and young. The second soil was the one we created in partnership with nature over the years—a high-nutrient clay loam exploding with organic matter. We explained that the restoration of organic matter to the soil was part of healing from colonialism.

As European settlers displaced Indigenous people across North America in the 1800s, they exposed vast expanses of land to the plow for the first time. It took only a few decades of intense tillage to drive over 50 percent of the original organic matter* from the soil into the sky as carbon dioxide. Rich prairie loam soils that once held 8 percent organic matter or more were reduced to less than 4 percent on average.

* Organic matter includes the living microorganisms in the soil ecosystem and the decayed remains of those organisms. Organic matter endows the soil with a healthy structure, nutrient richness, and water holding capacity.

There is a visible difference between the soils we have not amended with compost and organic matter (*left*) and those that we have stewarded for years (*right*).

The productivity of the US Great Plains decreased 71 percent during the 28 years following that first European tillage.[1] The initial anthropogenic rise in atmospheric carbon dioxide levels was due to that breakdown of soil organic matter. Cultivation and land clearing emitted more greenhouse gases than the burning of fossil fuels until the late 1950s.[2]

Restoring organic matter is a slow process, taking at least a decade to heal and only a year or two to undo. An acre (0.4 ha) of soil 6 inches (15 cm) deep weighs around 1,000 tons (900 MT), so increasing organic matter by 1 percent is a 10-ton (9 MT) change. It would be relatively easy if you could just add 10 tons of organic matter (straw, compost, leaves) and achieve that change, but the truth is that only 10 to 20 percent of what you add stays in the soil; the rest is broken down and released as carbon dioxide. So that amount of organic matter must be added year after year for it to return to precolonial levels.[3]

In this chapter we discuss how to care for the soil over the long term. We begin by exploring methods for testing the soil health, discuss how to make and use compost, share ideas for creating teas and inoculants that feed soil biota, and explore cover-cropping strategies that restore organic matter and nutrients to the soil.

Soil Tests

We can tell a lot about soil using only our senses and direct experience. Soil that is ready to receive our seeds is dark, moist, sweet swelling, crumbly, and teaming with biota. Soil that needs more love before we ask it to produce might be heavy with clay or depleted with sand, sour tasting, and devoid of life. One of the simplest low-tech soil tests is "texture by feel."

Soil texture refers to the size and type of mineral particles in the soil. Soil's inorganic portion is made

Adding organic matter to the soil is part of decolonizing our relationship to land. Photo by Jonah Vitale-Wolff.

One of the steps in the texture-by-feel determination is to form a ribbon with the soil and measure its length.

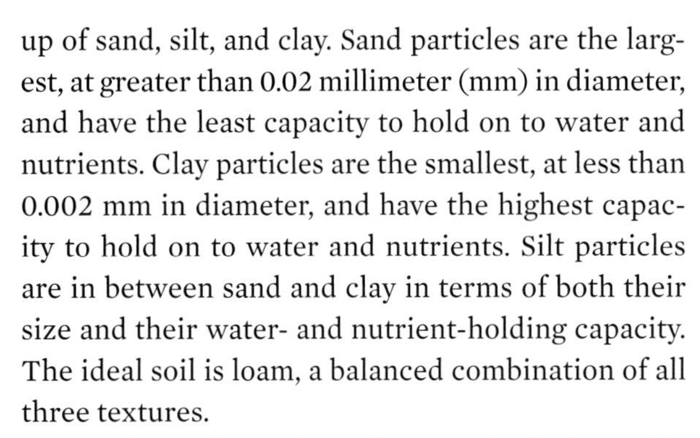

up of sand, silt, and clay. Sand particles are the largest, at greater than 0.02 millimeter (mm) in diameter, and have the least capacity to hold on to water and nutrients. Clay particles are the smallest, at less than 0.002 mm in diameter, and have the highest capacity to hold on to water and nutrients. Silt particles are in between sand and clay in terms of both their size and their water- and nutrient-holding capacity. The ideal soil is loam, a balanced combination of all three textures.

In order to determine soil texture by feel, you follow three basic steps that are reminiscent of making mud pies during childhood. First, see if a handful of the moistened soil can form a ball. Then determine if you can flatten the soil into a ribbon that can support its own weight, and measure the length of that ribbon. Finally, rub a bit of soil between your fingers and listen for its grittiness. The precise steps for texture by feel can be found on the Natural Resources Conservation Service website.[4] The end result is a determination as to which of the 12 soil textures you have. If your soil is not loam, you will need to improve the texture by adding organic matter, such as compost and mulch.

At Soul Fire Farm we have very heavy clay soils, which is a blessing in times of drought, but also a challenge in terms of water management. A few tips for managing clay soils:

- Add 1 to 2 inches (2.5 to 5 cm) of finished compost annually to keep organic matter above 8 percent. Concurrently, monitor the levels of phosphorus, potassium, magnesium, and calcium to avoid overnutrification. Cover crops can be used instead of compost once optimal nutrient levels are attained.
- Use permanent or semi-permanent raised beds so water drains into the pathways and infiltrates.

UPLIFT
Indigenous Soil Testing in Africa

For millennia farmers in Africa have developed intricate processes for testing soils and nomenclatures for soil classification. For example, the original name of Egypt was Kemet, meaning "alluvial dark and fertile soils," and the word *deshret* in ancient Egyptian was the name of desertic red soils.

Farmers in Africa classify their soils primarily based on productive capacity. Farmers test the soils by looking at environmental factors such as vegetation and fauna; topsoil morphology properties such as color, texture, density, and taste; and secondary factors including slope, workability, and stickiness.

Farmers rely on topsoil color and crop performance to assess nutrient levels. Topsoil's color reflects its fertility level, as dark soils are generally more fertile than red, white, or beige soils due to their higher organic matter content. Farmers in Niger distinguish three color classes and relate these to land degradation: *labu biri* (black soil), which is most fertile and contains relatively high levels of organic matter; *labu kware* (white soil), when through cultivation and erosion valuable nutrients are depleted; and *labu kirey* (red soil), the result of further degradation.

Texture is important to farmers to predict soil moisture-holding capacity. For instance, the Yoruba people in Nigeria rub soil between two fingers to tell whether it is *Yanrin* (sandy), *Bole* (clay), or *Alaadun* (loamy), or textures in between such as *Bole alaadun* (loamy clay). In the Mooré language (Burkina Faso), *Bissiga* means sandy soil, and is usually used for millet and vegetable crops; *Bolé* denotes clayey soils, suitable for sorghum and other crops that need more moisture. In terms of soil management, soils with coarse texture are easier to work with the hoe. They have a high infiltration rate but a low water-holding capacity.

Farmers use taste and smell to assess soil acidity and salinity. For example, farmers in Malaysia categorize soil on the basis of taste into *tanah payau* (sweet), *tanah tawar* (neutral), and *tanah masam* (sour) soil, relating fairly well to the Western concept of soil pH. Smell is used among a few of the Nigerian Yoruba people to determine "good" or "bad" soil.

Farmers use landforms to locate the topographic position of soils and estimate the risk of erosion. In the Mooré language the upper part of a slope is called *zegedga*, which denotes where erosion risks are high. In the Yemba language (Dschang, west Cameroon), soils of well-drained plains, called *tsa'a pepeuo*, are considered very fertile and are used for intensive agriculture. Poorly drained soils of inland valleys, *tsa'a ngui*, are used only for off-season crop production or grazing.

Farmers in Africa combine these soil characteristics to create complex soil classifications. For example, the Mamprusi people of the northeast of Ghana use the name *Kokua sabli* for dark soils with iron-manganese nodules and a general low fertility, especially in drought conditions. Dark-colored and sandy-textured soil found in the uplands are called *Bihigu sabli*. These soil categories correspond well with the soil classification of Western scientists.[5]

- Complete all bed prep in the late fall, since the spring is too wet to work the soil.
- Keep soil covered with cover crops or mulch at all times to prevent cementing.

We aspire to be able to determine the chemical composition of our soils by taste as farmers in Nigeria do, but are still practicing that skill. We acknowledge that we have the privilege to live and farm at the top of the watershed and in the relatively pollution-free mountain foothills where tasting soil is feasible; one should not taste urban soils that might be contaminated with harmful chemicals. In the meantime we rely upon laboratory soil testing to give us further insights into soil health.

Your local agricultural extension agent or agricultural university will offer soil testing, usually at a reasonable cost. The test comes with thorough instructions about how to gather the sample as well as a guide for interpreting the soil test results. We like to test our soil every fall, when it is most depleted from growing food all season, and use that information to decide what amendments are needed. As you can see from our soil test results summary (table 5.1), we have had an overall positive impact on the soil health over time. It is possible for human

Table 5.1. Soul Fire Farm Soil Testing Results (using Modified Morgan Extractable method)

Indicator	Optimal Range	Before Amending	After Amending
pH	6.0–7.0	5.0	6.6
Phosphorus (P)	4–14 ppm	0.6 ppm	24.1 ppm‡
Potassium (K)	100–160 ppm	128 ppm	490 ppm‡
Calcium (Ca)	1,000–1,500 ppm	720 ppm	2,657 ppm‡
Magnesium (Mg)	50–120 ppm	71 ppm	404 ppm‡
Sulfur (S)	>10 ppm	11 ppm	19.8 ppm
Organic matter	>5%	4.5%	11.8%
Lead (Pb)	<22 ppm*	0.8 ppm	0.8 ppm
Aluminum (Al)	<75 ppm	143 ppm	19 ppm
Boron (B)	0.1–0.5 ppm†	0 ppm	0.7 ppm
Manganese (Mn)	1.1–6.3 ppm†	3.9 ppm	9.1 ppm‡
Zinc (Zn)	1.0–7.6 ppm†	0.6 ppm	4.5 ppm
Copper (Cu)	0.3–0.6 ppm†	0.2 ppm	0.2 ppm
Iron (Fe)	2.7–9.4 ppm†	18.3 ppm	5.3 ppm
Cation exchange capacity	n/a	13.2 meq/100g	20.4 meq/100g

Note: Soil testing laboratories use acids to extract nutrients from the soil. The most commonly used acids are the "weak acid modified Morgan extract" and the "strong acid Mehlich 3 extract." Each of the aforementioned methods is valid, though it is best to maintain consistency in methods when comparing soil-testing results for the same area of land.

* Total estimated lead level should not exceed 299 ppm (state standard) or 400 ppm (federal standard.) This value represents the extractable lead, which is a fraction of the total lead.

† Optimal micronutrient values have not been established, so these values are the normal ranges found in Northeast US soils. Many farmers believe that these "normal" micronutrient amounts are inadequate, depleted by centuries of erosion and soil mining, and should be much higher.

‡ After amending, certain nutrients reached levels above optimal which could lead to problems with weeds, soil salinization, and nutrification of waterways. To moderate nutrient levels, we began using less compost and grew more cover crops.

beings to have a beneficial, mutualistic relationship with the Earth.

Depending on your soil testing laboratory, your results may come with recommendations to add inorganic chemical fertilizers to correct any "deficiencies." We strongly encourage you to use organic, natural amendments in your soil.

pH

Soil pH is the measure of the acidity or alkalinity of the soil and one of the most important soil properties in that it affects the availability of nutrients. When soils are closer to a neutral pH, around 6.5, microbial populations increase and convert nitrogen, sulfur, and other nutrients into forms that plants can use. When the soil pH is too low, too acidic, limestone can be added to bring the pH up. There are two common types of limestone: dolomitic and calcitic. Dolomitic limestone contains magnesium and calcium carbonates and is slow acting over time. Calcitic limestone contains only calcium carbonate and is fast acting, but short-lived. If your soil is deficient in magnesium and requires a higher pH, consider dolomitic limestone. Occasionally pH needs to be lowered. Elemental sulfur is an effective amendment to acidify soil.

Macronutrients

Nutrients that the soil supplies in relatively large amounts are called macronutrients, and include nitrogen, phosphorus, potassium, calcium, magnesium, and sulfur. While each has its unique contributions to plant metabolism, and specific natural sources, all can be found in rich compost. We recommend that you begin by trying to manage most macronutrients with compost. If deficiencies persist, you can move to using specific amendments:

Nitrogen contributes to plant growth, and to seed and fruit production, because it's a part of chlorophyll, the pigment responsible for photosynthesis. Nitrogen is also part of all proteins, thus essential for every metabolic process. Nitrogen lives in the soil for a short time, so we assume that it is almost completely depleted by the end of the season and amend accordingly. Nitrogen is supplied through leguminous cover crops (clover, vetch, peas), blood meal, composted manure, fish scraps, kelp meal, alfalfa meal, and bonemeal. In the northeast, where mineralization is slow in the spring, compost nitrogen release may be too slow to meet the needs of crops, making more soluble forms of nitrogen, such as blood meal, desirable.

Phosphorus is essential for photosynthesis and is involved in the formation of fats, sugars, and starches in the plant. It supports plant flowering and root growth. Phosphorus is supplied through rock phosphate, bonemeal, fish scraps, wood ash, and composted manure. Too much phosphorus can contaminate waterways and lead to eutrophication, a process of algae overgrowth that eventually deprives water ecosystems of oxygen.

Potassium supports the building of proteins and the process of photosynthesis in plants. It contributes to fruit quality and the plant's ability to fight disease. Potassium is supplied by wood ash, granite dust, seaweed, fish scraps, greensand, and composted manure.

Alfalfa meal is sprinkled on the soil as an organic source of nitrogen. Photo by Neshima Vitale-Penniman.

Calcium is an essential part of plant cell wall structure and facilitates the transport of elements throughout the plant. Calcium is supplied by limestone, gypsum, and seashell dust. Do not use limestone unless you wish to raise the pH.

Magnesium activates plant enzymes needed for growth and is essential to photosynthesis. Magnesium can be found in dolomitic limestone and Epsom salts.

Sulfur supports root growth, seed production, and cold resistance. It is supplied naturally through rainwater and is also found in gypsum and compost.

Micronutrients

Nutrients that the soil supplies in relatively small amounts are called micronutrients or trace nutrients. These include boron, copper, chlorine, iron, manganese, molybdenum, zinc, and many others. Micronutrient deficiencies are most likely to occur in sandy, low-organic-matter soils or soils with a high pH. If your soil test reveals deficiencies in micronutrients, we recommend that you first try adding compost and correcting your pH. If deficiencies persist, contact your local gravel quarry and purchase rock dust—the powdery waste product that results when rocks are crushed. Rock dust has trace amounts of dozens of micronutrients and can be added safely to soil. Seaweed, kelp meal, and azomite are other natural sources of micronutrients. We do not recommend purchasing micronutrient amendments like manganese sulfate or chelated iron as these alter soil chemistry too quickly, acting like an energy drink rather than a nutritious meal.

Cation Exchange Capacity

Cation exchange capacity (CEC) is the measure of the soil's ability to retain and supply nutrients. It is essentially the number of negatively charged binding sites on the soil that are ready to stick to the positively charged nutrients, including calcium, magnesium, potassium, ammonium, and others. The more clay and organic matter found in a soil, the higher the CEC. Soils with low CEC are more susceptible to losing nutrients through leaching. Forrest Lahens, a 2014 BLFI graduate, taught us to think of CEC in terms of a hip-hop metaphor in which Mos Def is like low-CEC soil that has few binding sites for nutrients because there's just one vocalist in the

Participants in the Black Latinx Farmers Immersion teach one another about macronutrients in a soil workshop. Photo by Neshima Vitale-Penniman.

project, while Wu-Tang Clan is like high-CEC soil that has more binding sites because there are more vocalists. CEC can be corrected through the addition of organic matter.

Lead

Lead is a dangerous element that can lead to birth defects and brain damage. If your lead test results are high, please refer to the previous chapter for remediation strategies.

Compost

Compost is proof of life after death. It is the original black gold and the primary way that our ancestors fed the soil. Among compost's myriad talents, it improves soil texture and structure, increases soil's moisture-holding capacity, adds macro- and micronutrients, controls erosion, protects against drought, buffers pH, stabilizes toxins, and controls weeds.

Compost is exceedingly easy to make if you use a passive, slow-composting method. It has only two categories of ingredients, "greens" and "browns." The greens are nitrogen-rich materials, such as fresh weeds, grass clippings, food waste, and manure. The browns are carbon-rich materials, such as dry leaves, newspaper, cardboard, hay, straw, and wood chips. Essentially, a compost pile is a receptacle into which you put alternating layers of browns and greens, with an occasional optional splash of wood ash and water. You may also sprinkle a little rock phosphate on your pile to reduce nitrogen loss. The microbes do all the work of decomposition if you are patient. Our compost pile comprises four shipping pallets tied together with twine. We put down a layer of small sticks or stalks at the bottom for airflow. We then add garden weeds or kitchen scraps (including vegetables, fruits, breads, meat, bones, and oil) one bucket at a time. Note that in urban areas, you may want to exclude meat and oil so as not to attract rodents. We keep a pile of leaves and straw nearby so that we can

UPLIFT

African Dark Earth

African dark earth is a very dark and fertile anthropogenic soil invented by women in Ghana and Liberia 700 years ago. The creation of dark earth involves the combination of several types of waste: ash and char residues from cooking, bones from food preparation, by-products from processing palm oil and handmade soap, harvest residues, and organic domestic refuse. African dark earth has a high concentration of available calcium and phosphorus, likely due to the addition of animal bones. The bones in combination with the char also make the soil more alkaline, so it has a liming effect on soil chemistry.

African dark earths store between 200 and 300 percent more organic carbon than other soils, reducing atmospheric greenhouse gases, as well as 2 to 26 times the amount of pyrogenic carbon than regular soil, a carbon compound that persists longer in the soil and contributes to fertility and climate mitigation. West African farmers continue to produce and use dark earth today. The elders in the community measure the age of their town by the depth of the black soil, since every farmer in every generation participated in its creation.[6]

cover each deposit of kitchen scraps. This keeps flies and odors down while adding the necessary carbon to achieve decomposition. Once our pile is full, we cap it with a generous layer of browns and let it rest for three to six months. For those who want to speed up decomposition, you can turn your pile over after one month to add oxygen and stimulate the growth of aerobic bacteria. In the interim we start a new pile. At the end of the resting period, the pile is finished compost and can be applied directly to soil as a topdressing. We don't mind the occasional eggshell or bone shard, so we don't sift our compost. If you would like a finer finished product, you can sift the rough compost through a wire mesh and throw the undecomposed items back into an active compost pile. Good finished compost has crumbly texture, dark color, and earthy smell; it teems with living organisms like worms and beetles.

The forest offers an even simpler composting model in the natural decomposition process on its floor. The forest deposits leaves and debris on the ground and microbes eat that detritus from below, forming new soil in the process. We mimic this type of "sheet composting" on our farm on the pathways between the beds. When the weather is hot and dry, we throw crop debris and weeds without seedheads right into the pathways while we work. (During cool, rainy weather weeds go in the compost pile to prevent them from re-rooting.) Over the course of the season this debris decomposes. We then shovel a 1-inch (2.5 cm) layer of pathway soil into the beds each fall, transferring that compost to the planting area.

Whatever your composting method, consider adding the following types of organic matter, depending on what is locally available and free:

Animal bones, innards[7]
Animal manure (livestock, not cats and dogs)
Autumn leaves
Grass clippings (no chemical spray)
Hay (only if hot-composting so that seeds will be killed)

Youth build a compost bin out of pallets at Soul Fire Farm.

Kitchen and table scraps
Pond muck
Straw
Weeds (not gone to seed, propagated by tubers, or otherwise invasive)
Wood ash
Wood chips

Many permaculture blogs would have you believe that more ambitious farmers make biochar rather than compost out of their organic debris. Biochar is the result of low-temperature controlled burning of wood and plant material, in a process called pyrolysis. Essentially, you dig a trench, place debris into the trench, set it on fire, and, once the smoke turns grayish blue, dampen the fire by covering it with about an inch of soil. The debris then smolders into charcoal chunks. This may seem similar to the process used to make African dark earth or Amazonian *terra preta*, but recent research shows that there are crucial differences. Making charcoal releases a substantial amount of carbon dioxide into the air, consumes vital oxygen, and drives deforestation. African dark earth is made through a process more like sheet composting, where waste ash from cooking fires is combined with other organic matter.[8] We do not recommend harvesting wood or other materials from nature just to burn them. Please only use ash if you already burn wood for cooking or campfires.

Soil Ecology

Soil comprises five ingredients: minerals, water, organic matter, air, and microorganisms. While microbial life makes up only 1 percent of the volume of soil, it is essential to soil's capacity to support plants. It is what makes soil alive. A common estimate is that 1 teaspoon of soil holds over 20,000 organisms, such as algae, fungi, actinomycetes, and other bacteria.[9] These organisms are decomposers of organic matter, consuming detritus, water, and air, and recycling it into nutrient-rich humus. Some

A balance of green and brown materials encourages healthy decomposition in the compost pile.

specialized microbes, like nitrogen-fixing bacteria, have symbiotic relationships with plants that allow them to sequester nitrogen from the atmosphere and convert it to an organic usable form. Other specialists, like mycorrhizal fungi, provide plant roots with water and minerals in exchange for sugars, made possible by their intricate web of hyphae spread wide throughout the soil matrix. I often reflect on these mutualisms in awe and wonder why we as humans struggle to collaborate when our nonhuman siblings combine superpowers to turn air into soil and dissolve rocks into plant food. We need to take a metaphorical seat and pay attention.

Soil wants to be alive, so the question is not how to add life to the soil, but how not to destroy it. Soil life is compromised by tillage and turning of the soil, leaving bare dirt exposed to the elements; by additions of chemical fertilizers and pesticides; and by lack of water. Simply using the no-till systems described in the last chapter will invite microbial life and earthworms to take up residence on your farm.

Children from the Black Lives Matter Toronto Freedom School took a field trip to a greenhouse on Six Nations Indigenous reserve, where they held worms, observed beehives, and learned about indigenous plants. Photo by Adabu Jefwa.

UPLIFT
Cleopatra's Worms

While there is some debate as to whether Cleopatra's regard for earthworms is legend or historical fact, many scholars believe that she declared the animal sacred. Recognizing the earthworm's contributions to Egyptian agriculture, Cleopatra (69–30 BCE) decreed that no one, not even farmers, was allowed to touch an earthworm for fear of offending the deity of fertility. Egyptians were not permitted to remove a single earthworm from the land of Egypt, and certain priests devoted their full-time study to the habits of earthworms.

We now know that earthworm casts are the highest grade of organic fertilizer yet analyzed by science. Along with making soil nutrients available, additional minerals from the worm's body are added to the castings. According to a study by the USDA in 1949, the great fertility of the Nile River Valley was the result of the activity of earthworms. The researchers found the worm castings from a six-month period to weigh almost 120 tons per acre (44 MT/ha), approximately 10 times the amount of castings on soils in Europe and the US.[10] In addition to caring for worms in the soil, ancient Egyptians also practiced composting by piling organic waste and spreading the decayed product on their soil.[11]

Additionally, you can "inoculate" your farm with native soil life by gathering partially decomposed autumn leaves from a healthy forest in your area and spreading them as mulch over your farm. Soil biota cling to these mineral-rich leaves and will likely make a new home on your land.[12]

Additionally, there is a low-tech method for giving soil and plants a boost of fast-acting beneficial microorganisms during the season. Our dear friends, elder Black farmers Demalda and Rufus Newsome of Newsome Community Farms, Tulsa, Oklahoma, shared with us their recipe for compost tea. In essence, compost tea is "a barrel of water, compost, and leaves that turns the color of piss, then we spray it to nourish the plant and run insects off."[13] Their compost tea recipe is as follows:

1. Fill a 5-gallon (20 L) bucket (or larger) with tap water. If the water is chlorinated, allow it to sit uncovered for 24 hours for the chlorine to evaporate.
2. Aerate the water for a couple of hours before adding the compost. This can be accomplished using a compost tea aerator or any air stone designed for aquariums. At the time of writing, prices for aerators were generally around $25. Aeration increases the oxygen level in the water, which is essential for growing the desirable bacteria.
3. The ratio of compost:water should be around 1:10. Measure out the desired amount of rich compost, either kitchen compost, vermicompost (worm castings), or the upper soil layer of the forest floor.
4. Transfer the compost into a nylon sock or pantyhose. Tie it off and submerge in the water like a tea bag. Suspend the nylon from a stick so that it is not resting on the bottom of the bucket.
5. Continue to aerate the bucket of water for another 48 hours, keeping the operation in a shaded area between 60° and 80°F (16–27° C). When finished the tea should have a brownish yellow color.
6. Apply within five hours using a watering can or sprayer, while the microbes are still alive.[14]

Collecting leaves from a healthy forest is one way to inoculate your soil with beneficial microbes and add organic matter. Photo by Jonah Vitale-Wolff.

Compost tea can be applied to the soil or as a foliar spray directly to plant leaves. It adds microorganisms to the soil, provides nutrients and minerals, and increases pest resistance.

Similar to compost tea, "purins" are liquid fertilizers derived from the fermentation of organic materials. To make a purin, mix about 2 pounds (1 kg) of healthy, freshly picked and chopped plants into a plastic bucket with about 3 gallons (12 L) of water. Comfrey, borage, alfalfa, clover, and yarrow are ideal plants for making purins due to their high nutrient accumulation, but almost any healthy plant can work. Keep the bucket of chopped plants and water covered and in the shade. Stir the mixture well two times per day. Funky smells will develop and then subside after two to three days. The fermentation will be finished in 7 to 21 days when bubbles no longer float to the surface during stirring. Strain out the solids and store the purin in the dark in a sealed container. To use, dilute the purin with irrigation water to 10 to 20 percent concentration.

Compost tea and fermented purins are liquid fertilizers made from natural, organic materials. Photo by Neshima Vitale-Penniman.

Cover Crops

If you pause in stillness, you can hear the honeybees dancing on the buckwheat crop all the way on the other side of the field. They busy themselves converting this meadow of nectar into honey sweetness and make stops at our vegetable and fruit crops on their way home. We planted the buckwheat to suppress weeds and pasture grass on the newly opened areas on the farm. Buckwheat is fast growing with a thick root system that outcompetes its neighbors. It produces enticing flowers that encourage pollinators to spend more time on your farm helping your crops reproduce. Buckwheat also has a beautiful, geometric, pyramid-shaped seed that is satisfying to hold and broadcast over the soil.

Buckwheat is just one of the plant allies that we partner with to feed our soil. Cover crops are plants that we grow not to feed the human community, but rather to alchemize air into soil nutrients, stabilize aggregates, and enhance soil health. At Soul Fire Farm we also use rye, oats, peas, bell beans, vetch, soybeans, sorghum sudangrass, sunn hemp, triticale, and clover as cover crops. In doing so, we pay homage to ancestor George Washington Carver, who was an early champion of the use of leguminous cover crops to fix nitrogen. He understood that the pink-colored nodules growing out of the roots of beans and peanuts were natural nitrogen factories that should be intentionally cultivated by land stewards.

We generally use cover crops in three ways: to prepare new land for cultivation, at the end of the season to restore soil over the winter, and as an understory to crops during the growing season. Here is a list of our favorite cover crops with their benefits and growing requirements.[15]

Buckwheat

Buckwheat is an annual that establishes quickly, and therefore is useful for weed suppression and the preparation of new growing areas. It can tolerate low-fertility soils and scavenges phosphorus and calcium, making these nutrients available for subsequent crops. It attracts beneficial insects, such as parasitic wasps, ladybugs, hoverflies, and honeybees.

In the Northeast, buckwheat is broadcast at 70 pounds per acre (13 kg/ha) in June or July. Buckwheat does not tolerate hard soil, flooding, or frost. Once the field has turned white with flowers, mow it

Buckwheat is a useful cover crop to attract pollinators, suppress weeds, and add organic matter to the soil. Photo by Neshima Vitale-Penniman.

down to prevent seeds from forming and resprouting. It is best to mow in the afternoon, when the honeybees have left off collecting the nectar. Buckwheat is a winter-killed cover crop, meaning that it will die during cold weather.

Red Clover

Red clover, a perennial, fixes nitrogen from the atmosphere, suppresses weeds, and breaks up compacted soil with its taproots. It attracts lady beetles, green lacewings, and hoverflies, all predators of aphids. Since its seedlings are slow growing, it benefits from being planted with a small grain as a "nurse crop" to help it along. Red clover prefers cool weather conditions and tolerates compacted and wet soils. In the Northeast broadcast at 10 pounds per acre (2 kg/ha) anytime from February until September. Clover can be overseeded into fall crops that are still maturing as a way to suppress weeds and enrich the soil. Clover can also be frost-seeded in late winter when the ground is freezing and thawing. The clover seed will work its way into the soil and germinate when the soil temperatures rise over 50°F (10°C). Clover survives the winter and is useful in preparing new growing areas. To plant with a nurse crop, mix two parts annual ryegrass with one part medium red clover and sow at 20 to 25 pounds per acre (3.5 to 4.5 kg/ha).

Dutch White Clover

Perennial white clover fixes nitrogen from the atmosphere, while maintaining a lower profile than red clover. It is very tolerant of wet conditions. We also use white clover as an understory for long-season brassicas in pathways between beds, a controversial practice. Some farmers shy away from planting understory cover crops because they can be slow to establish, retain moisture, limit airflow for disease control, and compete with crop roots. Note that white clover can be a host for root rot diseases, such as pythium and rhizoctonia, so it's best not to plant it with or right before other legumes. Broadcast at 7 to 14 pounds per acre (1.25 to 2.5 kg/ha).

Hairy Vetch and Rye

An annual, hairy vetch is one of the best nitrogen fixers and is good at weed suppression and improving soil texture. It does not tolerate compacted soil, but is comfortable with a wide range of pH. Plant in late August to September in the Northeast with a nurse crop of rye (another annual), sowing 20 to 30 pounds per acre (3.5 to 5.5 kg/ha) of hairy vetch with 70 pounds per acre of rye or oats (13 kg/ha). The vetch will overwinter, and most of the nitrogen fixing will happen in May of the following year. Vetch can get out of control as a weed if you let it go to seed, so it is best to mow it and incorporate it mechanically by early June. You can also roll and crimp fall-planted rye and plant directly into it the following spring, even in hand-scale farms.

Oats and Peas

Field peas (annual) are a source of nitrogen and organic matter, and are the only leguminous cover crop that thrives in early spring. Peas collaborate well with oats (also annual), with the oats providing a trellis and suppressing weeds while the peas contribute nitrogen. Some farmers like to add vetch to the mix to provide a cover over the winter after the oats and peas have died back. Sow peas in March–April or early September in the Northeast at 120 pounds per

UPLIFT
George Washington Carver

George Washington Carver (born c. 1864) was a pioneer in regenerative agriculture, one of the first agricultural scientists in the United States to advocate for the use of leguminous cover crops, nutrient-rich mulching, and diversified horticulture. Carver was dedicated to the regeneration of depleted southern soils and turned to legumes, such as cowpeas and peanuts, as a means to fix nitrogen and replenish the soil. He believed that the cowpea, a crop indigenous to Africa, was indispensable in a crop rotation, noting that a soil's "deficiency in nitrogen can be made up almost wholly by keeping the legumes, or pod bearing plants, growing on the soil as much as possible."

In addition to promoting cover cropping, Carver experimented with and advocated for compost manuring to address the deficiencies of the Cotton Belt soils. In 1902 he spoke at the annual convention of the Association of American Agricultural Colleges and Experiment Stations about his use of swamp muck, forest leaves, and pine straw to fortify soils. He believed that the forest was a "natural fertilizer factory" containing trees, grasses, and debris that produced "countless tons of the finest kind of manure, rich in potash, phosphates, nitrogen, and humus." Carver sought to persuade farmers to dedicate their autumn and winter to the collection of organic matter. He advised farmers to dedicate every spare moment to raking leaves, gathering rich earth from the woods, piling up muck from swamps, and hauling it to the land to be plowed under.

Carver believed that "unkindness to anything means an injustice done to that thing," a conviction that extended to both people and soil. A farmer whose soil produced less every year was being unkind to it in some way, and neglecting proper care. His belief in the moral imperative to care for soil was expressed in the Bible verses he invoked while teaching farmers, such as Proverbs 13:23: "Much food is in the tillage [fallow ground] of the poor; but that is destroyed for want of judgment." He would go on to explain how they should be taking advantage of "fallow ground" by gathering edible wild fruits and vegetables, feeding potato and pea vines for livestock, and gathering the abundant fertilizers rotting in the swamps and forests of Macon County.[16]

acre (22 kg/ha) alongside oats at 20 pounds per acre (3.5kg/ha). You can also harvest the top 6 inches (15 cm) of the plant and use the pea tendrils in salad without interfering with the efficacy of the cover crop.

Sorghum Sudangrass

Originally cultivated in northeastern Africa 5,000 to 8,000 years ago, sorghum is a high-biomass annual grass that is heat- and drought-tolerant. The grass provides abundant organic matter and suppresses knot root nematodes. It harbors greenbug, which in turn attracts lady beetles, lacewings, and other beneficial predatory insects. Plant at 30 to 50 pounds per acre (5.5 to 9 kg/ha) in July and August in the Northeast for a winterkill. You may also plant sorghum sudangrass earlier in the season and cut it several times using a flail mower. Either way, be sure to mow the sorghum in the late fall for easier spring management. Sorghum sudangrass can have an allelopathic effect on other plants, making it excellent for weed suppression and full-season cover cropping, but not conducive for interplanting or quick crop successions.

Sunn Hemp

Originally cultivated in India as a cover crop and livestock feed, sunn hemp, an annual, fixes nitrogen and tolerates low-fertility sandy soils. It is resistant to root knot nematodes and can be used to clean the soil of this pest. It requires 8 to 12 frost-free weeks of growth, achievable even in northern climates. Broadcast at a rate of 40 to 60 pounds per acre (7.25 to 11 kg/ha). Generally, sunn hemp does not have time to produce seed within the span of a growing season, so mowing is not essential. It will winter-kill and you can plant your spring crop right into the mulch.

Bird's-Foot Trefoil

This perennial legume is tolerant of soils with low pH, low nutrients, poor drainage, and fragipans. It can reseed itself and is resistant to phytophthora root rot. Plant in early spring at a depth of no more than 1⁄4 inch. Inoculate the seeds with *Bradyrhizobium lupini* bacteria to increase the nitrogen-fixing action of the plant. Seed at 8 to 10 pounds per acre (1.5 to 2 kg/ha) without a nurse crop.

Tree species can also be used as permanent cover crops, protecting soil, sequestering carbon, and providing windbreaks. In Northern Ghana farmers have used baobab (*Adansonia digitata*), neem (*Azadirachta indica*), shea nut (*Vitellaria paradoxa*), *Acacia* spp., dawadawa (*Parkia biglobosa*), mahogany, senya, mango (*Mangifera indica*), and gliricidia for generations as cover crops.[17] They plant the trees on the edges and borders of fields as a complement to annual crops. At Soul Fire Farm we allow the wild species of black cherry (*Prunus serotina*), ash (*Fraxinus americana*), sugar maple (*Acer saccharum*), beech (*Fagus grandifolia*), paper birch (*Betula papyrifera*), and wild berries to grow on all four sides of each crop field. Note that black walnut (*Juglans nigra*) is not desirable as a field-edge tree because of the root exudate juglone, which is allelopathic. These trees provide habitat for beneficial predatory insects and pollinators, a source of firewood and mulch materials, opportunities for maple sugaring and berry harvest, groundwater recharge, breaking of the wind, and aesthetic value. Being surrounded by the forest also means that we are visited by hawks, coyotes, deer, bears, porcupines, owls, wild turkeys, and other splendid beings regularly, reminding us of our place in the larger ecosystem.

Forest edges are useful for protecting soil, sequestering carbon, and providing windbreaks.

CHAPTER SIX

Crop Planning

A Small Needful Fact

Is that Eric Garner worked
for some time for the Parks and Rec.
Horticultural Department, which means,
perhaps, that with his very large hands,
perhaps, in all likelihood,
he put gently into the earth
some plants which, most likely,
some of them, in all likelihood,
continue to grow, continue
to do what such plants do, like house
and feed small and necessary creatures,
like being pleasant to touch and smell,
like converting sunlight
into food, like making it easier
for us to breathe.

—Ross Gay

Three hundred generations ago, during the hungry moon of a long winter, the people of this land were dropping from starvation. The Haudenosaunee people, original and rightful human stewards of much of the ground that we now call "New York," had no food to eat. Sky Woman, divine grandmother of the Universe, clothed herself as a beggar and came to the people, palms outstretched in supplication, asking to be fed. The people, generous of heart, cleaned the last seed and chaff from their baskets and fed Sky Woman. So touched by their abundant hearts, and true to the sacred law of reciprocity, Sky Woman honored the people with a gift. She offered to the community her three daughters—corn, beans, and squash—so they would never be hungry again. This small bundle of seeds contained a story of harmony, of three crops in synergy: the beans to grab nitrogen, the corn to grow tall and provide stability, the squash to shade the weeds. These three crops would combine to provide complete nutrition: proteins, starches, fats, and vitamins to the people. Over the next 8,700 years of genetic and cultural stewardship, these crops were named Sustainers of Life and maize herself named Mother of Life.[1]

The First Peoples of the West shared their gift with indigenous people across the planet. Maize came to Africa before the Portuguese and spread to all corners of the continent in under 500 years, becoming Africa's most abundant cereal crop and one honored in sacred ceremony.[2] The First Peoples of the West also freely gave their gift to the white colonizer, who disregarded its sanctity and turned that maize into a weapon against the givers. Maize was appropriated, exploited, commodified. Torn away from her sisters, beans and squash, maize was forced into a monoculture that would rape the soil of its carbon, driving climate change, driving hurricanes in Puerto Rico, Haiti, and Texas, and wildfires in

Maize was born on this land and has been cultivated by Indigenous people for over 8,000 years. Photo by Neshima Vitale-Penniman.

California. She was distorted into high-fructose corn syrup that was pumped into veins of our people living under food apartheid, driving diabetes, obesity, heart disease, and making it 10 times more likely for Black and Brown people to die from poor diet than from violence.[3] The USDA used her to justify policies of farmworker exploitation, pesticide exposure, child labor, legalized neo-slavery under the guestworker program, exile of Black farmers from 12 million acres of our land, and NAFTA-driven forced migration of Mexican farmers from their homes. They interrupted her lineage with GMO strains replacing native varieties, and terminator seeds blocking new life. Sky Woman weeps with us for the appropriation and distortion of her sacred gift.

We ask ourselves what it would mean to decolonize, to re-indigenize our relationship with seed and with crops. In this chapter we explore the process of deciding what, when, and how to plant crops in the earth. This is informed by an understanding of the cultural relationships between Black people and crops, as well as the relationships among the plants themselves. In doing so, we strive to honor the Indigenous communities who originated many of the crops we now cherish as Black people. Inspired by the intercropping strategies of our African-indigenous ancestors, we uplift crop varieties, combinations, and farm layouts that encourage mutually beneficial relationships between plant allies.

Annual Crops

"Pick it from the bottom, pick it from the bottom!" chanted Terressa Tate, 2015 BLFI graduate, as she waited for the Greyhound bus to take her back home after a week of intensive practice on the farm, collard greens arching out of the pockets of her overalls. She was playfully recalling the song we used to teach participants the proper method for harvesting these familiar leaves. She squatted slow and mimicked the pulling motions of the harvest, laughing all the while, deliberately ignoring the sideways glances of the other bus-station-goers who had not had the fortune of a recent immersion in Mother Earth and its side benefit of ecstasy. Collards are just one of the many crops sacred to the culture of Black people. Historian Michael Twitty and other Black scholars have worked hard to compile lists of our heritage crops and revive collections for us to steward. Below, we elucidate the stories and growing techniques for many of the plants we have come to love. These plants are organized by botanical family, just as we organize our human communities around the centrality of family. See the growing chart (table 6.1) for details on spacing and timing. Note that every variety has specific care requirements; seed catalogs are the best references for these details.

UPLIFT
African Rice and Rice Cultivation in the Americas

African rice, *Oryza glaberrima*, was domesticated between 2,000 and 3,500 years ago in West Africa, independently of *O. sativa* in East Asia. Along the "rice coast" of Senegambia and Guinea, the Wolof, Mandinka, Baga, Mende, and Tenne farmers developed sophisticated technologies for production and processing. The Carolina rice industry was built on the skills of enslaved Africans. These Black American farmers created embankments, sluices, canals, floodgates, and dikes almost identical to patterns of West African mangrove rice production. Their labor created a planter aristocracy wealthier than any other group in the British colonies. Consequently, the enslavers ramped up the forced importation of African farmers directly from the rice-growing areas of West Africa who possessed knowledge of the cultivation techniques, yet withheld credit or respect for these farmers.

Oryza glaberrima, also called Merikin Moruga Hill Rice, was brought to Trinidad by Black soldiers who fought for the British in the War of 1812 in exchange for their freedom. The soldiers were relocated to Trinidad, where they brought their favorite upland cultivars of rice. Gullah chef Benjamin "B. J." Dennis prepares Moruga Hill Rice with coconut milk, reviving the practice of his forebears. Of the recipe, he says, "It's up to us to tell our own story correctly. I feel our ancestors actually guide us and ask us to tell the story. And it makes my heart happy, chasing my ancestral roots through food."

Yet Black farmers are rarely credited for their contributions to rice production. The "denial of African accomplishment in rice systems," as Professor Judith Carney writes in her book *Black Rice*, "provides a stunning example of how power relations mediate the production of history." African rice farmers also made profound contributions to the musical and linguistic culture of Black people. Ironically, the song "Amazing Grace" was written by 1700s rice slaver Captain John Newton, who later repented of his participation in slavery and appropriated Senegambian musical style to create this song, which retains centrality in Black communities today.[4]

Cucurbitaceae Family: Gourds, Pumpkins, Squash, Cucumbers, and Melons

The cucurbit family of vining crops loves the heat and highly fertile soils. We start all cucurbits in the greenhouse in 50-cell trays or soil blocks. We transplant them outdoors when they are around three weeks old, taking care not to let the plants get "leggy" (overly long-stemmed) or to bury the necks of the plants too deep. To protect cucurbits from

Kaolin clay can be used as a pest and disease deterrent for cucurbits and other crops. Photo by Neshima Vitale-Penniman.

Table 6.1. Soul Fire Farm's Planting Chart

Crop	Rows per 44″ (1 m) Bed	Plant Spacing in Row (inches)	Outdoor Plant Date (NY)	Best Method	Seeds per 1020 Flat	Days to Maturity
Amaranth	2	10 (25 cm)	May	Direct seed	n/a	110–150
Arugula	6	1 (2.5 cm)	May–Aug	Direct seed	n/a	45–60
Basil	4	10–12 (25–30 cm)	May–July	Transplant	55	75–120
Bean	2	3 (7.5 cm)	May	Direct seed	n/a	45–75
Beet	5	3 (7.5 cm)	April–July	Transplant early season, then direct seed	110	50–75
Bok choy	4	10 (25 cm)	May–Aug	Transplant	55	60–80
Broccoli	2	18–24 (45–60 cm)	April–June	Transplant	55	60–80
Brussels sprouts	2	24 (60 cm)	May–June	Transplant	55	100–140
Cabbage	2	24 (60 cm)	April–July	Transplant	55	80–120
Carrots	6	1–2 (2.5–5 cm)	April–July	Direct seed	n/a	80–100
Cauliflower	2	18–24 (45–60 cm)	May–July	Transplant	55	90–115
Celery	4	8 (20 cm)	April–June	Transplant	55	115–135
Chard	3	8–10 (20–25 cm)	April–Aug	Transplant	55	50–60
Cilantro	4	½ (1.25 cm)	April–Aug	Direct seed	n/a	60–75
Collards	2	15 (38 cm)	April–July	Transplant	55	75–90
Cotton	1	24 (60 cm)	June	Greenhouse	40	110–130
Cucumber	2	12–18 (30–45 cm)	May–July	Transplant	50-cell tray	50–75
Dill	5	1–4 (2.5–10 cm)	May–Aug	Direct seed	n/a	65–75
Edamame	2	2 (5 cm)	May–June	Direct seed	n/a	75–110
Eggplant	2	18 (45 cm)	June	Transplant	40	110–120
Fennel	3	10 (25 cm)	April–Aug	Transplant	55	90–100
Garlic	5	5 (13 cm)	Late Oct	Direct seed	n/a	240
Gourds	1	16 (40 cm)	June	Direct seed	n/a	95–120
Hibiscus	1	24–36 (60–90 cm)	June	Transplant	55	90–100
Kale	2	12 (30 cm)	April–July	Transplant	55	55–65
Leek	4	5 (13 cm)	April–June	Transplant	200–300	120–150
Lettuce, cut	6	¼ (.6 cm)	April–Aug	Direct seed	n/a	20–30

pests and give them extra warmth at the beginning of the season, we provide a protective layer of floating row cover. For additional pest protection, we dip the young seedlings into a kaolin clay mixture, a practice that early Egyptian farmers developed. The clay irritates and deters pests, while also preventing disease and sunburn.

Our summer squash, zucchini, and cucumbers are planted into rich soil and left unmulched or planted into black plastic mulch. We found that straw mulch encourages the reproduction of the undesirable cucumber beetle (*Acalymma vittatum*) and the squash bug (Coreidae family). If you are short on space, cucumber will climb up a simple

Table 6.1. *continued*

Crop	Rows per 44″ (1 m) Bed	Plant Spacing in Row (inches)	Outdoor Plant Date (NY)	Best Method	Seeds per 1020 Flat	Days to Maturity
Lettuce, head	4	10 (25 cm)	April–Aug	Transplant	55	45–80
Maize	2	4 (10 cm)	Late May	Direct seed	n/a	80–100
Melon	1	12–18 (30–45 cm)	June	Transplant	50-cell tray	90–120
Mesclun mixes	6	¼ (.65 cm)	April–Sept	Direct seed	n/a	18–28
Millet	2	2 (5 cm)	May–June	Direct seed	n/a	60–90
Molokhia	2	10 (25 cm)	July	Transplant	55	60–70
Napa cabbage	4	10 (25 cm)	April–Aug	Transplant	55	60–80
Okra	2	15 (38 cm)	June	Transplant	40 holes/flat, 2 seeds/hole	55–60
Onion	5	4 (10 cm)	April–June	Transplant	200–300	95–120
Parsley	4	6–10 (15–25 cm)	April–Aug	Transplant	55	65–90
Parsnip	5	2 (5 cm)	April–July	Direct seed	n/a	100–150
Pea	2	1 (2.5 cm)	April/Sept	Direct seed	n/a	50–75
Peanut	2	2–3 (5–7.5 cm)	June	Transplant	55	100–150
Pepper	2	12 (30 cm)	June	Transplant	40	100–130
Potato	2	6–12 (15–30 cm)	June	Direct seed	n/a	60–90
Radish	5	1–2 (2.5–5 cm)	April–Aug	Direct seed	n/a	25–50
Scallion	5	5 plants per hole, 5″ (13 cm) apart	April–Aug	Transplant	55, 5 seeds per hole	60–80
Sorghum	1	8 (20 cm)	June–July	Direct seed	n/a	80–100
Spinach	6	½–1 (1.25–2.5 cm)	April/Sept	Direct seed	n/a	40–55
Summer squash	1	12–18 (30–45 cm)	May–July	Transplant	40	50–65
Sunflower	2	24 (60 cm)	June	Transplant	55	80–120
Sweet potato	2	18 (45 cm)	June	Transplant	n/a	90–170
Tomato in high tunnel	2	24 (60 cm)	April–May	Transplant	25	100–120
Tomato in field	1	18 (45 cm)	May	Transplant	25	100–120
Tomatillo	1	18–24 (45–60 cm)	June	Transplant	25	75–100
Turnip	5	3–4 (7.5–10 cm)	April–Aug	Direct seed	n/a	35–45
Winter squash	1	12–24 (30–60 cm)	June	Transplant	50-cell tray	60–110

trellis fence of 2-by-4-inch (5-by-10 cm) welded wire at least 4 feet (1.25 m) tall and supported with wooden stakes every 4 to 6 feet (1.25 to 1.75 m). The harvest of summer squash and cucumbers is based on the size of the fruits. These fruits mature quickly, so in peak season it's best to check them every 1 to 2 days.

We plant our winter squash and melons in black plastic mulch, which serves to heat up the cold mountain ground and suppress weeds. Winter squash are harvested all at once just before the frost when the vines start to die back. Winter squash need to be "cured" by resting in a warm, dry place for a few weeks before storing. This hardens the skins, concentrates the sugars, and reduces its tendency to rot in storage.

The harvest of melons is perhaps the most intricate of these cousins, as determining their maturity is something of an art. Larisa, our farm manager, taught us to look for three signs of ripeness: a brightly colored spot on the skin where the melon touches the ground, a resonant thump when we knock on the fruit, and the curly tendril closest to the fruit being brown and dry.

Cucurbit crops and varieties curated by Black people include:

African mini bottle gourd. The calabash or bottle gourd, *Lagenaria siceraria*, was one of the first domesticated plants, originating in Africa 11,000 years ago. It is used as a drinking vessel, musical instrument, fishing bob, medicine bottle, insect cage, and food source. DNA evidence shows that it drifted across the ocean from Africa to the Americas, where it was widely adopted.[5]

West India burr gherkin. The maroon cucumber or gherkin, *Cucumis anguria*, was originally brought from Angola to the Caribbean. It is used for pickling, cooking, and fresh eating and is an ingredient in cozido stew in Brazil.[6]

In upstate New York melons appreciate being grown on black plastic mulch. Photo by Neshima Vitale-Penniman.

Citron melon. The preserving melon, *Citrullus caffer*, is native to Africa and grows abundantly in the Kalahari desert. Its use dates back to 4,000 years ago in ancient Egypt. It is used as a source of water, food, and animal fodder. Its high pectin content makes it ideal for pickling.[7]

Georgia Rattlesnake watermelon. Watermelon, *Citrullus lanatus*, was first domesticated by the Bantu-speaking people of West Africa. It was cultivated widely in ancient Egypt, where it was buried in pharaohs' tombs to provide hydration for their journeys to the afterlife.[8] It was introduced to the US from Africa during colonial times. The Georgia Rattlesnake variety was grown by the Black community beginning in the 1830s. It is exceptionally sweet with a thick skin.

Haitian pumpkin. The joumou pumpkin, *Cucurbita moschata*, was introduced to enslaved Haitians by the Taino indigenous people of that island. This pumpkin is used to make Soup Joumou, the national dish that commemorates the Haitian Revolution of 1804, and marks the independence of our people from slavery. White enslavers did not allow African people to taste the joumou, so upon liberation we birthed the tradition of cooking and sharing the joumou on the anniversary of our freedom.

Green striped cushaw pumpkin. The sweet potato pie pumpkin, *Cucurbita mixta*, was domesticated by Indigenous people of North America sometime between 7000 and 3000 BCE and shared with enslaved Africans in Jamaica, who in turn brought it to the Chesapeake in the late 1700s. It is heat-tolerant, resistant to the squash vine borer, and stores well. Louisiana Creoles sweeten the squash for use in pies, puddings, and turnovers.

White bush scalloped squash. Scallop or pattypan squash, *Cucurbita pepo*, was domesticated by Indigenous people of the Americas. Enslaved Africans, who sold fruit to the Jefferson family in the 1790s, commonly cultivated this squash in their kitchen gardens. It produces pie-shaped white fruits that are mild and sweet.

Solanaceae Family: Eggplants, Peppers, Tomatoes, Potatoes, Tobacco

We affectionately call our 80-foot-long (24 m) high tunnel "North Carolina" because its climatic conditions mimic the South and are suitable for members of the cold-tender and nutrient-hungry Solanum family. We grow tomatoes inside of this plastic house, nourished by drip irrigation but spared of rain, so as to reduce fungal disease. With drip, the water goes directly into the soil around the roots without wetting the leaves. We start tomatoes in 1020 flats in the greenhouse and transplant out when they are about four weeks old. The tomatoes are trellised using plastic twine that hangs from the overhead supports in the high tunnel, secured with special clips. As the tomatoes grow, we prune them so that only one central leader remains and remove all the "suckers." Tomatoes tolerate bare soil, straw mulch, black plastic mulch, and even an understory of leguminous cover crops. We harvest tomatoes at 50 percent blush, and allow them to ripen in crates, as this reduces bruising and spoiling and increases overall yields.

One way to trellis tomatoes it to attach strong twine to the overhead supports in the high tunnel and use clips to support the laden plants. Photo by Neshima Vitale-Penniman.

Eggplants and peppers love the "North Carolina" house, but there is not always space for them, so we sometimes plant them outside in black plastic mulch. They appreciate a layer of floating row cover supported by hoops when the nights are cold. The floating row cover also protects them from flea beetles, which can turn eggplant leaves to fragile lace, greatly weakening the plants. It is important to remove the row cover once blossoms appear so they can be pollinated. We start eggplants and peppers in the greenhouse and transplant them outside after about six to seven weeks. Both crops are harvested when they reach peak coloration and desired size.

In northern climates solanums thrive on black plastic mulch or in a high tunnel. Photo by Neshima Vitale-Penniman.

Potatoes are one of the plants that we grow from the tuber, not directly from seed. The previous year's potatoes can be cut into pieces that each have at least two "eyes" and a minimum weight, around the size of a small chicken egg. Allow the cut pieces to sit and cure in a dry, well-ventilated area for a couple of days before planting. Then dig a trench 8 inches

Transplanting

Why should we start certain seedlings in a greenhouse rather than plant them directly in the ground? While some plants cannot survive transplanting, most thrive with the extra care provided during the transplanting process. When the plants are young and vulnerable, they experience controlled temperatures, precise watering, and protection from pests. Once they are bigger and more resilient, they make their way into the native soil, where they find themselves well ahead of the weeds and hungry insects. Our ancestors knew this well. The farmers in the Rio Nunez region of Guinea sowed their rice seeds in nurseries on higher and drier ground, allowing them to germinate. Once the seedlings reached the stage where they could withstand the inundation of water and salinity, the farmers replanted them on mounds and ridges in their rice fields.[9]

Wire mesh can be used to space out seeds when planting directly into a 1020 flat.

Starting seedlings indoors requires a location that receives 14-plus hours of direct sunlight or artificial light per day, temperatures

Participants in Black Latinx Farmers Immersion admire the newly seeded flats that they just worked on in the greenhouse. Photo by Neshima Vitale-Penniman.

above freezing, and access to water. For years we started our farm's seedlings in the window of our house on shelves we built with fluorescent lights hung directly over the plants. We now use a high tunnel with two layers of plastic for the skin, a propane heater, and fans and vents to keep the air moving. Additionally, we provide a seedling heat mat for crops that require more bottom heat.

We plant our seeds into open 1020 flats at a depth twice the length of the seed. We use organic potting soil that is free of weed seed that we purchase from a nearby farmer. We water the flats one or two times per day so that the moisture is that of a damp sponge throughout the entire soil media. Once the seedlings have two sets of true leaves, in contrast to the cotyledons that first emerge, they are ready to be moved to the outdoor seedling tables for "hardening off." For several days the seedlings experience the wind, rain, and temperature variability of the outdoors in preparation for transplanting into the soil.

To transplant our seedlings, we first supersaturate the flats with water. We use a hose to make the holes in the beds into which the seedlings will be transplanted. This extra water reduces transplant shock. We place the plants, one per hole, into the puddle we made, taking care to leave the soil on the roots. We then firmly close the soil around them at the level at which it touched the stem in the flats. If the seedlings are leggy, you may need to plant them all the way up to the growing point, the location from which the true leaves are emerging. Cucurbits will rot if buried too deeply. Newly transplanted seedlings require frequent watering while they get established.

Brassicas are transplanted directly into the soil beneath the straw mulch. Photo by Neshima Vitale-Penniman.

(20 cm) deep in which to place the potatoes. Cover with soil and as the plants grow, continually hoe new soil toward the potatoes such that only a few inches of green vegetation remain exposed. This process is called hilling and increases the yield of the potatoes. The potato pests that we deal with most often are the Colorado potato beetle, *Leptinotarsa decemlineata*, which we pick off by hand and drown in a small bucket of soapy water, and the leafhopper, *Empoasca fabaem*, that can cause devastating "hopperburn," which we attempt to control with floating insect cover material around the time the leafhoppers are taking flight. Rotations are important in growing potatoes; if you plant them in the same ground two years in a row, the Colorado potato beetles will come up with the emerging potato plants. By moving the potatoes to another place in your field, you give them a jump start on the beetles. Potatoes are ready to harvest when the green vegetation starts to die back. You can carefully dig with a potato fork and lift them from the soil or use a center buster plow or mechanical potato digger.

Solanaceous crops curated by Black people include:

Gbogname collard eggplant. This gbogname variety, *Solanum melongena*, was developed in Togo, West Africa, and unlike others in its genus, has edible leaves. Its tender, young leaves are cooked as one would prepare collard greens. The plant is heat tolerant and generates bitter yellow fruits.

Garden egg. The white garden eggplant, *Solanum gilo*, is indigenous to sub-Saharan Africa and one of the top three most consumed vegetables in Ghana. It has a small white fruit with a bitter flavor. It can be stored for months without refrigeration and retain its firmness.

Louisiana long green eggplant. Cultivated in Asia since prehistory, the eggplant, *Solanum melongena*, spread to Africa well before colonizers arrived. Africans introduced the Louisiana Long Green Eggplant to Southern and Creole cuisine, and enslaved Louisianians grew them in their gardens. It is a high yielding plant that generates green fruits with a rich, nutty flavor.

Black nightshade. The black nightshade, *Solanum nigrum*, is a Eurasian crop bred and cultivated across Africa and the Diaspora. Its native ancestor is toxic, but edible varieties produce palatable ripe berries and greens. In Ethiopia, the leaves are boiled in salt water and used in dishes until the maize ripens. In Ghana, the berries are called kwaansusuaa, and used in preparing palm nut soup. In South America, the greens are used in tamales, and in Central America to make chuchitos and pupusas. People in South Africa use the fruits to make purple jam. Black nightshade grows as a weed in most cultivated fields in the Northeast and can likely be found already flourishing on site.

Plat de Haiti tomato. Indigenous people in Mexico were the first to domesticate tomatoes. Enslaved Africans in Haiti cultivated the apple of Hispaniola, *Solanum lycopersicum*, since at least the 1550s. Creole refugees fleeing the revolution in Haiti brought the tomato to Philadelphia in 1793, convincing their neighbors that the tomato was not the poisonous ornamental most assumed it to be. This variety is compact, deep red, and balanced between sweet and acid tones.[10]

Moyamensing tomato. The Moyamensing tomato, *Solanum lycopersicum*, is an African-American heirloom that was grown by incarcerated people in the gardens of the Eastern State Penitentiary since the mid 1800s and popularized by the cook in that institution. It yields prolific orange-red fruits and is excellent in soups.

Paul Robeson tomato. The Paul Robeson tomato, *Solanum lycopersicum*, was developed in Russia, and named in honor of Black civil rights activist, opera singer, athlete, actor, and law school graduate, Paul Robeson (1898-1976). It produces dark, intensely sweet fruits with high juice over a short season.

Fish pepper. Cultivated peppers are descended from the wild American bird pepper, native to North and South America. The fish pepper variety, *Capsicum annuum*, was developed in the Caribbean and traveled to the Black American south where it became the "secret" ingredient at crab and oyster houses. The young, cream-colored peppers blended into white fish sauces adding invisible heat. The black folk painter Horace Pippin revived the variety in the 1940s.

Buena mulata cayenne pepper. The buena mulata pepper, *Capsicum annuum*, was passed on by the black folk artist Horace Pippin. It may have originated in Cuba. It is a highly productive hot pepper that changes color during ripening from violet, to orange, to brown, to deep red. Soilful City, a DC farm, makes a beautiful hot sauce from this pepper.

Scotch bonnet pepper, fatalii. The Caribbean red pepper, *Capsicum chinense*, is cultivated mainly in the Caribbean Islands, Central and South Africa. In Nigeria it is called *ata rodo*. It is about 100 times hotter than a jalapeño and gives Jerk its unique fruity, citrus flavor.

Mboga pepper. In Zanzibar where this pepper, *Capsicum frutescens*, was selected by Indigenous people, they call it *Pilipili Mboga*, in Swahili which means vegetable pepper.

Fabaceae Family: Black-Eyed Peas, Peanuts, Beans, and Sweet Peas

The legumes are one family that leaves the soil more enriched at the end of its tenure than it inherited at the beginning. Legumes provide nodular homes for nitrogen-fixing bacteria in their roots, thus facilitating the capture of atmospheric nitrogen. Consequently they can be intercropped with nitrogen-hungry crops like maize. They can also precede heavy feeders in a multiyear crop rotation.

Most legumes prefer to be direct seeded after the danger of frost. We dig a shallow furrow with the corner of the hoe and drop the seeds at the appropriate spacing. Beans and peas emerge quickly and outcompete weeds with more ease than other crop families. We apply straw mulch once the beans are about 6 inches tall. Bush beans and edamame do not require upright supports. However, sugar snap peas, shell peas, and pole beans need to climb. We install trellises made of 2-by-4-inch (5-by-10 cm) welded wire fence, 6 feet (1.75 m) high, supported by hardwood stakes every 6 to 8 feet (1.75 to 2.5 m). Beans and peas mature over time, so we pick them every two or three days during their harvest season.

Peanuts originated in Peru, where indigenous people cultivated them at least 3,500 years ago and used them as currency. While we are new to growing this crop at Soul Fire Farm, it is possible to attain a harvest in the North using short-season varieties like Early Spanish. The shelled peanut, still in its husk, can be started in flats in the greenhouse or planted directly outside in warmer climates. The peanut produces yellow, self-pollinating blossoms that, once fertilized, bend toward the earth and plant themselves in the soil to form a peanut pod. Peanuts are harvested when the leaves begin to yellow and the peanuts nearly fill out the pods.

Legume varieties curated by Black farmers include:

Brown crowder. The cow pea or black-eyed pea, *Vigna unguiculata*, originated in West Africa and was carried by Black farmers to the Americas during the slave trade. It is a drought-tolerant, disease- and pest-resistant nitrogen fixer that was promoted by Dr. George Washington Carver as a

Many seeds can be planted directly into a furrow in the soil. Photo by Neshima Vitale-Penniman.

tool for soil repair. The California Blackeye Pea variety is a prolific cultivar with mystical properties—attracting wealth, fertility, and luck when consumed on New Year's Day.

Goober nut. The peanut, *Arachis hypogaea,* is native to South America and made its way to Africa in the precolonial era, where it became a staple crop. During the era of enslavement, Africans brought peanuts back to the Americas, where they cultivated them for generations before Europeans adopted the crop.[11]

Bambara groundnut. The bambara nut or pindar, *Vigna subterranea,* originated in West Africa. It is one of the most important legumes in the semi-arid areas of Africa because it is tolerant of drought, heat, and marginal soils. It is highly nutritious and climate-adaptive. Its small white fruits form underground and can be eaten fresh, dried, or boiled, or ground into flour.

Sea Island red peas. The Geechee red pea was planted by African rice farmers in the Carolinas in rotation with rice to restore the soil. It is used in the popular African American dish Hoppin' John and the Gullah dish "reezy peezy," made with fresh peas and rice.

Bush bean, kebarika. The bush bean, *Phaseolus vulgaris,* is an heirloom from Kenya, used in soups, in baking, and dried.

Mbombo green. Drought- and heat-tolerant Mbombo green, *Lablab purpureus,* originated in Kenya where its name honors the Creator God Mbombo. It is a sacred crop, which brings fertility to the soil, prosperity to the people, and health to the nursing mother.

Malvaceae Family: Okra, Sorrel, Molokhia, Cotton, Cacao, Kola Nut

Without the mallow family, there would be no gumbo and there would be no sorrel tea—possibly a fatal blow to our Caribbean cuisine. These heat-loving crops can be coaxed out of northern soils with the aid of black plastic mulch or row cover. We grow okra in the hottest part of our southern-sloped soils, following a nitrogen-fixing crop. It's best to start okra indoors for five to six weeks and then set the plants out three to four weeks after the last frost date. We keep the area weed-free and harvest the okra when it is the length of an index finger.

Fortunately for us, the naturally acidic soils of the Northeast are perfect for hibiscus, though the cold temperatures mean that it grows as an annual, dying back each year, rather than persisting as a perennial. Hibiscus prefers to be transplanted into rich soil after the danger of frost, and mulched once it reaches 6 inches. The mature flowers can be harvested throughout the summer and made into a tangy, distinctive sorrel drink.

Where we live, cotton can only be grown indoors, and many Black farmers avoid it because of its association with slavery. The rise of "King Cotton" revitalized US slavery and drove a spectacularly devastating increase in the kidnapping of our people from 1787 to 1808, in addition to the internal slave trade that forced our ancestors from the border states into the Cotton Belt. Part of our collective healing from that trauma is the reconciliation with the cotton plant, and many Black farmers are reviving its growth on sovereign terms. Cotton can be transplanted or direct seeded, but requires 70°F (21°C) soil temperatures to germinate, and a long, hot season to mature. Keep weed-free and harvest when the bolls split open in late summer.

Here are a few mallow varieties that have been curated by Black farmers:

Levant cotton. The species of cotton native to semi-arid regions of sub-Saharan Africa is *Gossypium herbaceum,* a perennial shrub that reaches 2 feet (0.6 m) and produces short-fiber cotton. It is a sister to *G. hirsutum,* which is native to the Americas and was cultivated by the Incas since CE 1000. Colonizers brought it from the Caribbean to the US South, where enslaved Black farmers grew it.

Cowhorn okra. Okra or gumbo, *Abelmoschus esculentus,* likely originated in West Africa and was brought to Brazil by our ancestors in the 1600s and

Okra originated in West Africa and is a staple of dishes across the Diaspora. Photo by Neshima Vitale-Penniman.

Damaris, Soul Fire's assistant farm manager, holds the beautiful turquoise seeds of molokhia, a vitamin- and mineral-rich green eaten like spinach in soups and stews in the Middle East and Africa. Photo by Lytisha Wyatt.

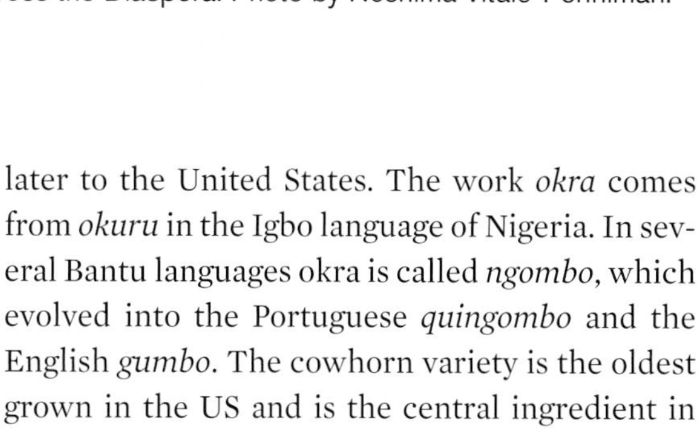

later to the United States. The work *okra* comes from *okuru* in the Igbo language of Nigeria. In several Bantu languages okra is called *ngombo*, which evolved into the Portuguese *quingombo* and the English *gumbo*. The cowhorn variety is the oldest grown in the US and is the central ingredient in gumbo (okra soup).

Molokhia. Egyptian spinach, *Corchorus olitorius*, is a staple food in Ghana and Burkina Faso and commonly eaten in Egypt. During the time of pharaohs, Egyptian royalty would drink molokhia soup to recover from illness. People in Ghana also use molokhia medicinally for fever, stomach issues, and loss of appetite. It has many names that refer to its texture: in Ewe, *Ademe* (slimy leaf); in Dagbani, *Salinvaa* (slimy leaf); in Vagla Dofila, *mwana* (grass okro or "grass okra"); and in Hausa and Twi, *Ayoyo* (slimy leaf).[12] It germinates in soils 77° to 86°F (25–30°C) and matures in only 60 days. It can be continuously harvested.

White sorrel. The Archer variety of roselle, *Hibiscus albus*, originated in Senegal and was grown throughout Jamaica starting in the 1700s. The more widely used red variety is *Hibiscus sabdariffa*. Its calyx is eaten as a vegetable and its flowers brewed as tea. It is one of the over 600 species of hibiscus, known as *bissap*, *zobo*, and *sobolo* in West Africa and sorrel in Jamaica and other parts of the Caribbean. Hibiscus is the national flower of Haiti.

Poaceae Family: Millet, Sorghum, Corn, Rice, Oats, and Rye

During initiation, practitioners of Ifa and Vodun receive individualized spiritual instructions through divination, known as *ita*. In my case I learned that I need to cultivate maize every year for the duration of my life. My adherence to this important mandate was lax one year and I realized with a panic midsummer that we had left corn out of the crop plan. Fortunately, nature had my back and self-seeded a little patch of maize where some chicken feed had spilled the previous fall. I made sure to tend and harvest that volunteer patch to fulfill my spiritual mandate.

Maize is an ancient and sacred crop, rich in protein, fiber, and minerals and relatively easy to grow among its cousins. You can try the passive method and hope for a magical crop with no effort, as I did, or intentionally cultivate this noble plant.

Patience is an important ingredient in northern maize cultivation, as early planting will result in rotten seed. Wait until the danger of frost has passed, the oak leaves on the trees are at least 1 inch long, and the dandelions are blooming to direct seed your maize. We plant our maize together with its sisters, beans and squash, in hills 4 to 5 feet (1.25 to 1.5 m) apart from one another. Each hill receives four or five maize seeds, followed 10 days later by four or five pole bean seeds and two of winter squash. Corn is wind-pollinated and needs at least four rows to attain adequate fertilization. To prevent corn earworm from burrowing into your ripe ears, insert a medicine dropper of mineral oil into the silk four to five days after it appears.[13] The maize is ready to harvest when the silk turns wilted and brown. Flour and dent corns can be left to dry on the stalk and stored over the winter for use as cornmeal, tortillas, or hominy. Sweet and flour corns can be harvested right away for roasting and fresh eating.

Millet and sorghum are staple grains domesticated by our ancestors in Africa. Millet was the primary crop grown by African farmers during slavery on their subsistence plots, or *kunuku*.[14] At Soul Fire Farm we primarily use sorghum as a high-biomass cover crop to remediate tired soils. It tolerates poor, dry soils and can be broadcast at 10 to 15 pounds per acre (1.8 to 2.75 kg/ha) to generate a thick forest of healing green. To grow sorghum as a grain crop, we plant it 8 inches apart when the soil has warmed to 60°F (16°C). It can be harvested once the grains no longer dent when pressed with your fingernail. Grains can be cleaned by vigorously whacking them against the inside of a bucket or rolling them over a screen with a collecting tub below. Millet is even more forgiving than sorghum and matures very quickly. Be prepared to share your millet harvest with the birds, who love this nutritious seed.

Coral sorghum. The majestic Coral Sorghum, *Sorghum bicolor*, was developed by the Shilluk people of South Sudan and cultivated across the savannas all the way to Mauritania. Its sweet canes can be pressed for juice to make sorghum syrup, and its large purple grains can be popped, boiled, ground into flour, and brewed as beer. It is tolerant of drought and marginal soil. Sugar Drip Sorghum is another variety popular in the South for its high sugar content.

African black rice. In rapid decline, *Oryza glaberrima* mainly persists for ritual use by the Jola people of Senegal. The farmers of the Niger River watershed domesticated it 2,000 to 3,000 years ago. It is superior to the Asian rice variety in terms of its climate adaptability and nutritional value.[15]

Finger millet, dragon's claw. Millet, *Eleusine coracana*, is native to the Ethiopian highlands and frequently intercropped with peanuts and cowpeas.

Distant Cousins

Many of the crops we have curated and cherished over millennia do not thrive in the soil to which we have been transplanted. For example, the yam (*Dioscorea* spp.), "King of Crops," sits tired looking on the shelf of the African market in town having endured a transatlantic journey. Our ancestors recognized that the sweet potato, of the Convolvulaceae family, native to the Americas, could supplant the nutritional and spiritual role of the yam in Black agrarian life. Similarly, some of the greens of our homeland struggle in cool temperatures, so we grow malabar spinach (Basellaceae family), collard greens, amaranth, and spinach to fill that culinary niche.

The following sections cover growing tips for Umbelliferae, Asteraceae, Brassicaceae, Liliaceae, Chenopodiaceae, and other crop families adapted to the northern climates that have become part of our cultural cuisine, even as recent adoptees by our people.

Umbelliferae Family: Carrots, Cilantro, Celery, Parsley, Parsnips

Carrots and parsnips are hearty root vegetables that can stand in for the tubers of our homelands in soups and desserts. Both of these crops have a long taproot that will not stand for transplanting, so we place the seeds directly into furrows in the ground. We recommend planting seeds at their precise spacing, rather than planting them thickly and thinning later, as it saves time overall. Weed prevention is exceptionally important, as carrots and parsnips are slow growing and easily overwhelmed by competitors. It is best to use a stale-bedding technique, which means weeding the area, waiting until another batch of weeds appears, weeding again, and then planting. This process depletes the weed seed bank in the top inch of soil so your roots can get a head start. Alternatively, you can use the tarping method described in previous chapters to smother the weeds in advance. Many farmers, including our fire-loving son Emet, like to flame-weed their carrots. Flame-weeding works by heat-damaging the growing point of the broadleaf weeds, but is less effective with grasses, which have a growing point below the soil line. One strategy is to plant a few fast-germinating radish seeds along with the slow-germinating carrots. When the radish seedlings come up, it means that the carrots are about one week behind. You can use a flame torch to burn the young weeds without damaging the pre-emergent carrots. It's ridiculously fun and quite effective. Another challenge for carrots and parsnips is heavy clay soil. In order to prevent stunted and gnarly roots, we use a soil fork to loosen up the ground before direct seeding. These roots can also be harvested with the aid of a gentle loosening by the soil fork.

Parsley and celery are prolific, cold-tolerant, disease- and pest-resistant herbs that are super packed with nutrients. They can stand in for most greens in soup and stew. Parsley can also be used in pesto. They are slow germinators, so be patient and keep hope when you start the seeds indoors. We transplant our parsley in late spring and throughout the season, interspersing it with lettuce and tomatoes as a pest deterrent. Parsley can be continuously harvested when you cut it so that about 2 inches (5 cm) of the plant remain for regeneration. Celery is harvested when it reaches desired size; the whole plant is removed at once.

Cilantro is a key ingredient in sofrito for our Afro-Latinx family. It can be direct seeded and grows quickly and abundantly. We find that we can get two to four cuttings of cilantro before it starts flowering. You can even plant the cilantro in early fall, cutting it once to leave ½ inch (1.25 cm) of stem and then allowing it to regrow. It will survive the winter and resume growth in the spring. After offering several rounds of tasty greens, the plant turns its energy toward making coriander seeds, which are a rich and

We manage to coax carrots out of our heavy clay soils, which brings us great joy. Photo by Neshima Vitale-Penniman.

potent spice. *Culantro* or *Recao* is also an essential ingredient in sofrito for many Dominicans, Puerto Ricans, Trinidadians, and other Afro-Latinx people.

Asteraceae Family: Lettuce, Dandelion, Chicory

Lettuce is a versatile, popular, fast-maturing, and lucrative green that can be grown year-round in the North. We seed lettuce every two to three weeks in succession to ensure that there is always a steady supply for the members of our farm share. Lettuce can be transplanted with a wider spacing and allowed to fully mature into heads, or it can be direct seeded with tight spacing to be harvested as baby greens. Recently, farmers have been switching to Salanova as an alternative to direct seeding lettuce mix. It is transplanted at close spacing to completely cover the ground and harvested by the fistful, saving weeding and harvesting time. We use all three methods throughout the season. It is best not to mulch lettuce, as it can introduce debris into your harvested product. In fact, lettuce itself can act as a living mulch when broadcast in the understory of brassicas or other long-season crops.

Eat your weeds! Dandelion is nutritious, flavorful, and naturally abundant. Photo by Neshima Vitale-Penniman.

Dandelion and chicory grow wild in our region and are commonly regarded as weeds, despite the fact that their nutritional value far outshines lettuce and other cultivated greens. Both were frequently used by our enslaved ancestors as vegetables. They are some of the first edible greens in early spring, and can provide essential nutrients in the cool season before other crops come in. The young leaves of both dandelion and chicory can be harvested for fresh eating, soup, or herbal infusions.

Brassicaceae Family: Collards, Turnip, Cabbage, Radish, Kale, Cauliflower, Mustards, Brussels Sprouts

Collards, turnips, and cabbage have become central to Black cuisine, despite their genetic origins in Europe. The brassica family thrives in the North and produces a diverse array of vegetables, from brussels sprouts to cauliflower. We grow both a spring and fall crop of most of the members of this family. We start collards, kale, cabbage, broccoli, cauliflower, and brussels sprouts in the greenhouse in 1020 flats and transplant them into beds once they have four true leaves. Brassicas appreciate a thick straw mulch. We have found it most efficient to lay down the mulch on top of the prepared bed and then transplant into the mulch, rather than mulching after planting. When brassicas are young, they are sensitive to pest damage from flea beetles (*Psylliodes chrysocephala*), so we install floating row cover to keep off the bugs and give some extra warmth. The row covers can be removed once the plant sizes up to about 10 true leaves. In the middle of the season, the cabbage worm (*Mamestra brassicae*) becomes the major threat to brassica thriving. We intercrop marigold flowers in the brassica beds because they host parasitic and predatory beneficial insects that consume the eggs and larvae of the cabbage moth. For kale and collards, *Pick it from the bottom* is the mantra for keeping the plant healthy while continually harvesting. Use a downward motion to remove the leaves evenly from the bottom, always leaving at least five leaves intact to generate new growth. Broccoli, cauliflower, and cabbage are harvested when they "head up"; use a sharp knife to remove the vegetable. One of the magical things about brassicas is that they get sweeter after the frost, continuing to give love to the farmer after the less hardy crops have quit for the season.

Turnips, radishes, arugula, and mustards can be direct seeded. We space turnips and radishes precisely, rather than seeding densely and thinning later. These roots are harvested based on size, and you can often get several harvests from the same bed. The greens of turnip are edible and delicious in stews. Some call turnip greens "the real greens of the Black South." We plant mustard greens as a complement to baby lettuce in our salad mixes. Mustards grow to size a few days faster than the baby lettuce, so we plant them a bit later with the goal of having the entire salad mix mature on the same day. We do not use mulch on mustards, arugula, turnips, or radishes.

William Alexander Heading collards. This heirloom variety of collard greens comes to us via Black farmer William Alexander. A group of African American sorority sisters and Master Gardeners in Winston-Salem, North Carolina, is keeping this variety alive for our community.

Green Glaze collards. This is the oldest known variety of collards, dating back to the late 18th century. Michael Twitty praises this variety as "pretty, waxy, crisp, tough against bugs, and extremely delicious." The consumption of collards, or "sprouts," was popularized by enslaved Black chefs in the American South. Ezra Adams, a formerly enslaved

Participants in BLFI tend to the collard greens at Soul Fire Farm. Photo by Neshima Vitale-Penniman.

chef, said, "If you wants to know what I thinks is de best vittles, I'se gwine to be obliged to [admit] dat it is cabbage sprouts in de spring, and it is collard greens after frost has struck . . . I lak to eat."[16]

East African kale. Amara, Ethiopian kale, has been grown in its homeland for over 6,000 years. *Brassica carinata* tastes like buttery young collards and can be grown like cutting lettuce, offering several harvests in a season.

Liliaceae Family: Onions, Garlic, Leeks, Scallions, Chives, Asparagus

The Liliaceae family gives us the profound gift of flavor. Onions, leeks, and scallions can be started from seed in the greenhouse in late winter and transplanted out as soon as the ground thaws. They are sensitive to weeds, so ensure that you prepared a weed-free area using tarping or stale bedding in the prior season. Some farmers like to plant onions and scallions in bunches to facilitate ease of weeding and harvest, rather than spacing them evenly. We have found good success with that strategy and plant scallions at five seeds per hole in the flat and then transplant those same five seedlings in a clump in the beds. Scallions, onions, and leeks are harvested based on size. For storage onions, wait until the stalks turn brown and die back to harvest. Then cure the onions by resting them in a warm, dry place for a few weeks before transferring to storage.

One garlic clove yields an entire bulb of garlic when planted directly into the ground in late fall. We plant our garlic around the time of Fet Gede,

Youth at Soul Fire Farm harvest and tie garlic into bunches for curing. Photo by Neshima Vitale-Penniman.

November 2, the Haitian holiday honoring the ancestors. Each garlic clove is separated, the papery covering intact, and placed in the ground with the pointy side up. We cover the planted ground with a thick layer of straw mulch and await spring, when the garlic shoots will penetrate the mulch covering and delight us with some of the first greenery of the season. Rather than discarding the very small cloves, you can plant them densely and cut the young leaves to make pesto. The leaves will regrow three or four times.

Egyptian walking onion. The perennial walking onion, *Allium cepa* var. *aggregatum*, originated in North Africa. It has a harvestable onion at the tip of its stalk. The onion is spreading and appears to "walk" across the field.[17]

Farmer Melony Edwards of Willowood Farm, Coupeville, Washington, harvests a giant chiogga beet that was overwintered. "We grow beets this big that are not woody and grow big from the result of LOVE." Photo by Kyle England.

Chenopodiaceae Family: Amaranth, Spinach, Beets, Swiss Chard

Callaloo is a Caribbean dish invented by enslaved Africans that features an indigenous green vegetable, sometimes in combination with African okra. For African Americans this green is usually collards or a member of the Chenopodiaceae family: amaranth, Swiss chard, or spinach. For Jamaicans, Belizeans, and Guyanese, *callaloo* almost always refers to amaranth. Amaranth is a high-protein, mineral-rich super green that tolerates poor soils. We direct seed amaranth in our *milpa* together with corn, beans, and squash. The leaves, seeds, and root are all edible.

We prefer to transplant Swiss chard, and to direct seed spinach. We transplant the first two successions of beets, and then direct seed the later season successions. Of these crops, only the Swiss chard receives straw mulch, since we continually harvest it throughout the season. The leafminer (*Pegomya hyoscyami*) leaves unsightly brown tunnels in the leaves of the Swiss chard. It is best controlled by removing the damaged leaves regularly and putting them in the garbage, and by keeping the area free of weeds. Spinach and chard are harvested using a sharp knife, such that the growth point remains intact for regeneration. Spinach can also be cut below the growth point and bunched for a "one and done" crop. Beets are harvested when they reach the desired size.[18]

Convolvulaceae: Sweet Potatoes

The Diasporic King of Crops (*Ipomoea batatas*) is a staple food for our people and a mainstay for Black farmers. It is the spiritual and culinary stand-in for the African yam (*Dioscorea cayenensis*). Unlike most crops that are grown from seed, sweet potatoes are grown from slips. You can order these or sprout your own indoors. To grow sweet potato slips, cut a sweet potato tuber into large chunks and suspend them in water, with half of the tuber exposed to air and the other half in the water. You can use toothpicks to hold it in place. After a couple of weeks, green sprouts will emerge. Carefully twist these sprouts off the potato and transfer them into a shallow basin of water, where they will develop roots. In the meantime, prepare the soil to receive the sweet potato slips. Fork it thoroughly, as loose soil is essential to the development of large tubers. Once the slips have roots 1 inch (2.5 cm) long, they are ready to be transplanted. We like to use black plastic mulch and plant the slips into soil through holes that we have cut in the mulch. The black plastic warms the soil and suppresses weeds. Take care to position your sweet potatoes away from the field edges where mice will have easy access to burrow under the plastic mulch to eat your tubers. Regular watering is essential for the first few weeks after transplanting.

Both the leaves and tubers of sweet potatoes are edible. We harvest both at the same time in late summer. A soil fork is useful to loosen the soil and liberate the tubers. Cure the sweet potatoes in a warm, dry place for a few weeks before transferring to storage. You can use the leaves fresh, sautéed like spinach with onion, garlic, and a splash of hot spice.

Pedaliaceae: Sesame

The oldest oilseed crop known to humanity originated in sub-Saharan Africa, with cultivation beginning around 3500 BCE. Egyptians called it *sesemt*, and included it in the list of plant medicines in the scrolls of the Ebers Papyrus over 3,600 years ago. Enslaved Africans brought sesame (*Sesamum indicum*) to the Carolinas. Sesame is known as the survivor crop because it grows in adverse environments with high winds, drought, high heat, and even flooding. Sesame requires 90 to 120 frost-free days and prefers highly fertile soils. Sesame seeds grow within a capsule that bursts open when the seeds ripen, a process called dehiscence. Harvest sesame seeds before full maturity and store them in an upright position to continue ripening until the capsules open.

Newly harvested sweet potato tubers and leaves at Soul Fire Farm. Photo by Neshima Vitale-Penniman.

Polycultures

Splashed with bits of soil kicked back from the spray of the hose, we added the finishing touches to our long beds of green cabbage by transplanting lemon drop marigolds down the middle every few feet. Their delicate yellow flowers understated the thick, sweet aroma that permeated the air and clung to our fingertips. Lemon drop marigolds (*Tagetes tenuifolia*) attract green lacewings, ladybugs, hoverflies, and other predatory insects that eat aphids and the eggs of pest insects. We laughed and joked, wondering if all of these synergistic interspecific love relationships were trying to give us some message about how we should take care of one another in the human community.

Marigolds and brassicas are one of our favorite polycropping combinations, for both practical and aesthetic reasons. In fact, planting beneficial flowers down the middle of beds or at the bed ends is the least intrusive way to get started with intercropping. Nasturtiums, marigolds, borage, sunflowers, zinnias, cosmos, and aromatic herbs all attract pollinators

UPLIFT
Polycultures of Africa

Just as the forest does not plant a monoculture of one species of tree, so have our African ancestors universally developed intercropping strategies on their farms. The Hausa farmers of Nigeria developed at least 156 systematic crop combinations, including no-till polycultures of grains, legumes, and root crops planted on ridges. The Abakaliki farmers of Nigeria used their hoes to construct mounds and then planted crops on distinct parts of the mound; they planted yams on top of the mound, rice in the furrow, and maize, okra, melon, and cassava on the lower parts of the mound.[19] They also intercropped egusi melon with sorghum, cassava, coffee, cotton, maize, and bananas because it confused the pests and blanketed the ground, suppressing weeds.[20]

Additionally, Indigenous people in Africa and the Americas developed a system called relay cropping, where a second crop is planted after the first crop is well established but immature. For example, maize can be established before seeding cowpeas and groundnuts. The cowpea's deep taproot mines the soil for water and nutrients and climbs up the stalk of the maize. This legume also fixes nitrogen and provides moisture-conserving shade, making the area more hospitable to other crops. Maize can also be relay-cropped with cassava and yam. The maize is shallow-rooted and relies on the cassava and yam to mine the soil for nutrients.[21] In sub-Saharan Africa the lablab ornamental flower is relay-cropped in rice fields following the rice harvest.

In Ghana farmers have developed several polycultures of vegetables and fruit trees. A few among the vegetable combinations are: okra, tomatoes, and peppers; carrots, papaya, moringa, and lettuce; cabbage, pepper, and onion; watermelon, garden eggs, and pepper; cucumber, okra, onion, and pepper. A few among the fruit tree combinations are: citrus, pigeon pea, sweet potato, platina, coco yam, and native timber species; mangoes, sweet potatoes, pigeon pea, banana, cassava, and native timber species; cashew, ground nuts, maize, and native timber species; cocoa, pigeon pea, coco yam, cassava, plantain, citrus, avocado pear, and native timber species.[22]

and predatory insects while adding a touch of color and joy to otherwise tedious rows of green. These flowers can be added at the end of virtually any crop row, but are especially suited to collaborate with brassicas, cucurbits, and solanums.

Another intercropping strategy that we employ is the use of one crop as the living mulch of its companion. For example, we broadcast lettuce seeds into a freshly prepared bed and then plant brassica seedlings into the same bed at their normal spacing. The lettuce smothers the weeds, yet does not compete with the brassicas, as their roots reach to different depths. We also use low-growing Dutch white clover as a living mulch in the pathways between mounds of winter squash, seeded at the time of transplanting. The clover fixes nitrogen and attracts spiders, which patrol the crops for damaging insects. Living mulches improve soil structure, control moisture, and prevent runoff.

Parsley, onions, basil, and other aromatic plants have pest-repellent properties and can be planted on bed edges and in the understory of tall crops. In our high tunnel the tomatoes are trellised up to the ceiling while onions and parsley fill in the spaces around their base. Occasionally we transplant a few beets in there as well, which love the warm, protected environment of the high tunnel and don't mind sharing space with other species. Similarly, we plant scallions, basil, and parsley along the bed edges of tomatillos, lettuce, and other crops out in the field.

Companion planting need not be simultaneous. One example of sequential companion planting is the use of wheat as a nurse crop for strawberries and other plants susceptible to wire worms. The wheat is direct seeded eight days before the strawberries are transplanted into the ground. Wire worms are drawn to the wheat and leave the strawberry crop alone. The wheat can be weeded out once the strawberries are well established.

Rye is another grain that can provide benefits to vegetable crops. It has allelopathic qualities, meaning that it secretes chemicals that prevent weed germination. We plant rye in the fall and use a mower to kill it before it goes to seed. The next spring tomatoes, broccoli, melons, and other crops can be planted right into the rye mulch. The rye protects the soil from erosion and drying out, in addition to leaching its 16 different allelopathic chemicals into the soil to control weeds.

The "three sisters," or milpa polyculture, is of course the queen of all intercropping systems. Developed by Indigenous farmers of this continent over 3,000 years ago, the brilliant interdependence of these crops provides complete nutrition to the people and higher productivity than any monocrop. Studies using the Land Equivalency Ratio—a way of measuring the agricultural productivity of land—found that intercropped fields often yield 40 to 50 percent more than monocropped ones. According

Maize and beans are grown together in a highly productive milpa polyculture. Photo by Neshima Vitale-Penniman.

to the UN Food and Agriculture Organization, an increase in "cropping intensity" could avert the need to clear an additional 153 million acres for crops by 2050, an area the size of France. H. Garrison Wilkes, professor emeritus at the University of Massachusetts, calls milpa "one of the most successful human inventions ever created."[23]

In the milpa, bean plants fix atmospheric nitrogen and help reduce damage caused by the corn earworm pest (*Helicoverpa zea*). Squash plants inhibit weed growth with their dense network of thick, broad leaves and retain soil humidity. Natural chemicals (cucurbitacins) washed from the leaf surface of the squash act as a mild natural herbicide and pesticide. For many farmers, the three sisters are just the foundation for a polyculture that may contain a dozen or more annual crops. In our milpa we integrate maize, pole beans, black-eyed peas, squash, gourds, amaranth, and sunflowers. We give the maize and sunflowers a head start before planting the legumes that will climb their stalks and the cucurbits that will spread out beneath them. Amaranth is planted on the edges of the beds so that it can get established before the vining squashes shade out the ground. All of these crops are fall-harvested, so the milpa gets to self-regulate throughout the season with only the occasional weeding. We harvest and dry all of the crops in autumn for storage and winter nourishment.

African farmers have experimented with our own versions of the milpa. Chris Bolden-Newsome

Beets, onions, and parsley share the precious space in the high tunnel. Soon tomatoes will join them in a synergistic intercropping. Photo by Neshima Vitale-Penniman.

of Philadelphia grows sorghum, sweet potato, and cowpeas as the African American "three mo' sisters" combination. These foods "are in our DNA," Chris explains. His garden doubles as a shrine and a learning space where okra, cotton, collard greens, taro, castor bean, and other culturally important crops thrive. The castor bean is notable, not only because it is allelopathic, repelling mice and voles, but because of its role in the biblical story of Jonah, a symbol of the gifts freely given to us by the Divine, regardless of whether we earn them.

In Kenya, Black farmers have developed a system where tick clover (*Desmodium* spp.) takes the place of squash in the "milpa" and serves a fourfold role. It suppresses striga, a devastating parasitic vine, fixes nitrogen, repels pests, and serves as a forage legume for livestock. The integration of *Desmodium* has doubled labor inputs, but tripled revenues for African farmers.[24]

Farm Layouts with Rotations

Colorado potato beetle (*Leptinotarsa decemlineata*) larvae are squishy, slimy creatures that devour potatoes, tomatoes, and eggplants. Not only can they pass through several generations in a season, with 300 offspring per female, but they also overwinter several inches down in the soil. In the spring they emerge ready to feed on your tender solanaceous seedlings as soon as you put them in the ground. Crop rotation is one way to "confuse" Colorado potato beetles and other herbivorous insects. Imagine that you have potatoes in a certain bed in the first year, and the beetles burrow down at the end of the season. They emerge in the spring hungry, only to find, not potatoes, but beans. They do not like to eat beans, so they have to spend days to weeks trying to find the potatoes. In the meantime the potatoes get bigger, stronger, and more able to withstand the hungry larvae.

Animal companions, such as our dog Rowe, are an important part of our pest management strategy. She scares off the deer, coyotes, mice, and rabbits. Photo by Jonah Vitale-Wolff.

Our ancestors gifted us with the concept of crop rotation with a fallow, or rest, period. The basic principle is to shift what is grown in a given area over time. Crop rotation has the benefit of reducing pests and diseases, and preventing soil depletion. There are several guidelines to keep in mind when designing your crop rotation:

- Avoid planting crops from the same plant family in the same place in successive years.
- Precede nitrogen lovers, such as brassicas, tomatoes, and corn, with nitrogen fixers (legumes).
- Crops with lower nitrogen requirements, such as root vegetables and herbs, can follow heavy nitrogen feeders.
- Organize your crop rotation around the plant families that will take up the most space on your farm. For example, the area of brassicas you plant may equal the combined area of alliums, solanums, and legumes.

Tables 6.2 and 6.3 are two example of crop rotation plans that we have used at Soul Fire Farm. The first plan (table 6.2) is organized by plant family. After the

UPLIFT

Swidden Agriculture

In sub-Saharan Africa, the dominant food crop production technology is swidden agriculture, also called the bush-fallow system, shifting cultivation, or "slash and burn." In the swidden system, short periods of agricultural production (1 to 2 years) are followed by long fallow periods (6 to 25 years). During the fallow period the forest regrows, sending its deep roots into the soil to recycle nutrients and build up organic matter. The roots prevent erosion, increase the infiltration of water, and reduce runoff. Further, the fallow period provides cooling shade to the soil, enhancing its moisture content and moderating its temperature. Farmers continue to interact with the land during the fallow period, gathering wood for building materials and fuel, and collecting herbal medicines and livestock feed. At the conclusion of the long fallow, the trees and shrubs are cut down and burned. This releases nutrients into the soil, making artificial fertilizer unnecessary. The burning also destroys weed seeds and eliminates pests, providing a suitable seedbed for food crops.

In 1957 the Food and Agriculture Organization pronounced swidden agriculture "a backward type of agricultural practice," and later Western theorists blamed it for climate change. However, recent studies by scientists at the Indigenous Knowledge and Peoples Foundation found that swidden practices sequester nearly 750,000 tons of carbon per 7,500 acres, while the burning only releases 400 to 500 tons, a ratio that puts industrial agriculture to shame. Soil formed under this system has more above- and belowground organic matter, and subsequently more carbon storage.

The challenge with swidden agriculture is not the method itself, but the reduction of the length of the fallow period due to land theft and overpopulation. As more people have been pushed onto smaller areas of land, the long fallow periods of 10 to 25 years have been reduced to just a few years.[25] The principle of intensive cultivation interspersed with fallow remains viable and sustainable.

Table 6.2. Crop Rotation by Plant Family

Area 1	Area 2	Area 3	Area 4	Area 5
Umbelliferae (carrot family) Asteraceae (lettuce family) Chenopodiaceae (beet family)	Brassicaceae (collard family) Malvaceae (okra family)	Fabaceae (bean family) Cucurbitaceae (squash family) Poaceae (corn family)	Solanaceae (tomato family) Liliaceae (onion family) Convolvulaceae (sweet potato family)	Fallow—cover crops

Table 6.3. Crop Rotation by Planting Date

Area 1 April/August Planting	Area 2 Early-May Planting	Area 3 Late-May Planting	Area 4 June Planting	Area 5 Fallow
Onion	Kale	Summer squash	Corn	Cover crops
Parsley	Broccoli	Celery	Tomatoes	
Beet	Collards	Flowers	Eggplant	
Chard	Cabbage	Cucumbers	Peppers	
Peas	Turnips	Melons	Okra	
Carrots	Lettuce	Potatoes	Winter squash	
Parsnip	Herbs	Beans	Basil	
Spinach			Sweet potato	

first year, the crops in area 1 move to area 2, the crops in area 5 move to area 1, and so on.

While the temporal crop rotation plan (table 6.3) does not perfectly follow the rule of not repeating by family, it is the most straightforward method I have devised and a great place to start if you are daunted by arranging your planting by families. I divide the planting area up into five or more sections. In the spring I start planting the seasonally appropriate crops in area 1, beginning with the nearest bed and working my way down the adjacent beds. In the following year, I start my planting in area 2, and the year after that in area 3. The result is that it takes five years for a given area to be cropped with the same vegetables. In real life it's not as neat and clean as the chart would suggest. We plant several successions of lettuce, turnips, radishes, carrots, and other fast-growing crops, so they end up planted in several areas. Still, I have had good success with arranging my crop rotation based on planting date.

CHAPTER SEVEN

Tools and Technology

The white man, preoccupied with the abstractions of the economic exploitation and ownership of the land, necessarily has lived on the country as a destructive force, an ecological catastrophe, because he assigned the hand labor, and in that the possibility of intimate knowledge of the land, to a people he considered racially inferior; in thus debasing labor, he destroyed the possibility of meaningful contact with the earth. He was literally blinded by his presuppositions and prejudices. Because he did not know the land, it was inevitable that he would squander its natural bounty, deplete its richness, corrupt and pollute it, or destroy it altogether. The history of the white man's use of the earth in America is a scandal.

—WENDELL BERRY

I am in love with the hoe. An ancient, versatile African tool that can open new ground, form mounds, plant, cultivate weeds, and harvest—this piece of flat iron was first put into my hands at age 16 at The Food Project in Boston. In the beginning it was awkward trying to work between the broccoli plants to remove the tender pigweed threatening from beneath. With mentorship and practice I got my stride and learned to dance with the hoe, employing a gentle rocking motion where the tool did most of the work and I just guided gently. I got the nickname "Alacrity" for moving quickly down the beds, outpacing my contemporaries twofold, little beads of sweat erupting on my nose.

When I first traveled to Ghana, West Africa, in my early twenties, I learned that the so-called hoe that I cherished was actually a flimsy lesser relative of the true African hoe. The farmers in Odumase-Krobo handed me a tool that weighed at least four times what a US-style hoe did, with a piercing blade and sturdy construction. They taught me to dig hills for beans, canals for irrigation, and terraces for erosion control, all with this single tool. Choice of technology drives the culture of use. In the case of the hoe, it works better when many people use it together in community. In moving to tractors, we begin to relinquish community. Now each time I go visit our sibling farm in Haiti, I purchase a few of these legitimate hoes and negotiate with the airline about my overweight, oddly shaped baggage. Used correctly, the hoe is an efficient technology to accomplish many farm tasks, and keeps us in direct contact with the land and our own physical power.

The farmers of Ghana also taught me how to build structures out of soil. Across the Dahomey region of West Africa, schools, homes, and places of worship are sculpted out of the earth. Inspired by this model, we constructed our home and educational center from local clay, straw, sand, ground limestone, local wood, and recycled materials. The primary building is a timber-frame straw-bale structure with passive solar design, interior thermal mass, earthen floor, and solar panels for heat and hot water. Choosing to build in harmony with nature is not the quick-and-easy path. We do not subscribe to the capitalist perspective that buildings should go up in a few months and be disposable after 30 years. It took us over three years to construct the durable primary building, and we endured many challenges along the way. For example, when we had almost finished digging the foundation using our undersized tractor and hand shovels through the hard, rocky clay of the mountainside, we realized that solar and magnetic south are about 13 degrees different in our area. This magnetic declination matters, because a solar home must face solar south. Needless to say, many tears of frustration were shed as we picked up the shovels to redig the foundation in the correct direction. We drew strength from the deep knowing that our ancestors faced greater challenges, and endured.

In this chapter we discuss the tools, technologies, and equipment needed for all parts of your vegetable growing operations, from bed preparation, to harvesting, to irrigation. We uplift the contributions of Black farmers in developing these technologies that make efficient, sustainable growing possible.

Bed Preparation

In 2012 Hurricane Sandy, the largest Atlantic hurricane on record, swept across the Caribbean and the entire eastern seaboard of the United States. We were alerted to Sandy's arrival on our farm in the middle of the night, when we heard a deafening and perplexing roar from the forest. Jonah and I sat straight up in bed, looked at each other wordlessly, and headed outside. The powerful sound was coming from a newly formed "river" cascading from the forest and headed right toward our crop fields. It was dark, windy, and cold, yet we knew that if we did not act, we might lose our fall harvest. We woke up the children, put shovels in their hands, and got to work digging a trench to divert the waters from the crops. After several muddy hours, we retired to our beds and hoped for the best. In the morning

UPLIFT

African Hoe

Egyptian farmers began forging and using hoes between 5,000 and 8,000 years ago, predating the first pharaoh. The hoe is versatile and can be used to plow the soil, dig planting holes, hill the earth into mounds, remove weeds, and harvest. The hoe is suitable for wetland and dryland farming, heavy clay soils as well as light, sandy earth. The hand hoe remains the most used tool in Zambia, Zimbabwe, Uganda, Senegal, and Burkina Faso.[1] The African style of hoe is distinct in its large, heavy blade suitable for primary tillage, in contrast with the lighter European hoe designed mainly for cultivation. Originally the African hoe had a short handle to ensure precision of work. However, it led to debilitating back injuries when applied to commercial agriculture with its relentless work hours and repetitive tasks. Cesar Chavez and others in the farmworkers' movement did essential organizing to replace the short-handled hoe, *el cortito*, with the long-handled version in the agricultural fields of the United States. Farmworkers were able to get the hand hoe banned in California in 1975.[2] The short-handled hoe continues to be used appropriately in small-scale, diversified agriculture across Africa.

news of devastation began trickling in. Both of the roads, Routes 2 and 22, that serve our town of Grafton were completely washed out, and residents were trapped. Neighboring farms had lost between 50 and 100 percent of their topsoil in the heavy rains. New York City was without electricity. And on our farm, we experienced erosion in one small section. In all of the other areas, the water was gently infiltrating in the pathways of our raised beds. We realized that our decision to "mound" the soil on our farm had quite literally saved our farm.

Bed Forming

We thank our ancestors for the invention of the brilliant technology of raised beds, which encourage water infiltration, prevent soil compaction, support weed management, and mitigate flooding. We are honored to carry on that legacy by growing all of our crops in mounds. To create a raised bed, you first need to smother the existing vegetation, which can be accomplished with tarps, fire, or tillage, as described in chapter 4. Next you need to remove soil from the pathway and place it onto the growing area for the plants. This work can be done with a shovel or an African-style hoe. In either case it is essential to keep your back straight and bend your knees to leverage the tool. Allow the weight of the tool and the leverage points on your body to do the work, and alternate the side you work on to balance out the strain on your muscles.

It is best to lay out the beds with a long measuring tape, placing stakes or other markers to show where the pathways should be dug. We dig pathways that are 12 inches (30 cm) wide and beds that are 44 inches (1 m) wide at the top and 48 inches (1.25 m) wide at the bottom, for a total of 5-foot (1.5 m) spacing between the centers of the beds.

New beds can also be formed by using tractor implements. If a potential crop area is still in pasture, we begin by flipping the sod with a moldboard plow. We wait several weeks to allow the sod to break down a little, then make a few passes with our 7½-foot (2.25 m) disc harrow. This slices up the sod

Farmers at Soul Fire Farm demonstrate proper back position while using a hoe. Photo by Neshima Vitale-Penniman.

and allows us to plant a cover crop to further build the soil. The moldboard plow and discs are very disruptive to soil structure and are only used to break new ground. Historically, the plow is responsible for much of the devastation of our soils, causing erosion and compaction, and destroying soil life.

Our plow and disc are International brand, and we found them in the reject pile of a neighboring farmer's barn. Older pieces of equipment are often made of higher-quality steel and are consequently more durable. Once our cover crop has established, we mow it down with our 5-foot (1.5 m) Bush Hog brush cutter. After that we come through with a bed former to create our crop beds. Our tractor is a reliable 1997 Ford New Holland 2120, and we have been able to purchase many of its implements secondhand from other farmers in our networks.

Tillage, or Not

When our farm was under ½ acre (0.2 ha), we were able to do all of our tillage by hand. After digging the pathways with a shovel, we would work down the growing area with a hoe to break up clods and then use a garden fork to aerate the soil without turning it over. Standing in the pathway to prevent compaction in the bed, it's best to step on the top of the fork to insert it completely into the soil, then use leverage to push the handle down and draw the tines up through the soil. After aerating the soil, you can create a smooth surface with a hard metal rake. Hand raking is a fine art that involves getting a smooth back-and-forth motion via long, even strokes. Some farmers like to leave a little "lip" on the beds on the downslope side to encourage water infiltration during the dry season, and a slightly raised center of the bed to shed water during the wet season.

As our farm grew, we thought our only option was to switch over to tilling with a tractor, as the conventional farmers in our county were doing. At the end of the season, we used the bed former to toss pathway detritus into the beds and make them accessible to the tiller. We then used a 5-foot (1.5 m) rotary tiller to turn under and break up the

Cheryl DeSanctis operates a Ford New Holland tractor to shape up the beds at Soul Fire Farm.

plant debris from the season. For the final pass, we formed the beds again and applied annual cover crop or thick straw mulch. While we had short-term gains with this aggressive method of tillage, after a couple of years we developed a tiller pan, which is a layer of compaction below the soil surface. We also attributed the spread of perennial weeds, such as thistle, grasses, and yellow nut sedge, and the increase in pests, such as the leaf miner, to the use of this equipment.

For love of the climate and soil, we are experimenting with ways to achieve no- or low-till methods once again. We use large, 24-by-100- or 50-by-100-foot (7.5-by-30- or 15-by-30-meter) silage tarps to smother weeds and encourage microorganisms to do their own "biological tillage." The weeds germinate in the dark and, unable to find light, die off. The warm, moist environment is perfect for worms and other soil life to consume detritus and form aeration tunnels throughout the soil. After removing the tarps, we minimally disturb the soil to avoid turning up latent weed seeds. Thus, it is essential to thoroughly prep the beds before tarping. In very hot weather, silage tarps have the benefit of being able to be flipped so that the white side is exposed, lessening the heat that could kill beneficial microbes.

Tool Care

Tools require regular care and maintenance. At a basic level this means keeping track of tools and putting them back in an organized storage area. I recommend counting the tools that you bring out to the field and establishing a practice of leaving them in a wheelbarrow whenever they are not directly in your hands. Tools need to be stored in a dry place. Each time you are finished, hose off the soil or brush it away using a hard-bristle brush. Several times per season you will need to sharpen your tools using a metal file. Tools with moving parts also need tool oil or mineral oil applied regularly. For any moving harvest tool, such as Felcos, you need to use a food-safe oil.

Tools need a well-organized, dry storage place for their own longevity and to ensure farmer safety. Photo by Neshima Vitale-Penniman.

Tractor Tips

A tractor is a lifetime investment, and we have come to love our 1997 Ford New Holland 2120 with New Holland 7309 loader arms. It is a compact four-wheel-drive tractor that is just the right size to navigate the tight corners on our modestly sized mountainside farm. This tractor has a wide wheel base that makes it safer on uneven ground. We have a rollover bar (ROPS: rollover protection structure) and always wear a seat belt when driving. Tractors can be extremely dangerous, causing over half of farmworker fatalities in the US. Using ROPS and a seat belt brings your chance of surviving a rollover up to 99 percent.[3] In several states there are programs to help farmers pay for installation of a ROPS if the tractor does not have one.

We installed a skid-steer quick-attach on the front loader as well as quick-attach couplings for all of the hydraulic hoses, which allows us to easily swap out attachments. For example, just recently we were moving the chickens to their winter home so we first attached the bucket loader to help us clean up brush from the new area, and then easily switched that out for the pallet forks to pull the chicken housing into place.

Participants in the Black Latinx Farmer Immersion learn how to swap out implements on the tractor. Photo by Neshima Vitale-Penniman.

Our loader is a powerful New Holland 7309. We added extension hoses from the rear hydraulic remotes so that we can use hydraulics with equipment mounted on the front loader. The essential attachments for the front include a loader bucket and pallet forks. The essential attachments for the rear include a brush cutter and bed former. It is important to size the brush cutter, and any other PTO-driven equipment, to the horsepower of the engine. (A PTO or "power take off" pulls power from a running engine and transmits it to an attached implement at a power loss around 10 percent.)

We have a Bush Hog SQ600 that is 5 feet (1.5 m) wide. This smaller cutter works well for cutting pasture on our slopes and narrow pathways. It is also useful for chopping up green manures and crop debris at the end of the season. We also have a PTO-driven tiller, a 5-foot (1.5 m) Kuhn that we used more extensively before we shifted toward no-till methods. Our bed former is a simple model from the Buckeye Tractor Company. Most of the other tractor equipment we have bought from local farmers or the classified listings in our area.

Something to keep in mind for those blessed with heavy clay rocky soils like ours is finding tractor equipment that is durable enough to withstand a thorough beating by the stones. One of the great things about our two-bottom plow is that it has a trip mechanism that allows it to just pop out of place when it hits an immovable rock, rather than breaking. Other plows have similar mechanisms: for example, shear bolts or a PTO with a slip clutch. What you do not want is a rigid piece of equipment that aims to pit its strength against that of the earth—it will lose.

Propagation

Extending our season by harnessing the greenhouse effect allows us to enjoy our cultural crops in the environments to which we have been transplanted and to reduce the environmental footprint of our food miles. When our operation was smaller, we were able to start all of our seedlings on wooden shelves supported on vertical brackets in a large south-facing window in our home. We installed daylight fluorescent lights over each level so that they hung just above the emerging plants on an adjustable chain. While daylight fluorescent bulbs do not perform quite as well as full-spectrum bulbs, they save a lot of money. As our farm grew and the seedling flats crept onto the floor and kitchen table, we realized it was time to build a proper greenhouse.

Our greenhouse is a plastic hoop house that we constructed from scratch. Its skeleton is made of 10-foot-long (3 m), 1⅜-inch-diameter (3.5 cm) fence rail sections that we bent using a hoop bender and joined together with 1-inch (2.5 cm) self-tapping metal screws to attain an overall 30-foot (9 m) hoop length. We drove 1⅝-inch-diameter (4 cm) hollow fence posts into the ground at 2 feet (0.6 m) deep, taking care not to directly hit the pipe end lest they "mushroom" and be ruined. You can use a block of wood or a specialized cap to protect the pipe from your striking sledgehammer or pneumatic driver. We then inserted the bent hoops into the verticals in the ground. The supports are 4 feet (1.25 m) apart on center and reinforced at the peak with two rows of purlins. The double plastic skin is a four-year lifetime ultraviolet-resistant greenhouse poly plastic held in

The heated greenhouse has double plastic walls, a polycarbonate south end with a wide door, forced-air propane heat, fans, and ventilation. Photo by Neshima Vitale-Penniman.

We added struts to our Haygrove high tunnel for additional strength in the heavy snow and to allow us to trellis tomatoes from above. Photo by Neshima Vitale-Penniman.

place using wiggle wire. We installed two layers of the plastic for insulation and use a blower to keep an air pocket between the layers. Our greenhouse has plywood on the end that does not receive sun, and clear polycarbonate panels on the south side. One side has a standard-sized door and the other side has wide doors that allow carts and seedling tables to enter and exit, as well as material deliveries from the tractor. To maintain a warm environment in the greenhouse, we installed an 80,000 BTU propane forced-air heater. There is also a thermostatically controlled active vent that opens and closes to maintain the temperature set by the farmer, and a passive louver vent that opens and closes based on the temperature. There are two hanging fans at opposite ends of the greenhouse that are on at all times to keep the air circulating. Air movement is essential to climate control and disease prevention. At the height of the season, we install a shade cloth over the exterior of the greenhouse to reduce solar gain and maintain cooler temperatures.

We also use an unheated high tunnel to extend the season in either direction. Spinach, lettuce, onions, radishes, and parsley all survive over the winter in the tunnel and provide nourishment in the early spring before the ground outside is even thawed. Our high tunnel is an 80-foot-long (24 m), 25-foot-wide (7.5 m) Haygrove that we built from a kit, and modified along the way. It utilizes simple construction that involves no wood and no carpentry. The plastic is tensioned by ropes rather than wiggle wire, which allows for the sides to be manually pulled up for ventilation in the hot months. The kit came with limited structural supports for hanging, so we added struts between the collar tie and arch. This offered structural integrity to withstand wind, rain, and snow. We also made purlins from 1⅜-inch-diameter (3.5 cm) fence rail to allow us to trellis the heavy tomato plants from above.

Low tunnels are temporary structures that can be used for climate regulation and pest protection. Photo by Neshima Vitale-Penniman.

These modifications were all put together using chain-link-fence parts from the local hardware store.

In addition to the high tunnel, we also use low tunnels and mulch to achieve season extension out in the field. A low tunnel is simply a series of ½-inch (1.25 cm) electrical conduit bent into hoops the width of the bed. We push these hoops firmly into the bed at 4-foot (1.25 m) intervals in the winter and 8-foot (2.5 m) intervals in the summer. Since the hoops have a tendency to collapse toward the center of the bed, we drive two pieces of 18-inch (45 cm) rebar at a 20- to 30-degree outward angle at the end of each bed. We then install the end hoops over the rebar for extra reinforcement. Additionally, for our winter low tunnels we bunch and tie the ends of the plastic and firmly tie them to a 2-foot (0.6 m), 2- to 3-inch-diameter (5- to 7.5-cm) hardwood branch driven into the ground. Then we install 6-mil greenhouse plastic (not construction plastic) over the hoops, weighing it down with stones or long lengths of rebar on the sides and stones or stakes at the ends. For example, if we plant a healthy crop of kale in the late fall, we can install a low tunnel over it and it will be protected through the winter and resume growth in the spring. During the warm season, we can use the same hoops to support floating row cover for pest protection and climate regulation.

Overwintering of root crops, strawberries, and greens can also be achieved with the simple use of mulch. Cover the crops with a "blanket" of 6 to 8 inches (15 to 20 cm) of straw or leaves. For extra protection you can add a layer of clear greenhouse plastic on top of the mulch, and secure it on the edges with stones, sandbags, or rebar. In the spring remove the mulch and allow the vegetables to continue growing.

Whether you start your seeds in your window or a greenhouse, there are a few essential supplies. First of all, it's best to sort and store your seeds in waterproof

Simple seedling tables can be built from 2-by-4 lumber and welded wire fencing. Photo by Neshima Vitale-Penniman.

containers in a dark, dry place with minimal temperature fluctuations. We like to plant our seeds in open 1020 flats, with the exception of cucurbits, lettuce, parsley, celery, beets, and basil, which prefer plug trays. It is necessary to have plastic or wooden labels and a permanent marker to identify what is growing in each flat along with seeding date. You can water seedlings with a watering can; this is often the most efficient solution in the winter when hoses freeze. Fill a giant trough with water in your greenhouse and dip the watering can into the trough to fill. In the warmer seasons hook up a hose to a spray wand that generates a gentle, even rain for the seedlings.

The best height of a worktable is just above your thumb knuckle when you are standing with your arms at your sides, hence the saying *rule of thumb*. For many people this height is 33 to 36 inches (84 to 90 cm), so we build our seedling tables accordingly. Consider building worktables at several heights to accommodate people with various body sizes on your team. They have a simple 2-by-4-inch (5-by-10-centimeter) wooden frame with 2-by-4-inch welded wire fence stapled on the frame to form the tabletop. We have brackets installed along the edges so that we can put galvanized steel hoops over the tables. These hoops can be skinned with greenhouse plastic or floating row cover to provide pest, rain, and wind protection when the tables are used outdoors.

Suppliers

When you're sourcing equipment and supplies, it is best to find local suppliers. Talk to the farmers near you and support Black-owned businesses whenever possible. We get most of our supplies through our local farmers exchange and local stores. However, we order certain items from the suppliers listed in the resources, page 321, and have been satisfied with the quality of the products.

Seeds

- Southern Exposure Seed Exchange
- Truelove Seeds
- Johnny's Selected Seeds
- Fedco Seeds and Fedco Trees

- High Mowing Organic Seeds
- Seed Savers Exchange
- Sow True Seed
- Homowo African Heritage Seed Collection at Old Salem Museums and Gardens
- Lakeview Organic Grain
- Baker Creek Heirloom Seeds
- Indigenous Seed Initiative
- Hudson Valley Seed Company
- Fruition Seeds

Supplies

- Rain-Flo Irrigation
- Nolt's Produce Supplies
- Griffin Greenhouse Supplies
- Carts Vermont

Transplanting and Direct Seeding

As soon as the dandelions bloom and the maple trees leaf out, the soil shows its readiness to receive our warm-season crops. As the climate changes, calendar dates become less reliable predictors of seasonal change than phenological events like flowers blooming and the migration of birds. When the soil is warm and the wild plants are reviving, it is time to bring out your hoe and seed container to plant your fields.

We direct seed vegetable crops by hand rather than using a mechanical seeder. On our uneven, clay soil, it is challenging to get an even distribution of seeds mechanically. So we train our hands and our eyes to be able to determine seed spacing. At the beginning of the season, everyone takes a ruler and measures the length of one of their knuckles, fingers, the full span of their hands, and the length of their stride as a reference point for plant spacing. We carve shallow furrows into the beds using PVC pipe sections attached to a seeding rake to ensure consistently spaced and straight rows. Into this furrow, we place or scatter the seeds. We then use the hoe or a rake to gently push the elevated soil into the furrows to cover the seeds.

Transplanting seedlings from flats into the soil requires its own careful technique. We ensure that the bed is already prepared, complete with irrigation line and straw or plastic mulch, if applicable. We then use a hose with spray nozzle to "dig" holes for the seedlings by using the water to spray a hole at the appropriate plant spacing. We like to use a siphon-mixer to add fish or seaweed emulsion (approved by the Organic Materials Review Institute, or OMRI) into the transplanting water in order to give the plants a boost to help with the transition to a new environment. Into these puddles of water and nutrients, we transplant the seedlings, pressing the soil firmly around their moist roots. Some farmers also add a beneficial bacterial drench, like Actinovate, as a root shield for cucurbits. In order to prevent the hose

UPLIFT

The Oldest Tool in the World

The digging stick originated in Africa in pre-agricultural times as a tool used by foragers to dig out wild tubers, and was later adapted for farming. It is a stick with a pointed end, used for making holes into which seeds can be dropped. Subsequently, farmers began adding iron tips to be able to work denser soils. Over time these iron ends became flatter and wider to resemble the first long-handled hoes. In the early 1900s farmers added perforated stone weights to the top of their digging sticks to increase the penetration of the pointed end. In Cameroon, Gabon, and Congo, farmers developed some of the first planting knives, called *couteau de culture*. They were leaf-shaped blades used for planting and the basis for the modern trowel.

Transplant drench can be added to irrigation water to reduce transplant shock and support the transition from greenhouse to field. Photo by Neshima Vitale-Penniman.

from dragging across other beds of crops, it's best to insert a pitchfork, shovel, or wooden stake into the end of the bed and thread the hose through the tines so that it stays in the pathway.

Irrigation

Water is life. If we have the good fortune of a favorable climate, rainfall may provide enough moisture to enliven the soil biota, nourish the crops, and fill our drinking wells. In fact, most farmers in the world still rely on rainfall as their only source of agricultural water. In a rural community outside Odumase-Krobo, Ghana, where farmers depend on rainfall, the community chief keeps a record of the dates of the rains scratched into his doorpost. When I was living there in 2002, he showed me how climate chaos was making the rains come later each year and end sooner, with the consequence that not all of the crops could fully mature. Climate variability and climate change mandate that we think about supplementary water sources for our farms. Irrigation is a survival strategy.

For the small-scale grower, it may be reasonable to simply hand-water the garden using a watering can or a hose with a spray nozzle. Once you scale up, however, the amount of time it takes to stand there holding the hose becomes untenable. For crops that are direct seeded, a sprinkler system works well. We pay homage to Black farmer Joseph Smith for his invention of the lawn sprinkler. We use mini sprinklers that feature adjustable distance and droplet size to nourish our direct-seeded crops that would have trouble germinating without an even penetration of water. The sprinklers sit on risers about a foot above the ground and can be adjusted to cover just one or two 100-foot (30 m) beds at a time. These mini sprinklers hook up to a garden hose and utilize its water pressure to operate. They have disadvantages in that they waste much of the water through evaporation and need to be continuously moved to new areas in order to cover the farm. Still, these low sprinklers do conserve more water than those with large arcs and high spray volumes.

Our preferred irrigation method is a drip system, the most water-conservative and easiest to use of any we have tried. We use a Rain-Flo irrigation kit and have added components over time. A 1-inch (2.5 cm) lead pipe carries water from our well that goes through a screen filter to remove grit, then to a pressure reducer to lower the pressure to 12 PSI (83 kPa). After the pressure reducer, the water flows to a check valve where we can divert it to a fertigation line; there it can pick up a specific concentration of compost tea or other organic nutrients to be delivered to the plants along with the water. Next the water travels out to the crop area, where it flows into four arteries that correspond to the four main sections of our growing area. From there we attach two lines of drip tape per bed off the main header lines or arteries. The whole system mirrors the circulatory system in our bodies or the roots of a tree in its branching pattern. The drip tape is midweight and has a flow rate of 0.45 gallon (1.7 L) per minute per 100 feet (30 m). The drip tape attaches to the artery by a reverse-threaded valve. The far end of the tape is folded over and secured with an end sleeve. We also install irrigation staples at the ends and middle of the bed to prevent the drip line from blowing out of place. Should someone accidentally puncture the tape with overzealous use of a garden fork, it can easily be repaired with coupling hardware.

UPLIFT
Irrigation Systems of Africa

Ancient Egyptians developed the first irrigation systems in the world, over 5,000 years ago. Their water management technology, known as basin irrigation, involved the construction of a network of earthen banks near the Nile River that formed basins of diverse sizes. The farmers built sluices that directed floodwater into the basins, where it was allowed to percolate until the soil was saturated. Then the excess water would be drained to another basin or canal, allowing the farmers to plant their crops. In this manner Egyptian farmers cultivated nearly 2 million acres for millennia.[4]

In arid areas devoid of floodwaters, African farmers developed climate-adapted irrigation technologies. In the Sahel desert of North Africa, farmers dug underground channels through bedrock to transport water from deep aquifers to the surface. The channels were 100 to 14,000 yards (90 to 12,800 m) in length with aeration wells 8 to 40 yards (7 to 36 m) deep at 10 to 500 points along the channel. These *foggaras* were built through collective labor and shared ownership. Each owner received a share of the water commensurate with their contribution to its development and maintenance. The foggara is the oldest irrigation technology still in use today, with channels in Algeria over 1,000 years old that continue to irrigate thousands of acres of date palms.[5]

African farmers have long used bunding and canals to manage surface water. In dry areas, farmers build mounds of stone or soil (bunds) along the contours of the slope in order to slow the rare storm waters. The bunds have periodic outlets at half their height to release excess water; that water is further retained with mulch and shelterbelts. Bunds are essentially the opposite structure to canals, used by farmers in the Rio Nunez region of Guinea for managing waterlogged soils. These farmers dug an intricate irrigation system of dikes and canals to move fresh water. They planted seedlings on the mounds and rice in the trenches.[6]

The heart of the drip irrigation system: grit screen, pressure reducer, and fertigation line. Photo by Neshima Vitale-Penniman.

Drip irrigation tape installed in a bed, ready to deliver water directly to the plant roots. Photo by Neshima Vitale-Penniman.

Drip irrigation can also be used with a gravity system. We are currently constructing a method to pump water from our pond up to a 1,500-gallon (5,700 L) poly storage tank at the top of one of the fields. The drip irrigation system that we will connect to the storage tank will essentially be the same as the one currently connected to our pressure tank, less the pressure reducer. The one downside of drip irrigation is that is does generate a fair amount of plastic waste once the lines are spent after a season or two.

These systems do not preclude the need for hoses and direct watering. Even with the sprinklers and drip irrigation, we still use hoses for transplanting, spot watering, and cleaning crops and equipment in the field. We have hundreds of feet of hose carefully laid out across the farm such that it can reach even the farthest corner of the crop beds as well as avoid being mowed by the tractor or impaled by a soil fork. At the barn, we have a hose manifold with shutoff valves to control each of the four to six main hoses departing from that junction. We then have a series of splitters allowing hoses to further branch off into the field. Spray nozzles are installed at the end of each hose.

Weeding and Crop Maintenance

The sprawling cucurbits and tidy brassicas stay low to the ground and are able to support their own weight into maturity. However, plants whose wild ancestors were accustomed to vining up trees and grasses require supports in order to thrive, as do those crops that we have bred to be so laden with heavy fruits as to bend. Trellising can be as simple as gathering three long sticks from the forest, setting them into a tripod shape, lashing them together at the top, and planting your climbers around the base. For a small-scale garden, this method is more than adequate, but once you scale up, a more streamlined trellising approach is needed.

For our crops in the field, we like to set a trellis fence of 6-foot-tall (1.75 m) 2-by-4-inch (5-by-10-centimeter) welded wire down the middle of the bed, supported with 7-foot-tall (2 m), rot-resistant white oak garden stakes woven in every 6 to 8 feet (1.75 to 2.5 m). We then plant the peas, pole beans,

A fence made of welded wire and garden stakes is useful for supporting pole beans, peas, and other climbing crops. Photo by Neshima Vitale-Penniman.

Tomatoes and tomatillos can be trellised outdoors using a basket weave of garden stakes and twine. Photo by Neshima Vitale-Penniman.

cucumbers, or tomatoes on either side of the trellis fence, such that there are two crop rows in the bed. As the plants grow, we reinforce the trellis by weaving a length of twine around the plants and looping it to the fence every 8 feet (2.5 m) or so. Once the plant is fully mature, it may have three to five lengths of twine holding it up against the fence.

For tomatoes and tomatillos grown outdoors, we use a basket-weave trellis. First we pound garden stakes into the row of established tomatoes every 4 to 6 feet (1.25 to 1.75 m). We then weave twine around the plants, so that they are sandwiched between the supportive string on either side. A pass of twine every 18 inches to 2 feet (45 to 60 cm) will provide adequate bracing, and you should continue to add passes as the plant grows. The Immokalee workers, a coalition of Mexican, Guatemalan, Haitian, and Caribbean tomato pickers, also invented a basket-weave trellising tool. According to Larisa Jacobson, the tool is a 2- to 4-inch-long (5- to 10-inch-long) wooden dowel with two screw eyes on one side of it. You run the trellis line through the screw eye guides and tie to the end post. Then you

We use specialized clips to attach tomatoes to the trellis string. Photo by Neshima Vitale-Penniman.

can scoop up plants and basket-weave very quickly and much more tightly. You clip the tomato twine box to your belt with a carabiner or tie it to a belt loop with a loop of twine. You use a gloved left hand (if you're right-handed) to tension the line, scooping up wayward branches with the taut line itself, jam the end of the dowel against each stake as you move down the row, tension the line with other hand, and let the dowel spin freely on its own around each stake to rotate back to you, then repeat on down the row and back down the other side.

Inside the high tunnel, we find it simpler and gentler on the tomato plant to hang trellis twine from the overhead purlins or wires of the tunnel. The twine hangs from the ceiling almost to the earth; as the plants grow we secure them to the twine using specialized tomato clips.

In addition to providing structural support to the plants as they grow, it is crucial to weed out their competitors. Many of these "weeds" have edible and medicinal properties, to be discussed in a later chapter, and the process of removing them can double as harvest. Of course, weeds can be removed by hand: Grab them firmly from the base and offer a smooth tug to dislodge the roots. A hand cultivator can assist the hand-weeding process when the roots you are trying to remove are more resilient. However, hand weeding is time consuming and not feasible for market gardens and farms. If you time your workflow to catch weeds when they are young, at two true leaves

Selecting the right tool for the job is essential. Here the farmer demonstrates the incorrect tool selection for weeding, but the correct one for looking fly. Photo by Jamel Mosely.

or fewer, you can use a J-hoe, collinear hoe, or stirrup hoe to quickly scuffle them away. These hoes employ a short-stroke back-and-forth motion that cuts the roots right below the surface. They are compact tools that allow you to work between plants, even when tightly spaced.

Harvest

I admit that we might have been a little too informal in our harvesting procedures in our early days. We did not have access to a refrigerator or cooler for our produce, so we harvested into clean bedsheets doused in water and allowed the evaporative cooling to preserve the freshness of the leaves until distribution. We left the washing to the consumer, only hosing off soiled roots before packing them into boxes.

Because we are comparatively small and distribute all of our produce through a CSA, we are exempt from many of the rules under the Food Safety Modernization Act (FSMA), but we are updating our procedures around the farm to comply with these important health and safety standards. The produce rule for FSMA ensures that water and soil that come in contact with the vegetables are uncontaminated, that we as workers wash our hands and don't work when sick, and that the storage vessels we use are clean and sanitary. We recommend that you spend some time reading these rules and signing up for a training course or webinar.

UPLIFT
US Black Farmer Patents

Black farmers have made significant technological contributions to agriculture in the United States. Many of us are familiar with the story of Eli Whitney's invention of the mechanized cotton gin (1794), which catalyzed an industrial boom and led to the expansion of slavery. What is less often told is the fact that the person who gave Whitney the idea for the cotton gin was an enslaved farmer known to history only as "Sam," and that the basic principle of Sam's gin was rooted in African technologies dating back to the 5th century CE.[7] Enslaved people were not granted patents by the government, so records of their myriad antebellum inventions are lost. In 1834 Henry Blair, a free Black man, became the first African American to be identified on a US patent application. He was awarded a patent for his invention of the corn planter, which combined plowing, placing the seeds, and covering the seeds with soil.[8] A slew of inventions by Black farmers followed, including: Norbert Rillieux's sugar refiner (1846), Alexander P. Ashbourne's coconut oil refiner (1880), Joseph Lee's kneading and bread-crumbing machines (1890s), George W. Murray's eight agricultural patents for planters, cultivators, and fertilizer distributors (1890s), Peter Smith's potato digger (1891), John T. White's lemon squeezer (1896), Joseph Smith's lawn sprinkler (1897), William H. Richardson's cotton chopper (1899), John A. Burr's lawn mower (1899), and Leonard Julien's sugarcane planter (1966).[9]

Prior to the onset of Western patenting, Black farmers invented myriad hand tools for agriculture, including the cutlass (sub-Saharan Africa), the banana cutter (Uganda), the socketed ax (sub-Saharan Africa), the transpierced ax (sub-Saharan Africa), the curved sickle (Sahel region), the lateral sickle (Mali), and the plow (Ethiopia).

We harvest all of our crops by hand, using a harvest knife or scissors for the crops that cannot be picked directly. We bring a garden cart out to the field with us, stacked with hard plastic harvest crates and carrying a small bucket with the knives and rubber bands we will need. It is best to harvest directly into these clean crates, close them, and put them in the cart under a clean damp cloth while we continue with the harvest. Certain crops, like tomatoes, do not tolerate being stacked in deep containers, so we harvest them into shallow open crates that are stackable. Once the cart is full of packed crates, we transport the crops to the barn where we label them with the date and contents on the exterior and leave them at ambient temperature. This small step of labeling saves a lot of time when it comes to distributing the produce later on. Crops like broccoli benefit from being dunked in cold water and "hydrocooled" right

Most crops are harvested directly into these durable, stackable bins.

Dunking certain vegetables in cold water immediately after harvest helps preserve freshness and increase storage life. Photo by Neshima Vitale-Penniman.

Farmers crowd into the cooler to see how the CoolBot tricks an air conditioner into maintaining low temperatures. Photo by Neshima Vitale-Penniman.

after harvest, which greatly increases their freshness and storage potential.

Most of our crops are stored in a homemade walk-in cooler, which is an 8-by-8-foot (2.5-by-2.5-meter) insulated box with an insulated door and wooden shelves. It stays cool via a CoolBot device, invented by farmers in the Hudson Valley, New York. The CoolBot "tricks" a standard off-the-shelf digital air conditioner into maintaining the desired low temperatures needed for storing vegetables. The system is much less expensive than a commercial walk-in cooler and has needed no maintenance for us in its 5-plus years of continuous use on our farm.

When it's time to distribute our produce to the community, we divvy it out into 1 1/9-bushel cardboard boxes that can easily stack in our van. We assemble and lay the boxes out in the barn and then pack them, with the heavier items on the bottom. Some of our crops, like beans, peas, and lettuce mix, are bagged before we box them. We use grocery-store-brand gallon-sized plastic storage bags and brown paper lunch bags as containers. We also have both a hanging scale and a bench scale to help us accurately and fairly divide the produce among the number of shares.

Apparel and Gear

The sum total of all the equipment and supplies mentioned in this chapter does not exceed the value of your labor as the farmer. Taking proper care of your health and well-being and setting yourself up for success with protective work gear are both essential to the success of your farm. Experienced farmers can recognize one another right away by what they are wearing, and it's not generally skinny jeans and a fedora.

While there are a few tasks where it is workable to be barefoot, more often than not the task at hand involves navigating sharp tools, traversing muddy soil with hidden sticks, or working with moving equipment. A high-quality pair of waterproof boots is essential. Many farmers swear by the Muck brand, which is breathable and waterproof. Additionally, farmers should have a waterproof rain jacket and rain pants, work gloves, insulated waterproof gloves for cold, wet days, and a sun hat for hot, sunny days. I like to don a fanny pack to carry my phone, to-do list, water bottle, and small items like seeds or fasteners. Don't hate—fanny packs are the new slay!

Most of these young farmers are appropriately dressed for a workday in the rain, with impermeable jackets and rubber boots. Photo by Neshima Vitale-Penniman.

Farm Tools and Equipment Checklist

Bed Preparation

- ☐ Shovel
- ☐ Garden hoe
- ☐ Hard rake
- ☐ Tractor with bed former, plastic mulch layer, disc harrow, power harrow, Perfecta, and/or spader (tractor and implements optional for smaller operations)
- ☐ Silage tarps for no-till
- ☐ Measuring tape—100 or 300 feet (30 or 90 m)

Propagation

- ☐ Waterproof seed container
- ☐ Seedling table
- ☐ 1020 flats and plugs
- ☐ Plastic labels
- ☐ Permanent marker
- ☐ Hose and watering spray wand
- ☐ Watering can
- ☐ Ventilation
- ☐ Heat source
- ☐ Watering trough

Transplanting and Direct Seeding

- ☐ Hose and spray nozzle
- ☐ Trowel and/or Hori Hori
- ☐ Pitchfork
- ☐ Garden hoe
- ☐ Seeding rake

Weeding and Crop Maintenance

- ☐ Colinear hoe, J-hoe, stirrup hoe, garden hoe
- ☐ Hand cultivator
- ☐ Row cover
- ☐ Trellising, such as white oak stakes
- ☐ Galvanized steel hoops
- ☐ Weedwacker
- ☐ Lawn mower and/or tractor-drawn mower

Irrigation

- ☐ Sprinkler *and/or*
- ☐ Drip irrigation system—drip tape, header line, shutoff valves, valve hole tool, drip tape repair couplers, pressure reducer, inline filter, irrigation staples
- ☐ Hose, manifold, splitter, spray nozzle
- ☐ Siphon for transplant drench

Harvest

- ☐ Harvest knife and scissors
- ☐ Soil fork
- ☐ Harvest crates
- ☐ Plastic bags
- ☐ Scale
- ☐ Washing bin
- ☐ Cold storage
- ☐ Hand clippers
- ☐ Garden cart, wheelbarrow
- ☐ Boxes

Gear

- ☐ Notebook
- ☐ Waterproof rain jacket and rain pants
- ☐ Tall waterproof boots
- ☐ Insulated waterproof gloves
- ☐ Work gloves
- ☐ Sun hat
- ☐ Hip pack/water bottle holder
- ☐ Basic first-aid kit
- ☐ Hammock, for after-lunch siesta ☺

Tool Care and Maintenance

- ☐ Metal files
- ☐ Tool oil (such as food-safe mineral oil)
- ☐ Dry storage location
- ☐ Hard-bristle brush

CHAPTER EIGHT

Seed Keeping

My grandmothers were strong.
They followed plows and bent to toil.
They moved through fields sowing seed.
They touched earth and grain grew.
They were full of sturdiness and singing.
My grandmothers were strong.

My grandmothers are full of memories
Smelling of soap and onions and wet clay
With veins rolling roughly over quick hands
They have many clean words to say.
My grandmothers were strong.
Why am I not as they?

—MARGARET WALKER, "Lineage"

As the family story goes, my grandmother's grandmother's grandmother, whose true name was forgotten and whose enslaved name was Susie Boyd, was kidnapped from the shores of West Africa around the year 1800. She and other mothers of the community had been witnessing the kidnapping and disappearance of their community members and experienced a rising unease about their own safety. As insurance for an uncertain future, they began the practice of braiding rice, okra, and millet seeds into their hair. While there were no report-backs from the other side of the transatlantic slave trade, and rumors abounded that white people were capturing Africans to eat us, they still had the audacity of hope to imagine a future on soil. Once sequestered in the bowels of the slave ships, they continued the practice of seed smuggling, picking up grains from the threshing floor and hiding the precious kernels in their braids.

Though we do not need anthropologists and historians to confirm that water is wet, it is notable that these experts have verified our family legends of seed keeping. The Djuka Maroon communities of Cayenne possessed varieties of rice specific to the Baga people of Guinea, brought by enslaved women who smuggled them in their hair. A separate Maroon group, the Saramaka of Suriname, independently corroborated this seed transport story in a conversation with anthropologist Richard Price. In South Carolina the earliest rice varieties were the African *Oryza glaberrima*, grown mainly by women in their provision gardens. Enslaved Africans relied on informal networks of seed exchanges within and between plantations to maintain their cultural crops.[1]

At Soul Fire Farm, when we grow and save the seed of ají dulce, Eritrean basil, callaloo, sorrel, Moyamensing tomatoes, fish peppers, bee balm, Ashwagandha, Hopi blue maize, Tuxpan and Tatume squash, and watermelon, we do so with the intention of honoring the grandmothers who braided seeds in their hair as an act of resistance. In this chapter we uplift the work of the Haitian Peasant Movement,

Alliance for Food Sovereignty in Africa, and the seed keepers of the US to conserve our genetic and cultural heritage through seed keeping. We explore practical methods for planning a seed garden, harvesting seeds, and exchanging seeds within the community.

Why Save Seed?

"You've got a few companies that want to control all the seed stock of the world, and they've just about got a handle on marketing three of the main commodities: corn, soybean, and cotton. [For us], it's hard to find seeds that aren't treated with the Monsanto-manufactured [herbicide] Roundup Ready. I've tried to find cotton that wasn't treated, but I couldn't. Now they're working on controlling wheat and rice," explained Black farmer Ben Burkett, in an interview with the blog *Black Left Unity*.[2] His experience underscores the importance of the work to save heritage seed in our own communities. Just 60 years ago, seeds were largely stewarded by small farmers and public-sector plant breeders. Today the proprietary seed market accounts for 82 percent of the seed supply globally, with Monsanto and DuPont owning the largest shares.[3] In our work with sibling farms in Haiti, we learned about Monsanto's insidious practice of making "donations" of seed for a few seasons, until the native seed stock was depleted, and

UPLIFT

Haitian Peasant Movement, and Alliance for Food Sovereignty in Africa

The Haitian Peasant Movement, also known as the Group of Four (G4) or "4 Eyes Meet," is a coalition of the four largest social movements in Haiti: National Congress of Papaye Peasant Movement; Peasant Movement of Papaye (formed in 1973); Heads Together Small Producers of Haiti (Tet Kole); and Regional Coordination of the Southeast. G4 brings together over 25 million rural farmers on a united platform of food sovereignty and conducts survival work, such as pig repopulation, tree planting, and native seed exchanges.[4] G4 contributes significantly to La Via Campesina, the global peasant movement that has more than 200 million rural and peasant members in 79 countries. As part of their international work, they formed the Dessalines Brigade, named after the Haitian independence leader Jean Jacques Dessalines. In 2007 the Dessalines Brigade cooperated with South American farmers, agroecologists, and activists to share ways of protecting native seeds and local farming practices. In 2013 the G4 won the Global Food Sovereignty Prize for their rejection of a substantial donation of hybrid seeds by US megaproducer Monsanto following the 2010 earthquake, choosing instead to protect their native seed heritage.[5]

Farmers are also organizing for their food sovereignty on the continent of Africa. Thirty African farmer collectives, grassroots organizations, and consumer movements came together to form the Alliance for Food Sovereignty in Africa. They work to protect seed diversity in situ, from the native grasses of the savanna to fruit tree crops.[6] One of the member organizations, the National Coordination of Peasant Organizations of Mali, with around 2.5 million members, convinced its country to amend the constitution to name food sovereignty as a basic right.

then charging farmers unreasonable prices for the company's proprietary seed in subsequent seasons. In this way, Monsanto could ensure dependency of the farmer on their corporation indefinitely.

Beyond simply preserving the genetic heritage of the seed, it is also crucial to our survival that we preserve the stories of our seeds. Owen Taylor and Chris Bolden-Newsome, of Truelove Seeds, encourage us to think about the work of preserving seeds as "seed keeping" rather than "seed saving," in that our obligation is to keep the stories of the farmers who curated the seeds alive along with the plant itself. It matters to know that roselle is from Senegal and that the Geechee red pea is an essential ingredient in

Saving the seed of Sea Island red pea and other heritage varieties preserves genetic diversity and cultural memory. Photo by Owen Taylor.

A meeting of the peanut-seed-saving cooperative of Bigonet, Haiti. Photo by Jean Moliere.

the Gullah dish known as Hoppin' John. In keeping the stories of our seeds alive, we keep the craft of our ancestors alive in our hearts.

In addition to the ethical reasons for seed keeping, there are economic and ecological considerations. Small farmers spend hundreds to thousands of dollars per year on seeds, and seed keeping can help cut those costs. Additionally, allowing plants to go to flower and then seed increases pollinator habitat. The Food and Agriculture Organization estimates that we have lost 75 percent of the world's crop varieties in the past 100 years, and 22 percent of the wild relatives of certain staple crops. This puts us at great risk in a climate-unstable future. By saving heritage seeds, we can contribute to biodiversity preservation and global food security.[7]

The Seed Garden

Even after a decade of farming, I was intimidated by the idea of saving seed. I imagined that the people who packed seeds into those tiny paper envelopes and mailed them to me each year had some secret inaccessible magic to make those seeds viable. I did not trust myself not to ruin the life force potential of these tiny beings of possibility. I did not trust the plants growing on my farm to make progeny as healthy and vibrant as those I could purchase in the store. Of course, I realize that I had internalized the corporate messaging of agricultural megacorporations, which try to get us to believe that we are not good enough to steward our genetic heritage. I was wrong. As Chris Bolden-Newsome affirmed, "Our ancestral grandmothers did not have a seed catalog, a life raft, a backup plan. They didn't know nothing about isolation distance and all that stuff. They had to do and they did, and we are here."[8]

We started saving seed bit by bit, first garlic and potatoes, then calendula and lettuce, then tomatoes and beans. The key is to choose healthy, vigorous plants with the ideal characteristics of that variety. Rather than eat the very best, we save the very best for the future. While we are by no means expert seed savers and owe much of the wisdom in this chapter to our sibling farmers, we have discovered the accessibility and ease of keeping our own seed. In planning your first seed garden, there are a few things to keep in mind.

Open-Pollinated Versus Hybrid Seeds

Open-pollinated varieties retain their distinct characteristics when they mate with others of the same variety. They share pollen via insects, birds, wind, or other natural mechanisms. Open-pollinated heirlooms are in danger of becoming inbred if we don't take measures to increase their genetic diversity by maintaining large populations and trading seeds with other seed keepers. Hybridization is a method of human-controlled pollination in which the pollen of two different varieties is intentionally crossed to yield a desired trait. Due to a phenomenon known as hybrid vigor, the first generation of hybridized seed, called F1, tends to produce higher yields than the parent varieties and have higher genetic diversity. However, any seed saved from a hybrid plant will not breed "true to type" and will be less vigorous. Seed keepers are best advised to choose open-pollinated seeds to start. Once you are an advanced seed keeper, you may try stabilizing a hybrid variety after about seven years of breeding.

Species and Isolation Distances

Broccoli, brussels sprouts, cabbage, and cauliflower are all members of the same species, *Brassica oleracea*, which means that they are able to reproduce together. On the other hand, peas (*Pisum sativum*) and beans (*Phaseolus vulgaris*) are members of different species and cannot reproduce together even if planted in the same row. Certain species, like legumes, lettuce, and tomatoes, are able to self-pollinate, which means they can transfer pollen from the anther to the stigma of their flowers without the aid of insects or wind. Other species, like corn, collards, okra, and squash, cross-pollinate, meaning they rely on the transfer of pollen from the anther of a flower on one plant to the stigma of a flower on a different plant of the

Plants that cross-pollinate, like maize, need to be isolated from other varieties in order to breed true to type.

same species. Understanding the genetic relationships between crops allows us to prevent unwanted cross-pollination between different varieties of the same species. This can be accomplished by maintaining prescribed isolation distances between varieties or by spacing out their cultivation temporarily. Some growers also use hand pollination or pollination barriers to achieve isolation.

Annuals, Biennials, and Perennials

Some crops grow, flower, set seed, and die all within a single growing season. These crops are known as annuals and are the easiest plants to grow for seed saving. Annuals include legumes, lettuce, tomatoes, peppers, okra, and maize. Other crops require a cold period in order to flower, a process called vernalization. These biennials include carrots, onions, leeks, beets, and some brassicas. Other plants, such as asparagus, raspberries, and apple trees, are perennials, surviving and flowering for many years. It is easiest to start with annual, self-pollinating crops that require little to no isolation and a small population size to produce seeds. The easiest crops for beginning seed keepers are peas, lettuce, beans, and tomatoes.

Hand Pollination

In order to ensure that open-pollinated varieties are not contaminated with genetic material from other varieties, advanced seed keepers can use hand pollination. With plants that produce unisex flowers, the pollen from the male flowers is manually transported to the unpollinated stigma of the female flower. Then the female flower is covered to prevent

Table 8.1. Seed Garden Planning Chart

Crop	Species	Pollination	Life Cycle	Isolation Distance (feet)	Population Size (viable seed, long-term maintenance)
Bean	*Phaseolus vulgaris*	Self or cross	Annual	15 (4.5 m)	1, 6
Beet, chard	*Beta vulgaris*	Cross	Biennial	800 (240 m)	5, 25
Brassicas	*Brassica oleracea*	Cross	Biennial	800 (240 m)	5, 25
Butter beans	*Phaseolus lunatus*	Self or cross	Annual	15 (4.5 m)	1, 6
Callaloo	*Amaranthus* spp.	Self or cross	Annual	800 (240 m)	1, 25
Carrot	*Daucus carota*	Cross	Biennial	800 (240 m)	5, 25
Celery	*Apium graveolens*	Cross	Biennial	800 (240 m)	5, 25
Cilantro	*Coriandrum sativum*	Cross	Annual	800 (240 m)	5, 25
Corn	*Zea mays*	Cross	Annual	800 (240 m)	10, 100
Cucumber	*Cucumis sativus*	Cross	Annual	800 (240 m)	1, 10
Eggplant	*Solanum melongena*	Self or cross	Annual	300 (90 m)	1, 10
Gourd	*Cucurbita* spp.	Cross	Annual	800 (240 m)	1, 10
Leeks	*Allium ampeloprasum*	Cross	Biennial	800 (240 m)	5, 50
Lettuce	*Lactuca sativa*	Self	Annual	10 (3 m)	1, 10
Melon	*Cucumis melo*	Cross	Annual	800 (240 m)	1, 10
Mustard	*Brassica juncea*	Cross	Annual	800 (240 m)	5, 25
Okra	*Abelmoschus esculentus*	Cross	Annual	500 (150 m)	1, 10
Onion	*Allium cepa*	Cross	Biennial	800 (240 m)	5, 50
Parsley	*Petroselinum crispum*	Cross	Biennial	800 (240 m)	5, 50
Peas	*Pisum sativum*	Self	Annual	20 (6 m)	1, 10
Peanut	*Arachis hypogaea*	Self	Annual	20 (6 m)	1, 10
Peppers	*Capsicum* spp.	Self or cross	Annual	300 (90 m)	1,20
Roselle*	*Hibiscus sabdariffa*	Self	Perennial	20 (6 m)	1, 20
Sorghum	*Sorghum bicolor*	Self or cross	Annual	200 (60 m)	1, 20
Southern peas	*Vigna unguiculata*	Self	Annual	20 (6 m)	1, 20
Spinach	*Spinacia oleracea*	Cross	Annual	800 (240 m)	10, 75
Squash	*Cucurbita* spp.	Cross	Annual	800 (240 m)	1, 10
Tomatoes	*Lycopersicon* spp.	Self or cross	Annual	20 (6 m)	1, 10
Turnip	*Brassica rapa*	Cross	Biennial	800 (240 m)	5, 50
Watermelon	*Citrullus lanatus*	Cross	Annual	800 (240 m)	1, 10

Sources: "Seed Isolation Distances," Vegetable Seed Saving Handbook, http://howtosaveseeds.com/table.php; *Seed Saving Guide* (Decorah, IA: Seed Savers Exchange, 2017), https://www.seedsavers.org/site/pdf/Seed%20Saving%20Guide_2017.pdf; *Seed Saving Chart* (Emeryville, CA: Seed Matters, 2015), http://www.seedsaversalliance.org/uploads/4/3/9/8/4398404/seed_matters-seed_saving_chart.pdf.

* Roselle is a perennial crop that is grown as an annual in cold climates.

A farmer in Kenya hand-pollinates maize. Photo by F. Sipalla/CIMMYT.

additional pollen from entering. In the case of corn, seed keepers place a bag over the male tassels at the top of the plant to collect the pollen. They also cover the young female ears before the silks emerge as these can be wind-pollinated by varieties on neighboring farms. The pollen from the tassels is collected in the bag and manually placed on the silks once they emerge. The ears are then rebagged until maturity. In the case of squash, the female flowers are distinguishable by the presence of a large ovary below the flower that resembles an immature fruit, something lacking in the male flowers. When both the male and female flowers show a yellow blush of color, they are mature and will likely open the next day. At this point, seal the flowers closed with flagging tape. The next day, remove the male flowers and use them to brush pollen onto the female flowers at the rate of three male flowers per one female flower. Then gently reseal the female flower and allow it to mature.

Population Size

Certain plants require several individuals in close proximity in order to reproduce. Others can reproduce with just a single individual. However, for long-term preservation of genetic diversity and its associated vigor, it is best to keep seed from a large population of individual plants.[9]

The Seed Harvest

I could hear the sound of the maize shelling before I could see it, cresting the hill to visit Mr. Kwabla's farm in Oborpah-Djerkiti, Ghana. The whole family was engaged on the threshing floor. They had spread tarps and feed bags over the ground of their home and outdoor kitchen. The cobs of maize waited in a thick layer on top of the tarps. Elders, adults, and children arranged themselves amid the maize with thick sticks in their hands, beating the cobs to release the grains. It was a rhythmic, joyful cacophony. When the maize was finished, we moved onto the beans. In this case, we didn't beat the crop with sticks, but rather danced on it with bare feet. This practice of dancing and stomping on dried pods made its way to the Diaspora, where farmers in Trinidad's Palo Seco hamlet put rice plants into large cloth sacks laid out in the sun. Once dried, they separate the rice seed by dancing and stomping on them barefoot and later remove the hulls in a mortar and pestle.[10]

The process of harvesting and processing seeds for storage can be as straightforward and joyful as the methods of the farmers of Ghana and Trinidad. Seed harvesting and processing are best accomplished in community accompanied by a great soundtrack.

Many seeds need nothing more than to be gathered in their mature state from the field and stored in their pods in a dry place, such as a hoop house. For beans, peas, lettuce, peppers, okra, brassicas, spinach, leeks, corn, and most herbs and flowers, this simple dry harvest method can be used. When the seedpods are fully mature, harvest them before the rain, then allow them to dry further indoors before threshing the seeds and winnowing. Here are more details for major crops:

Beans and peas. Harvest the pods when they begin to turn brown and leathery. Bring them to a warm, dry location and spread them out on a single layer to dry for about one week. Once the seeds rattle in the pods, put the pods into a pillowcase or basin and stomp on them to remove the seeds. Winnow to remove the chaff by pouring the contents of the

Thorough drying is an essential step in the seed saving process. Photo by Owen Taylor.

Wear gloves when processing hot peppers for their seed. Photo by Owen Taylor.

basin into a large container with a fan blowing off the falling chaff. Note that screens can also be used to separate seed from chaff. Once the seeds have been fully winnowed and cleaned, lay them in a single layer out of direct sunlight and allow them to further dry for about a week.

Lettuce. Allow the lettuce to send up its flower stalk rather than harvesting it before bolting. When half of the seeds appear feathery, like dandelion fluff, cut the stalks at the base and allow them to continue to mature on tarps for a couple of weeks. Then knock the seeds into a bucket by shaking the plants back and forth and knocking them against the side of the bucket. Winnow by blowing on the seeds gently or using a fan on low speed. Spread the cleaned seeds in a single layer out of direct sunlight and allow them to further dry for about a week.

Okra. Harvest the pods once they are large, brown, and splitting open. Allow them to rest in a warm, dry place for about a week. Open the pods and remove the seeds. Dry the clean seeds in the open air for a few days before transferring into storage.

Brassicas. After the winter cold the plants will bolt and produce seedpods. Harvest when the pods and the seeds inside of them are brown. Lay them out to dry in a single layer for two weeks. Then thresh by whacking the stalks against the interior side of a large bucket, allowing the ripe seeds to fall while not forcing the unripe pods open. Winnow on low speed. Allow the cleaned seed to dry for another few days out of direct sunlight.

Alliums. The plants will make flowers in the second year. When the seeds are brown and hard, harvest by shaking the heads into a bucket. Then allow the seeds to dry out of direct sunlight in a single layer for two additional weeks.

Corn. When husks are brown, dry, and brittle and the silks are dark and dry, harvest the ears and allow them to dry further in their husks for 1 to 2 weeks. Then shell the corn using your hands or a blunt object. Take care with the latter method as it can crack the kernels, reducing their storage life.

Peppers. When the fruits are completely ripe, having turned red, orange, yellow, or whatever color is their final stage, cut them open to reveal the seeds. Scrape the seeds onto a paper plate in a single layer and allow them to dry for two weeks out of direct sunlight. With hot peppers, use gloves during processing.

Other plants, such as eggplant, tomatoes, cucumber, squash, and watermelons, rely on fermentation in order to complete their natural ripening cycle. In nature

Tools of the Seed Keeper

Seed Cleaning

- Large bowls and buckets
- Flat tub or large bucket (for dancing and stomping on pods)
- Screens mounted on wooden frame, openings of various sizes (for removing seed from chaff)
- Wire strainers and colanders (for cleaning wet and dry seed)
- Electric fan (for winnowing)
- DIY seed cleaner/aspirator (optional; you can get the design free from the Real Seed Catalogue)
- Gloves (for processing hot peppers)
- Serrated knives, spoons
- Glass jars with loose-fitting lids (for fermenting seed)
- Blender (for processing husk cherries, tomatillos, and Ashwagandha)
- Paper plates or paper towels (for drying seed)

Seed Storage

- Paper bags
- Sharpies
- Coin envelopes
- Glass or plastic jars with screw-top lids (to keep out bugs and moisture)

A homemade seed aspirator uses vacuum suction to separate seed from chaff. Photo by Owen Taylor.

Saara from Added Value Farm, Brooklyn, New York, threshes seeds by whacking the stalk against the inside of a large bucket. Photo by Owen Taylor.

certain fruits drop to the ground, where hungry bacteria get to work fermenting the sugars and cleaning the seeds. A similar process happens in the digestive tract of animals. When we harvest fermentation-dependent seeds, we can mimic that natural process indoors with the help of a little water. Fermentation separates seeds from pulp, reduces the risk of disease, and increases germination. When you let the pulp sit in water, the seed generally sinks, making it possible to pour the pulp off the top. It is important to keep a cover on the fermenting slurry to prevent insect infestation. Here are details for each major crop:[11]

Children scoop seeds from mature tomatoes with Owen Taylor of Truelove Seeds. Photo by Owen Taylor.

Dry cleaned seeds thoroughly out of direct sunlight before storing. Photo by Owen Taylor.

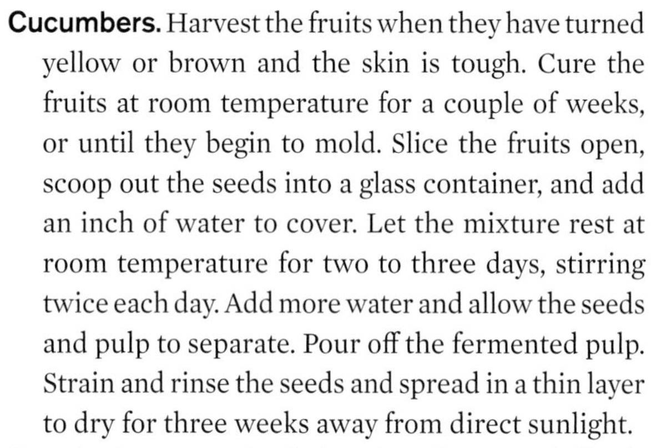

Cucumbers. Harvest the fruits when they have turned yellow or brown and the skin is tough. Cure the fruits at room temperature for a couple of weeks, or until they begin to mold. Slice the fruits open, scoop out the seeds into a glass container, and add an inch of water to cover. Let the mixture rest at room temperature for two to three days, stirring twice each day. Add more water and allow the seeds and pulp to separate. Pour off the fermented pulp. Strain and rinse the seeds and spread in a thin layer to dry for three weeks away from direct sunlight.

Eggplant. Harvest the fruits when they turn dull yellow or brown and the skin becomes tough. Cut the fruit into quarters or eighths, exposing the seeds in the middle of the fruit, and squish the seeds and pulp through a screen. Allow the pulp and seeds to ferment for two to three days. Remove the seeds from the pulp and spread them in a thin layer on a paper plate or cloth and allow to dry for three weeks.

Watermelons. Harvest the fruits when they are fully ripe or slightly overripe. Scoop out the flesh and push it through a fine mesh over a bucket. The seeds will remain on top of the screen. Rinse the seeds through a strainer. Spread them in a thin layer to dry for three weeks.

Squash, zucchini, cantaloupe. Leave the fruits on the vine until they are full-sized and the skin is very hard. Harvest the fruits before the first frost and store for one or two months at 50° to 60°F (10–16°C) to cure. Open the fruit and scoop out the seeds. You can clean them immediately or soak them in water for 24 hours, stirring occasionally, then rinse. Spread the clean seeds on a paper plate to dry for three weeks.

Tomatoes. Harvest the fruits when they are at peak color and fully ripe. Squish the whole fruit through a screen, or halve the tomatoes first along the equator to expose all seed cavities and then scoop the seeds into a jar with a pointed vegetable peeler. Cover the seedy pulp with about 1 inch of water and cover the jar with a paper towel or piece of cloth. Label the jar with the variety and date, let the mixture ferment for four to five days, stirring occasionally. Remove the large pieces of flesh, then rinse the seeds in a strainer. Spread in a thin layer to dry for three weeks.

Seeds are alive! They are embryos encased in a protective shell of nourishment. Seeds have evolved resiliency and maintain their vitality for several years, even in harsh conditions. However, we can keep them alive longer through proper storage until the following season's planting time. Seeds need to be stored in a location that is dry, dark, and cool. We keep our seeds in carefully labeled paper envelopes that are stacked in plastic Tupperware on a shelf in a cool, dark closet. The best-case scenario to avoid insect damage and the introduction of moisture is to put envelopes in jars with screw-top lids. For seeds that will be stored for five years or more, the freezer is the best location. Just remember to allow frozen seeds to acclimate to room temperature before opening the package. Table 8.2 is a list of crops with the respective longevity of their seeds.

Table 8.2. Seed Longevity Under Proper Storage Conditions

Seed Type	Longevity Under Proper Seed Storage Conditions (years)
Arugula	3–5
Beans	3–4
Beets	4–6
Broccoli	3–5
Brussels sprouts	4–5
Cabbage	4–5
Callaloo	7+
Carrots	3
Cauliflower	4–5
Celery/celeriac	5
Chard	4–6
Collards	5
Corn	4–6
Cucumbers	5–8
Eggplant	4–6
Gourd	4–6
Kale	4–5
Leeks	1–2
Lettuce	2–3
Melons	5–7
Mustard	4–5
Okra	3–4
Onions	1–2
Peas	3–4
Peanut	1–2
Peppers	2–4
Pumpkins	4–6
Radish	5
Roselle	1–3
Sorghum	4
Spinach	2–4
Summer squash	4–7
Tomatoes	4–5
Turnips	5
Watermelon	4–6
Winter squash	3–4

Source: O. L. Justice and L. N. Bass, "Principles and Practices of Seed Storage," in the *USDA Agricultural Handbook 506* (Washington, DC: US Department of Agriculture, 1978).

Seed Exchange

Back in the day, while organizing community gardeners and youth urban farmers in Worcester, Massachusetts, we would convene an early-spring seed and plant swap for the soil stewards of our city. We booked the large community room for free at the public library and invited everyone to bring the seeds they saved personally and their purchased surplus seeds to exchange with other growers. These were the days before social media took off, so we spread the word through door-to-door knocking, flyers, and phone calls. The excitement was palpable. The snow melted outside and mild winds hinted at the coming thaw of the earth and return of green. The seed swap was our testament of faith that another season of sowing and reaping would soon arrive.

We learned a few key logistical parameters for seed swapping by trial and error. It is important to let the growers know what they are expected to bring and what they can expect to take away. Make sure they know that seed collected from hybrid plants will not reproduce the same characteristics as their parents and should not be exchanged at the swap. Also, share a seed viability chart so people know that very old seeds may not retain their vitality. Generally, you can ask people to take only as much seed as they bring. However, we like to have extra seed on hand so that if someone does not have seed to exchange they

UPLIFT
Black Seed Keepers Today

Ira Wallace is a seed keeper and worker-owner at Southern Exposure Seed Exchange, a cooperative that sells over 700 varieties of heirloom seed grown by 50 farmers. She grew up gardening with her mother in Florida and went on to study native plants domestically and internationally. She was one of the co-founders of the Acorn community (1993), where she helps farm 60 acres of land organically, and co-manages Southern Exposure Seed Exchange. She authored *The Timber Press Guide to Vegetable Gardening in the Southeast* and serves on the Organic Seed Alliance. Her personal mission is to "create sustainable, regional food systems built on cooperative self-reliance that provide safe, healthy food for everyone—starting with home and community gardens."

Clifton Slade is another Black farmer who grows seed for Southern Exposure Seed Exchange. He is a third-generation farmer and extension agent at Virginia State University. Formerly a "big tractor" farmer growing GMO corn, he switched to organic seed growing after he observed wildlife avoiding the GMO crops. It took five years for Ira to convince Clifton to become a seed farmer, because he was concerned about being able to make a living. As a seed grower, he can now make the same income on 2 acres that his neighbors make on 300. One of his specialty crops is sweet potato slips. Virginia Baker, the heirloom that he grew up growing with his father, nearly drifted into obscurity. He helped recover this variety to preserve the legacy for his four children and future generations.[12]

Tierra Negra Farm, North Carolina, convened Seed Keepers, a people-of-color-led seed-keeping collective that works to keep alive knowledge, culture, and ancestry through the preservation of seeds. In addition to saving and exchanging seeds, Seed Keepers pass on the stories of the seeds and those who stewarded them. There are six farms and 16 farmers along the Atlantic Coast engaged in seed-keeping projects inspired by their time organizing together.

can still take a few packets. The host should supply some items to make the swap go smoothly:

- Tables for seeds, plants, and handouts
- Pots, trowels, and potting soil for separation of perennials
- Blank labels and Sharpies
- Instructional signs that indicate how you want seeds organized and labeled
- Coin envelopes and scoops for dividing bulk seed into smaller amounts
- Seed and plants (have extra so no one leaves empty-handed)
- Charts including seed viability, seed saving how-to, and gardening how-to
- Sign-in sheet, print or digital

Of course, seed exchanging may already be happening in your community. As expert seed keeper Ira Wallace advises, "Meet people who are already doing it in your area." She recommends looking up the Organic Seed Alliance and Seed Savers Exchange

We honor our ancestors through keeping their seeds and seed stories alive, as with this okra pod descended from West African varieties. Photo by Owen Taylor.

to find others already engaged in your community. Seed Savers Exchange is part of the Community Seed Network, an alliance that provides tools and advice to community groups interested in creating their own seed exchanges and seed libraries. The network compiled an interactive map of seed libraries across the world. To jump-start your heirloom seed collection, you can purchase seed from Seed Savers Exchange, Truelove Seeds, Baker Creek Heirloom Seeds, Indigenous Seed Initiative, Hudson Valley Seed Library, Southern Exposure Seed Exchange, Fruition Seeds, or another heirloom, organic provider (see the list of seed companies in chapter 7, page 138, and the resources, page 321). You can also talk with other farmers in your area and ask them to share seed from their cherished varieties. These locally adapted strains are likely to thrive in your microclimate. However you get started, taking the leap to stewarding seed honors the work of your ancestors who braided seeds in their hair before being forced to board transatlantic slave ships, believing against the odds in a future on soil.

CHAPTER NINE

Raising Animals

Ngai is the creator of everything. In the beginning, Ngai was one with the earth, and owned all the cattle that lived on it. But one day the earth and sky separated, so that Ngai was no longer among men. The cattle, though, needed the material sustenance of grass from the earth, so to prevent them dying Ngai sent down the cattle to the Maasai by means of the aerial roots of the sacred wild fig tree, and told them to look after them.

—JENS FINKE, *Traditional Music and Cultures of Kenya*

We have never had a television in our home, partially because "chicken TV" is more entertaining and life affirming. For years we raised chickens in the backyard when we lived on Grand Street in Albany until the city changed the ordinances and outlawed our flock. Our laying hens preceded us as full-time residents at Soul Fire Farm and will likely be a permanent part of the soil fertility management on this land. We have always had at least one rooster in the mix with the hens. While the roosters have often been loud, showy, and even intimidating, we never had one outright mean until the summer of 2017. One tiny rooster with dull gray plumage must have had some early trauma, because he went out of his way to inflict bodily harm on any farmer who entered the pasture. He hid behind the wheels of the chicken coop where we couldn't see him, waited for us to turn away, and then sneak-attacked our ankles with claws and spurs engaged. He pecked at our hands when we gathered eggs, and hurled his body toward our faces. We love all beings and tried everything to get this rooster, whom we named Rocky Balboa, to come around. One of our farmers even carried him around in a baby sling for a few days, having read that scared animals just needed swaddling. This was not successful, and it got so bad that a few of the Soul Fire Farmers refused to take care of the chickens altogether, while the remaining willing farmer went in with makeshift armor: padded gloves and double pants in the hot weather.

One day after making it through chicken chores alive, this farmer mused, "Universe, can you arrange some type of rooster swap? I pray for a kind rooster to join our flock." The next day a large, majestic rooster with white plumage appeared at the edge of the forest, then retreated. We called the neighbors, none of whom raise crops or livestock, to see whether anyone had lost a chicken. No one had. Over the next several days, this magical rooster made wide rounds at the perimeter of the field, observing but not coming too close. After a week it went right into the chicken paddock, attacked Rocky Balboa, and left him huddling in the corner nursing his wounds. We named the new chicken Apollo Creed. Creed continued to beat Balboa from time to time until Balboa retreated into the forest. Creed happily adopted Balboa's 25 hens and ample home, and showed gentleness and restraint with the farmers. While we do not celebrate Balboa's demise, we do muse at the intricate lives of chickens and the magic of prayer.

Animals have long been part of holistic, balanced agricultural systems in the Diaspora. In Haiti the Creole black pig was the main source of protein and the ritual offering made at the beginning of the revolution of 1791–1804. In the African Sahel farmers developed a rotational grazing system to maintain the health of the soil and their livestock. This chapter explains how to raise poultry and pigs in a sustainable, humane manner, including slaughter and packaging processes. A discussion on strategies for sustainable meat consumption in the era of human overpopulation is included, rooted in African ancestral wisdom.

Raising Chickens for Eggs

At Soul Fire Farm the chickens make gentle happy noises and produce eggs with sunflower-yellow yolks. The key to these outcomes is the use of the rotational grazing practices perfected by our ancestors. We once raised a flock of 200 chickens for eggs, but now partner with a neighboring farmer to supply eggs to our Ujamaa Farm Share members. We maintain an educational flock of just 25 laying hens in addition to about 200 chickens for meat each year. Except when they are too young to endure the rigors of outdoor living, we raise our birds on pasture.

Chicks

We purchase day-old laying hens from McMurray hatchery and day-old meat birds from Freedom Ranger hatchery. We make a point to order rare and heritage breeds, like Dominiques and Andalusians, to do our part toward conserving agricultural biodiversity. The babies come directly to the post office in a specialized cardboard crate, and need to be picked up and fed right away. The nursery can vaccinate your birds for Marek's disease, which we recommend. (Do *not* use medicated feed for baby chicks that are vaccinated.) Make sure the brooder is set up before

UPLIFT

Rotational Grazing of Africa

Rotational grazing systems have long been a part of traditional African agriculture. When colonizer James Watt came to Guinea in 1794, he observed the Fulbe of Labé in Futa Jallon, "rotating their upland rice fields with pasture lands." He noted that the farmers pastured their cattle on the fields before converting those areas into rice cultivation. As a result, they created "exceedingly rich land" with high quality soil and a good rice crop.[1] Additionally, these African farmers gathered up manure, heaped, and burned it in the rice nurseries for additional fertility. Likewise, in Senegal farmers alternate rice fields and pasturage, which fortifies the soil and reduces tsetse-fly infestation. Farmers in the Madiama Commune, Mali, are preserving the rotational grazing practices of their forebears in the West African Sahel. They raise cattle by dividing the pasture into four to seven paddocks that are intensively grazed for limited amounts of time such that all of the plant species are consumed. When the animals are rotated to the next paddock, the original pasture recovers. As a result, the density of plants on the grazing land increases and the health of the calves and adults likewise improves.[2]

the chicks arrive, as they may be hungry and chilled from the journey.

Baby birds live in a brooder until they are large enough to move out to pasture at around three to four weeks old. A brooder can be any protected, nondrafty space with solid walls and floors for pest protection. A box 2 feet (0.6 m) wide by 3 feet (1 m) long and 18 inches (45 cm) deep will house 25 chicks. Adjust the size based on the number of chicks you are rearing. Place clean, dry wood shavings, dry sand, or other absorbent material in the bottom of the box. Brooder lights (heat lamps) are used to keep the chicks warm, 1 light per every 25 birds. For the first week, keep the temperature at 90°F (32°C). Lower the temperature to 85°F (30°C) for the second week, then 80°F (27°C) the third week, and 75°F (24°C) the fourth. More important than the temperature is to monitor the chicks' behavior. If they are panting and located far from the light, they are too hot. If they are "piling" under the light, they are too cold. On hot, sunny days we open and screen the door to the brooder, then close and bolt it at night.

Young chicks need a heat lamp or heating plate to stay warm. Photo by Neshima Vitale-Penniman.

The brooder is constructed to be predator-proof, with a layer of check wire lining the walls and floor. Photo by Jonah Vitale-Wolff.

Chicks are at a tender stage of life, and should be checked and given food and water twice daily. At first feeders and waterers are placed directly on the ground so that the baby birds can reach. As the birds grow, prop the feeders onto wooden blocks and then eventually hang them, always at the height of the average bird's back. Add fresh bedding every one to two days to keep conditions sanitary in the brooder. When the chicks grow full feathers and reach about four weeks old, they may be ready to go to pasture, but do not mix new pullets with older hens. The youngsters may not be able to compete for food and the hens may injure them seriously. When it is time to move the birds, do so at night when they are sleeping. Otherwise you may learn the hard way that hens do not herd well.

Pullets and Hens

Once hens have their full feathers and are big enough to be contained by an electric fence, they can move out to pasture housing. Our laying hens live in a coop on wheels that is suitable for both summer and winter dwelling. It is equipped with roosts for night perching as well as nest boxes for laying. We place straw on the floor of the coop to help absorb moisture from the chicken manure and compost the bedding for use on crops. Hens are quite flexible about the design of their home, so long as the housing provides sufficient floor space, protection from the weather and predators, ventilation without drafts, a place to roost, and nest boxes for laying eggs. Table 9.1 shows the floor space requirements for different life stages. Use this to plan the minimum space for your chicken housing. We believe that providing considerably more than the minimum recommended living space per bird is best for their health, aiming for 2.5 to 3.5 square feet (0.23 to 0.27 square meters) per bird inside the weather-tight coop and an additional minimum of 4 to 5 square feet (0.37 to 0.46 square meters) per bird in the fenced, outside area.

Table 9.1. Floor Space Requirements for Chickens

Type of Bird	Minimum Floor Space per Bird (square feet)
Chicks (0–10 weeks)	0.8–1.0 (0.08–0.09 sq. m)
Chicks (10 weeks to maturity)	1.5–2.0 (0.14–0.19 sq. m)
Brown egg layers	2.0–2.5 (0.19–0.23 sq. m)
White egg layers	1.5–2.0 (0.14–0.19 sq. m)

Additionally, laying hens require 8 to 10 inches (20 to 25 cm) of roost space per bird and one nest box for every four to five birds. Our nest boxes are 12 inches (30 cm) cubed and are located in the dark eaves of the coop. A board on the front about 4 inches high is necessary to hold the shavings and straw in place. If possible, include a hinged door accessible from the outside of the coop to gather eggs and clean the nests. The roosts can be constructed out of 2-by-2-inch (5-by-5-centimeter) lumber or hardwood saplings from the forest. Allow at least 14 inches (36 cm) of horizontal distance between the perches. The birds will sleep on the roosts at night, affording additional predator protection.

The laying hens are moved to fresh pasture every three weeks. Their grazing area is fenced in with electronet fencing with a perimeter of 164 feet (50 m). To install the fence properly, make sure it's tight from bottom stake to bottom stake. Reinforce the corners with posts inserted at a 45-degree angle away from the fence that are then tied with twine to the fence post itself. The end result will be a straight rectangle with no gaps or sagging. The fence is electrified with a solar electric energizer. Remember to turn the charger off whenever moving the fence and to turn it back on before leaving the area.

Feed can be grown right on your farm or purchased from a local supplier. If store-bought, for the first eight weeks the hens should eat turkey starter, which is 27 percent protein. From 8 to 21 weeks old, they switch to turkey grower, which is 21 percent protein, and when they start laying eggs, they switch to layer mash, which is 16 to 18 percent protein. They continue to eat layer mash for the rest of their lives at the rate of about ⅓ pound (0.15 kg) of food per laying hen per day. We use one 3-gallon (12 L) feeder per 25 birds and fill it all the way to the top every morning. Hens also may be fed table scraps and garden products. To

An electronet fence protects the chickens from predators and supports rotational grazing. Photo by Neshima Vitale-Penniman.

The hens live in a movable coop with nesting boxes and roosts. Photo by Neshima Vitale-Penniman.

avoid spoilage and rodents, give them only as many scraps as the hens can consume in under 30 minutes.

At the same time, we rinse and refill two 3-gallon rubberized watering pans to the top. In the winter our water freezes and a submersible heater must be used to keep the hens healthy.

We recommend collecting the eggs in the late morning (approximately 11 AM) and early evening (approximately 4 PM). Frequent egg collection prevents pecking. Keep the nest boxes lined with hay, straw, or other bedding to prevent eggs from breaking. Wash eggs with warm water (cold water pulls in bacteria), air-dry, pack large-end up into cartons, and place in the refrigerator. Properly stored eggs last for about one month.

Most chicken illnesses can be prevented with good sanitation and proper housing. Watering pans should be washed with a soap, water, and vinegar solution every couple of weeks or when visibly grimy. Bedding should always be fresh enough that you enjoy visiting the chickens. Chickens are social animals and should never be left without other chicken companions or they will decline. Chickens naturally spend almost half of their day scratching and pecking for food, so they require adequate forage area. If denied this natural behavior, they will instead peck at one another. In the winter you can support this natural behavior by hanging a cabbage from the ceiling of the coop for them to peck at and eat. If one of your chickens does become sick, isolate her in separate housing and add a splash of apple cider vinegar to her water. If necessary, manipulate her beak manually to help her eat and drink. Almost all of our sick birds have recovered with isolation and extra care.[3]

Raising Chickens for Meat

"How many of you eat chicken?" we asked the group of youth gathered around the movable coops in the field. Almost all of the hands went up. "How many of you would raise a chicken and kill it yourself?" No hands went up. The youth protested that killing a chicken would be cruel and gross. I shared that I voluntarily raise and kill hundreds of chickens every year, even though I am a vegetarian, because I know our community eats meat and I want that meat to be humanely and sustainably raised.

At Soul Fire Farm we raise Freedom Ranger meat birds, a variety bred to exhibit natural behaviors on pasture, unlike the ubiquitous Cornish Cross broiler chicken. The Cornish Cross hybrid variety is bred solely for weight gain and broad breasts, and becomes so obese by the time it is six weeks old that it cannot bear the weight of its own body, collapses, and rubs its belly raw from friction with the ground. After experiencing that inherent cruelty firsthand, we searched for a breed of chicken that would still grow to a good size over time, while retaining the ability to scratch, peck, and move about naturally. Freedom Ranger, Bresse, Jersey Giant, and Orpington all emerged as good options.

We raise four to five batches of 50 pastured meat birds per season, for a total of 200 to 250 birds. Just as with laying hens, the chicks spend their first few weeks of life in a brooder under a heat lamp. For the meat birds, our "brooder" is simply a 100-gallon (380 L) rubberized watering trough that we leave in the barn. When they outgrow that space in about 12 days, we move them to a larger coop, then out to the field. Since space is tight on the farm and we have so many rotations, we create a chart to manage the movement of these birds and make sure they have adequate housing. We also have to plan ahead for the slaughter date to make sure that our volunteers, equipment, marketing plan, and delivery are all lined up in advance.

When the birds are about one month old, they are ready to move out to pasture. The meat birds live in movable hoop coops, sometimes called chicken tractors. These structures are open on the bottom so the chickens have access to grass and insects and can deposit their manure onto the earth. The chicken tractors are rotated daily to fresh pasture using a dolly. I have used several models of chicken tractors, none of which I particularly like because of their bulky construction. I decided to design my own movable coop to be lighter and simpler. It consists of a 12-by-6-foot (3.5-by-1.75-meter) wooden frame at the base, and metal conduit bent into hoops for the roof covered by

UPLIFT
A Legacy of Raising Fowl in the Black Community

Guinea fowl are among the oldest of the gallinaceous birds, predating chickens, turkeys, pheasants, and grouse. They are endemic to Africa, with first archaeological evidence suggesting origins 3.6 million years ago. Footprints found in the Laetoli flats of Tanzania of what is believed to be guinea fowl were probably made in damp volcanic ash and then buried and hence preserved. Guinea fowl are foragers, surviving on insects and weed seeds and producing an all-dark gamy meat.[4] Fowl is a staple food in the West African diet, served with a starchy root vegetable or grain along with palm oil stew. On the other side of the continent, Egyptian farmers likewise raised fowl for thousands of years. Textual records indicate that chickens were already present in Egypt by the time of the Third Dynasty of Ur (c. 2113–2006 BCE).[5]

The interdependent relationship of Black farmers and fowl continued during the time of enslavement. Fearful that enslaved Africans could buy their freedom from profits made by selling animals, the Virginia General Assembly in 1692 made it illegal for slaves to own horses, cattle, ducks, geese, or pigs. Chickens, though, weren't considered worth mentioning. Black farmers—both free and enslaved—built their farm businesses on the raising of chickens.[6] Black farmers sold chicken and eggs to white households, including that of President Thomas Jefferson. On September 29, 1805, the Jefferson kitchen purchased (among other items) 47 dozen eggs and 117 chickens from the Black farmers at Monticello. To put that single purchase in context, the kitchen bought 564 eggs and about 527 pounds (240 kg) of poultry. That's more than a quarter ton of chicken (at an average weight of 4.5 (2 kg) pounds per bird).[7]

Table 9.2. Soul Fire Farm's Pastured Meat Birds Schedule

Batch #	Birth, Sending	Arrival, Brooder	Date to Coop	Date to Field	Days to Maturity	Maturity, Slaughter
1	4/18	4/20	5/2	5/23	78	7/5
2	5/2	5/4	5/16	6/6	78	7/19
3	5/16	5/18	5/30	6/20	78	8/2
4	5/30	6/1	6/13	7/4	78	8/16

wire fencing and a tarp. Both ends are covered with chicken wire, but on one side the chicken wire is held shut using metal spring clamps so that the farmer can easily open and close the entrance to care for the birds.

Our hoop coops are constructed using the following materials:

(3) rot-resistant 12-foot (3.5 m) 2x4s (for base)
(1) 12-foot 1x6 for cross bracing (for base)
(5) 10-foot pieces of ½-inch (1.25 cm) EMT galvanized metal conduit (bend for hooped roof)
(20) two-holed straps for ½-inch pipe (to attach conduit to base)

A simple movable coop can be constructed with lumber, conduit, fencing, and a tarp.

(1) 10-by-12-foot (3-by-3.5-meter) piece of 2-by-4-inch welded wire fencing (attach over the conduit, staple to the base)
(1) 9-by-11-foot (2.75-by-3.5-meter) tarp (place over the fencing, zip tie)
(2) 4-by-6-foot (1.25-by-1.75-meter) pieces of chicken wire (for the ends)
Nails or screws (for base)
Construction staples (to attach fencing to wood)
Zip ties (to attach metal to metal)
Metal spring clamps (for the end that will be accessed to feed birds)

Meat birds are moved to adjacent fresh pasture daily using a dolly. Be careful not to injure birds by running over them with the back end of the structure. You may need to walk around to the back and bang on the fence to get them to move out of the way, then return to the front and resume the move. After a few days, the chickens learn the routine and eagerly walk along with their moving home toward the fresh pasture.

Meat birds consume turkey grower at 21 percent protein for their entire life cycle. To care for the meat birds, fill two 3-gallon (12 L) feeders per 50 birds once daily, or one 3-gallon feeder per 50 birds twice daily. Since meat birds grow rapidly, keeping the feeders full is important. Also, fill two 3-gallon watering pans to the top each morning. Hanging feeders are best and should be continually adjusted to be at the level of the backs of the average-sized chickens. To prevent disease, watering pans should be washed with a soap, water, and vinegar solution every couple of weeks or when visibly grimy. The day before slaughter, the feed containers need to be removed from the meat birds. Fasting cleans out their bowels and prevents contamination of the meat with feces.

From Field to Freezer

Taking life should not be taken lightly. It is important to learn slaughter and processing through hands-on practice with an experienced mentor before attempting the work on your own. We also recommend that

Participants in the Black Latinx Youth Immersion rotate the meat birds on to fresh pasture daily. Photo by Neshima Vitale-Penniman.

you check out the New York State *On-Farm Poultry Slaughter Guidelines* published by Cornell Cooperative Extension.[8] What follows are the steps we use at Soul Fire Farm.

First we gather the necessary supplies, most of which we rent from a neighboring farmer, but some of which we purchase from Cornerstone Farm Ventures. The equipment needed includes:

- Killing cones
- Large metal pot
- Floating thermometer
- Gas burner
- Plucking machine
- Plastic tables
- Cutting boards
- Sharp knives
- Needle-nose pliers
- Three or four hoses
- Trays or racks for draining birds
- Hanging scale
- Buckets with lids

Minimize fear for the chicken by holding her firmly and gently. Photo by Neshima Vitale-Penniman.

- Garden cart—chicken wire, sheet
- Towels
- 100-gallon (380 L) trough for use as chill tank
- Ice (10 bags for 50 birds)
- Three bowls for organs
- Vacuum-seal bags
- Masking tape
- Sharpies
- Dish soap
- Spray cooking oil
- Spray bottles with 10 percent bleach solution
- Aprons or designated clothing

We begin by saying Jewish and Haitian Vodou prayers for transitioning life and honoring death.

Hebrew Prayers

Kol han'shama, hallelluyah, t'halayl Yah, halleluyah
Halleluyah, praise Yah with every breath, halleluyah

Kee hu Ayl zan oom-far-nayse la-kol oo-may-teev l'kol oo-may-cheen mazon l'chol b-ree-yotav asher bara
For You are Almighty, feeding and sustaining all and doing good to all and preparing enough food for all the creatures You created

Baruch atah Adonai hazan et ha-kol
Blessed are You, Nourishing One, who feeds all

Baruch ata ado-nai elo-heinu melech haolam asher kideshanu bemitsvotav vetsivanu al hashechitah
Blessed are you Adonai, Sovereign of the Universe, you have sanctified us with your commandments and instructed us regarding shechitah (ritual slaughter)

Vodou Prayer to the Lwa of Death

Gedeviyawe gede o
Gedeviyawe gede nibo se lwa

Gede Yahway (God)
Gede Nibo (smooth swaying rhythm of Gede) is a lwa

Yoruba Prayer for Sacrifice of Hen or Rooster

For use in cases when the blood offering is designated for orisa. Note that the heart, liver, feet, and other organs may also need to be prepared and offered to orisa. Ask your spiritual elder for guidance.[9]

Call: *Agbe bo adie* (hen)/ *Aku ko adie* (rooster) **Response:** *Sara yeye gbo nku lo, Sara yeye*	Hen/Rooster Shake your feathery body and carry away Death, Shake your feathery body
C: *Agbe bo adie / Aku ko adie* **R:** *Sara yeye gbarun lo, Sara yeye*	Hen/Rooster Shake your feathery body and carry away Illness, Shake your feathery body
C: *Agbe bo adie / Aku ko adie* **R:** *Sara yeye gbofo lo, Sara yeye*	Hen/Rooster Shake your feathery body and carry away Loss, Shake your feathery body
C: *Agbe bo adie / Aku ko adie* **R:** *Sara yeye gbegba lo, Sara yeye*	Hen/Rooster Shake your feathery body and carry away Paralysis, Shake your feathery body
C: *Agbe bo adie / Aku ko adie* **R:** *Sara yeye gboran lo, Sara yeye*	Hen/Rooster Shake your feathery body and carry away Big Trouble, Shake your feathery body
C: *Agbe bo adie / Aku ko adie* **R:** *Sara yeye gbepe lo, Sara yeye*	Hen/Rooster Shake your feathery body and carry away Curse, Shake your feathery body
C: *Agbe bo adie / Aku ko adie* **R:** *Sara yeye gbewon lo, Sara yeye*	Hen/Rooster Shake your feathery body and carry away Imprisonment, Shake your feathery body
C: *Agbe bo adie / Aku ko adie* **R:** *Sara yeye gbese lo, Sara yeye*	Hen/Rooster Shake your feathery body and carry away All the Others, Shake your feathery body

Birds are slaughtered in killing cones, allowing the blood to drain into a receptacle below. Photo by Neshima Vitale-Penniman.

It is significantly less traumatic for birds to be transported a few hundred feet from the pasture to the place of transition than it is to box them into a vehicle and transport them miles to a slaughterhouse. At all stages of the process, we strive to make the work as gentle as possible. Still, we cannot prevent suffering. In the raising of animals (and plants for that matter), there is always pain and loss. We encourage you not to gloss over this or numb yourself to it, rather to feel the emotions it engenders and move from a place of compassion for these beings you steward and for yourself.

On transition day we first set up all of the equipment. The killing cones are erected and sprayed with cooking oil to prevent the blood from sticking. A large pot of water is warmed to 150°F (66°C) on the gas burner with a floating thermometer inside. The plastic folding tables are erected and sanitized, cutting boards and knives arranged neatly atop and buckets waiting underneath for offal. The 100-gallon (380 L) chilling tank is filled with cold water and ice.

Once our transition area is completely ready and the scalding pot hot with steam, we gather the first batch of 16 birds from their housing, holding each one firmly with wings against the body. The birds are transported in a garden cart that we retrofit with some chicken wire and a sheet to make it into a dark and secure carrier. They rest quietly in the dark garden cart at the processing station until their turn to be transitioned.

We place the birds upside down into the killing cones, belly facing out. We then pluck the feathers on one side of the neck to allow unobstructed access to the carotid artery. A very sharp knife is used to make a firm, singular stroke through the artery but not through the bone, which would dull

Scalding water loosens the feathers and makes their removal easier. Photo by Neshima Vitale-Penniman.

the knife. We hold the neck open for a few moments to ensure blood flow, then wait several minutes for blood to drain and all electrical impulses and muscle spasms to stop.

The next step is to remove the dead bird from the killing cones by the feet. Hose off the bird to remove blood, dirt, and any feces. Place it upside down into the hot-water pot at 150°F (66°C). Depending on the size of the pot, you may be able to put two or three birds in the hot-water bath at the same time. Submerge the bird up to the knees, swishing back and forth, not in and out. Keep the animal in the hot water for approximately 60 seconds, until a large wing feather pulls out with a light tug.

Once the feathers are loose, you can place the bird into the plucking machine. The claws of the chicken can tear its skin in the plucker, so it is best to remove the feet before turning on the machine. Remove the

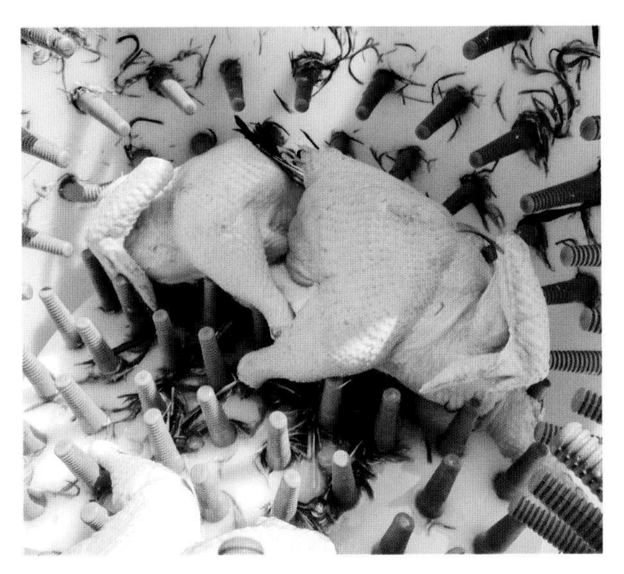

The plucking machine removes most of the feathers using water and rubber "fingers." Photo by Neshima Vitale-Penniman.

feet by cutting through cartilage, not bone. Place the feet in a covered bucket. Run the plucking machine until almost all of the feathers are removed. Before we were able to rent a plucking machine from a neighboring farm, we did our plucking by hand. If you choose to use the hand-plucking method, pull the feathers off in handfuls, tugging in the opposite direction of the feathers' natural orientation. Remove the stubborn feathers with tweezers or needle-nose pliers.

Once plucked, the birds move to the sanitary evisceration zone of the operation. Wash hands thoroughly before proceeding. All table surfaces, cutting boards, and knives are sanitized with bleach in this area, and workers do not cross back and forth between the slaughter area and the sanitary evisceration area. At this station all chickens and their parts should be covered with towels to prevent flies.

The steps of evisceration are as follows:

1. Cut off the oil gland on top of the tail.
2. Remove the neck by cutting the muscle at the base and then pulling off; do not cut through bone.
3. Open the body cavity by making a narrow triangular cut from under the breastbone down to either side of the cloaca; cut through skin and fat but don't nick organs.
4. Reach into the bird with the back of your hand against the inside of the belly, break through connective tissue until you reach the heart, and pull all the organs out in one go. Be careful not to pierce the gallbladder and spill bile on the meat.
5. Scrape the lungs off the ribs against the back.
6. Clean up any esophagus or trachea remaining in the neck area, and trim the neck skin if ragged.
7. Save the heart, liver, neck, and any other desired organs for consumption. Place the remaining offal in a covered bucket for composting later.

Rinse the bird thoroughly inside and out with a hose and then place it in the tank of ice water to chill. If you have many birds to do, it is best to add ice intermittently to ensure that the water remains cold. In between batches of birds, scrub equipment with soapy water and rinse with 10 percent bleach solution.

After all the birds are processed, drain them in a drying rack. Place a neck in each bird and a bird in each vacuum-seal bag. Squeeze the air out of the bag and seal it with its designated clip. Dip the bag into

Evisceration is the process of removing the internal organs from the bird. Photo by Neshima Vitale-Penniman.

Once evisceration is complete, birds are cooled immediately in an ice-water bath. Photo by Neshima Vitale-Penniman.

water at 180°F (82°C) to achieve a vacuum seal. Trim excess plastic from the bag. Weigh the bird, label with the weight, and transfer to the freezer immediately.

To clean up, compost all offal deeply into a compost pile or bury it in the earth. Wash all equipment and surfaces with soapy water, spray with bleach solution, and leave in the sun to dry and further sanitize.

In addition to cleaning up the physical space after slaughter, we have experienced the importance of cleaning ourselves emotionally and spiritually. Even having transitioned hundreds of birds, I personally feel a heavy and powerful energy cling to me each time I participate in slaughter. This energy is necessary for the work, but not something I want to carry into play with my children or dinner with my friends. At Soul Fire Farm we use spiritual baths to cleanse ourselves and mark the transition between the time for taking life and the time for existing in life. (See "Herbal Baths" on page 64.) Prayer, song, ritual, journaling, and contemplative silence all may be used to help you make this shift in a manner that is life affirming.

Raising Pigs

I pulled up on my bicycle as the sun began to rise and evaporate the dew. My 6:30 AM chore at Many Hands Organic Farm in Massachusetts was to care for the pigs before the crew workday began at 7 AM. Most days the task was simple, to check that the automatic waterer was flowing, refill the feed cans, and offer some kind words of affection to the sounder (family group of pigs). Once in a while, though, the bright and courageous sows burrowed under the fence and set themselves free, eagerly devouring crops and weeds nearby. I learned quickly that trying to chase and tackle a pig is not a good strategy. Instead I would lure the errant sow with treats and song back to the pen. Our pigs were expert at turning new ground and clearing old crops in preparation for planting. We rotated them around the farm, where they rooted out weeds and left their rich manure to compost in place. The following season we planted vegetables in the renewed earth.

As our family in Haiti knows well, pigs are adaptable scavengers, soil fertilizers, and quick-growing protein sources. They are also relatively easy to raise. Each pig requires a minimum of 50 square feet (4.5 square meters) of space in their outdoor pen and a simple shelter to protect from sun and rain. Ideally, this is a three-sided shelter, open on the fourth side, about 3 to 4 feet (1 to 1.5 m) tall. Leave openings under the eaves for ventilation and provide fresh bedding material regularly.

Pigs are incredible escape artists. Rhyne Cureton, an African American pig farmer in North Carolina, shared his love story with this powerful animal. He was in Texas managing his first sow, an English Large Black named Louise. She escaped her paddock by lifting a 7-foot (2-m) chain-link fence 5 feet (1.5 m) up in the air. Prior to that moment he did not like pigs much. Seeing her brute strength and sincerity, he said to her, "I am going to love you so hard that I'll begin to understand and appreciate you, and not get frustrated with you." He looked her right in the eye and told her, "I want to understand you." He has kept to that pact, letting go of pride and resistance in favor of connection. He fell in love with pigs.

Since pigs are natural diggers and can escape from any pen made of mesh wire, electric fencing with a strong energizer is best. Many farmers even put an extra strand of electric wire inside the pen at ground level to discourage the pigs from digging underneath the barrier. Biosecurity is also very important. The diseases brucellosis and trichinosis can be transmitted to pigs through wild boars. Rhyne recommends using two layers of fencing at least 1 foot apart from each other to prevent the nose of the boar from touching the nose of the domestic pig and spreading disease.

You can buy your 50-pound (23 kg) piglets in spring or early summer. Select either gilts (females) or barrows (castrated males). Always raise pigs with at least one companion, as they are social animals. Regarding varieties, Rhyne recommends Tamworths, which are easy to raise, are talented foragers, and produce lean meat. He also likes Ossabaw Island hogs, Guinea hogs, Mangalitsa, Large Blacks, and Gloucestershire Old Spots for fattier cuts of meat.

UPLIFT

Creole Pig

Haitian farmers bred the perfect pig for their unique climate and culture. The Haitian black pig, or Creole pig, became the center of the rural peasant economy. They were a hearty breed that ate readily available waste products and could survive for days without food. Nearly 85 percent of rural households raised pigs, which served as a "piggy bank" to pay for school fees, emergencies, funerals, marriages, and medical expenses. The Creole pig also played an important role in maintaining the fertility of the soil.[10] In 1982 the US government initiated a campaign to eradicate the Creole pig under the pretense that they might become sick with African swine flu, which was present in the Dominican Republic, and spread that disease to pigs in the US. Convinced to cooperate, the Haitian government helped to destroy all 400,000 Creole pigs in the nation over a 13-month period. They made promises to replace the pigs with "better" pigs. These "better" pigs from Iowa required clean drinking water, expensive imported feed, and roofed pigpens, none of which were attainable for peasant farmers. The Haitian farmers called the new pigs *les princes a quatre pieds* (four-footed princes).[11] In the year after the devastation of the Creole pig, school enrollment dropped between 30 and 50 percent in rural areas because families could no longer afford tuition fees and school uniforms. All told, Haitian peasants endured half a billion dollars' worth of damages.[12] Further, the Creole black pig was central to the practice of the island's religion.[13] The 1791 ritual at Bois Caïman that launched the Haitian Revolution involved the sacrifice of a black pig. Houngan Dutty Boukman led the Vodou ritual, and Mambo Marinette, who was possessed by the Iwa Erzulie Dantor, a Vodou goddess of love and warrior mother, performed the pig sacrifice. This offering forged a bond of unity and pact of victory between the enslaved Haitians and Iwa Erzulie Dantor to rise up together and defeat the French.[14]

Pigs need access to food 24 hours per day, which is best accomplished with a self-feeder. They start on a 16 percent protein feed and can substitute a 14 percent protein feed once they reach 125 pounds (57 kg). Additionally, pigs can eat surplus produce. Throughout their lifetime each pig will consume about 700 pounds (318 kg) of feed. Pigs also need a continuous supply of clean water, best attained through an automatic waterer that the pig can turn on and off. Be sure to bury the hose running to the pen under an inch of soil to prevent the sun from overheating the water along the way. The food can be set up adjacent to the shelter. It is best to keep the water supply on the far end of the pen, since pigs tend to defecate near their water. According to Rhyne, it is also important to build a bond with your pigs, by staying with them for at least 15 minutes each day at chore time to play and walk with them and observe their behavior.

He explained, "I like to spend time with them to observe their behavior because the more you spend time with them, the more you're able to detect or understand when something is wrong. If you have pigs that have a habit of getting out, most likely it's because they either are bored or lack resources. As a pig farmer I have to make sure that my pigs aren't bored, so I rotate them regularly to prevent boredom;

Pigs require a three-sided shelter and plenty of space to forage.

Farmer Rhyne Cureton enjoys a “kiss” from one of his pigs. Photo courtesy of Rhyne Cureton.

it’s called enrichment. If they escape their paddock or enclosure, it might have to do with what I’m doing wrong. Do the pigs eat enough or are they escaping so that they can forage? Do they have enough clean water or did they flip it over in the morning and didn’t have water for the rest of the day? Do they have enough shade throughout the entire day or are they directly in the heat by high noon? If a pig farmer doesn’t take that into account, then it is very easy to get frustrated and blame the pig for the farmer’s lack of pig husbandry. Pig husbandry goes beyond just feeding, water, and shelter. That’s why it’s important to spend time with your pigs. It’s like having a relationship with someone. The more time you spend with someone, the more you notice the little details about what they enjoy, what they hate, their range of emotions, and how to avoid unnecessary conflict.”

Pigs can also be rotated through paddocks in the forest, orchard, or field to eat the harvest of the land.

Dr. George Washington Carver lamented that a "great quantity of acorns produced in our oak forests" had "been hitherto practically a waste product . . . and the feeding values of this natural product [had been], in great measure, lost sight of." He observed that acorns not only were free to impoverished Black farmers, but also provided an overlooked economic justification to preserve the forests from the axes of timber companies.[15] Finishing pigs on acorns imparts a rich, desirable flavor to their meat.

With proper feeding, pigs will reach market weight of 250 pounds (113 kg) in approximately 100 days. Rhyne shared his method for "weighing" pigs without a scale. He measures the girth of the pig around its chest cavity. The heart girth correlates to weight regardless of the breed. When the hog measures 44 inches around the heart, it weighs around 240 pounds (109 kg).

To move pigs to another pen or to move them into a horse trailer to bring to the butcher, patience is key. Pigs prefer to be lured with food treats rather than pushed. Someone should walk behind the pigs holding a piece of plywood, called a pig hurdle, so they do not escape by turning back.[16]

The blessing of having pigs on the farm extends beyond their value for meat and manure. Jon Jackson, veteran of the Iraq and Afghanistan wars and Georgia Black farmer, started Comfort Farms to help fellow veterans recover from PTSD. He has observed firsthand the therapeutic benefits of raising pigs. "Animals don't care about your bad day," Jackson explained. "They're going to come up, and they're like, 'I want you to pet me.' And you're like, 'Okay, I'm feeling really mad right now, but I'm petting you.' Man, [these animals] don't know the amazing stuff that they're doing for our vets who come through."[17]

Meat and Sustainability

Intensive, industrial livestock production is an environmental justice disaster, adversely impacting communities of color. In the eastern part of North Carolina, a billion-dollar hog industry is clustered in the African American community. Hog farms collect billions of gallons of untreated pig urine and feces in cesspools, and dispose of the waste by spraying it into the air. The droplets of waste permeate the

UPLIFT

Collard Greens with Ham Hocks, Fatback, or Turkey Neck

Our enslaved ancestors were denied prime cuts of meat, allowed only to eat the leftover and undesirable cuts and the offal of the animal. We combined these meager rations with the vegetables that we could grow ourselves in provision gardens. Our ancestors used vegetables from the brassica family—collards, mustards, and turnips—as stand-ins for the greens we ate in our homeland. We flavored these greens with bits of ham hock, fatback, and turkey neck to impart taste and nutrients to the pot. While this practice was born of necessity, not necessarily desire, it provided a model for the sustainable consumption of meat. Using meat as a "spice" and not a "slab" may represent the correct ratio of animal to plant foods in the sustainable modern diet. Similarly, our ancestors and contemporary cousins around the world use meat as a special-occasion food. Chickens were slaughtered on Sundays and larger animals on special occasions, such as holidays, weddings, and funerals. Outside of these sacred days, our diets were rooted in plant foods.

homes, clothes, and bodies of the Black residents of the area. Five hundred primarily African American residents have sued Murphy-Brown, the state's largest hog producer, for compensation. Many had lost access to their wells because the hog waste imparted unhealthy colors and odors to the water.[18]

Further, animal agriculture is possibly the most environmentally impactful human activity. About 30 percent of the world's ice-free surface is used not to raise fruits, vegetables, and grains, but to support livestock for human consumption. Animal agriculture devours a full third of the world's fresh water. In industrialized nations it takes 75 to 300 kilograms (kg) of feed to produce 1 kg of animal protein. In sub-Saharan Africa that number is much higher, as a cow might eat 500 to 2,000 kg of dry matter to produce 1 kg of protein. Livestock produce at least 18 percent of the anthropogenic greenhouse gases according to the Food and Agriculture Organization. Meat is arguably destroying the planet.[19]

At the same time, meat is part of our cultural heritage and ancestral cuisine. The offering of animals is even central to many of our traditional religious practices. Many people experience that their health is dependent on the consumption of animal protein. How are we to reconcile these contradictions?

Collard greens with turkey neck provide the answer for some of us. We can think of meat as a spice rather than a staple, something we add in small amounts to flavor a dish. We can reserve the consumption of meat in quantity for special occasions: religious ceremonies, holidays, and life-cycle events. Further, we can focus on the most climate-friendly meats, pigs and poultry, which account for only 10 percent of total livestock greenhouse gas emissions even as they provide three times as much meat globally as cattle. Poultry and pork also require five times less feed per kilogram of protein than cows, sheep, or goats. I personally have found a home in the flexible vegetarian diet. Plant foods are the center of our family's cuisine, but we are open to eating Grandmommy's chicken soup or the goat stew prepared for us by friends in Haiti. We participate in the cultural cuisine of our people with joy and also keep love of the planet and the sanctity of life at the center of our consciousness.

Pigs and poultry have a smaller environmental footprint than cows, sheep, and goats, especially when raised on forage.

CHAPTER TEN

Plant Medicine

Self-care and healing and attention to the body and the spiritual dimension—all of this is now part of radical social justice struggles.

—ANGELA DAVIS

At the end of morning "hands on the land" teaching block during Black Latinx Farmers Immersion, we traveled to each work area to share lessons on what we learned. Team Onyx spent their morning transplanting Eritrean basil and other medicinal herbs with sweet peppers as part of an intercropping system. This basil is native to Eritrea, as its name suggests, and offers an attractive spicy aroma that is central to East African cuisine as well as myriad health benefits. We challenged each team to come up with a creative way to impart their new knowledge to the other teams. Onyx chose a call-and-response song that undoubtedly deepened our reverence for this sacred plant.

Feeling anxious and sad?
  Eritrean basil!
Got phlegm in your lungs?
  Eritrean basil!
Itchy swelling skin?
  Eritrean basil!
Need to pass a kidney stone?
  Eritrean basil!
Bad breath got you down?
  Eritrean basil!
Eyes feeling fatigued?
  Eritrean basil!
For your spiritual needs?
  Eritrean basil!

Earlier that year a few members of our team traveled to my ancestral homeland of Haiti for a delegation of Ayiti Resurrect, a solidarity project with farmers in Komye, Leogane. While the central work of the trip was to complete the projects identified by the farmers of Komye—the installation of irrigation systems, digging of a new well, and planting of mango trees—we also managed to fit in a spiritual pilgrimage. We traveled to the town of Ville-Bonheur for the annual festival for Ezili Dantor, lwa of motherhood and healing. Ever since she appeared at the Sodo waterfall in the mid-1800s and began healing the sick, tens of thousands of Haitians have journeyed to these waters every July 14 through 16 to ritually bathe in the waterfall and pray for blessings. When we arrived in the crowded and peaceful town, the first local we met sold us a large bunch of bush basil to use for washing ourselves in the waterfall. Basil anointed us with its healing energies and spiritual purification.

Basil is just one of the thousands of medicinal herb species used by African people across the Diaspora. Our people's relationship with plant medicine extends

During the annual pilgrimage to Sodo, Haiti, to pay homage to Ezili Danto, devotees bathe with basil and other herbs in the sacred waterfall.

even before our first written account in 1500 BCE on the Ebers Papyrus, on which ancient Egyptians listed the recipes for over 850 herbal medicines. Our knowledge of plants traveled with us in the bowels of slave ships and was kept alive in the root and conjure work of the Black American South, in Harriet Tubman's deft use of wild plants to keep her Underground Railroad passengers healthy, and in the natural pharmacies of orisa worshipers. In this chapter we explore how to cultivate, wildcraft, and use herbs that have been important to Black health. In an elegant interplay of spiritual folk wisdom and Western science, this chapter helps the reader understand how plant allies can assist in our physical, emotional, and spiritual health.

Species Accounts of Cultivated Plant Allies

Our medicine cabinet is lovingly planted in stone-lined raised beds right outside of our home and education center. The plants that thrive there were chosen as allies for the specific needs of our human community. Like so many Black and Brown survivors, members of our family struggle with anxiety, depression, and insomnia. Lemon balm, vervain, codonopsis, skullcap, chamomile, and lavender are here to support us. We're a multigenerational collective with children among us, so colds and flus abound. Wormwood, bee balm, yarrow, thyme, echinacea, and elderflower provide the immune support and symptom relief to keep our children well. Some of us lean on rosemary, oregano, and sage to help manage high blood sugar and diabetes. Others harvest delicate calendula flowers to make healing salves. The children enjoy eating the fresh peppery nasturtiums and mild sweet borage flowers that sneak up between the perennials.

Herbs That We Cultivate

The accounts in this section are of just a few of the medicinal species that we have firsthand experience cultivating or wildcrafting. All of the species are

UPLIFT
Traditional Medicine of Africa

According to the World Health Organization, over 80 percent of people on the continent of Africa rely on traditional medicine as their primary health therapy. This percentage is over 90 in Ethiopia. While Western medicine views the isolation of phytochemicals and their single-use chemical entities as superior for targeted remedies, traditional African medicine values the synergy of phytochemicals as collectively surpassing the individual constituents. For example, a single plant may have bitter substances that stimulate digestion, phenolic compounds that act as antioxidants, anti-inflammatory compounds that reduce swelling, antibacterial and antifungal tannins, diuretics to support waste elimination, and alkaloids to enhance mental well-being. Not to mention, this same plant may also attract pollinators, stabilize soil, and fix nitrogen.[1]

Africa is blessed with astounding biodiversity resources, estimated to host up to 45,000 plant species of potential medicinal value, of which 5,000 are in active use. Western pharmaceutical companies have extracted and capitalized on that biodiversity, patenting and marketing medicines from many African species including *Acacia senegal*, *Aloe ferox*, *Artemisia herba-alba*, *Aspalathus linearis*, *Centella asiatica*, *Catharanthus roseus*, *Cyclopia genistoides*, *Harpagophytum procumbens*, *Momordica charantia*, and *Pelargonium sidoides*. *Aloe ferox* is native to South Africa and has been used by the local population since time immemorial. Its use is depicted in the ancient San rock paintings. Today the export of aloe is central to the South African economy and supplies the cosmetic and pharmaceutical industries with a powerful ingredient that is anti-inflammatory, antiviral, anticancer, soothing, and healing. Similarly, *Artemisia herba-alba* or wormwood, a plant native to North Africa, is now used worldwide to treat hypertension, diabetes, and malaria.[2] African herbalists have also provided the world with rooibos, honeybush, castor bean, and myriad other plant allies.

included because they have deep and enduring relationships with Black people in the Diaspora. I recently had the honor to ask Wande Abimbola, the Awise Awo ni Agbaye (Spokesperson of Ifa in the Whole World), about his perspective on the use of New World plants for spiritual medicine, when so much of our ancestral knowledge is connected to African plants that may not grow in the US. He affirmed the expansion of our apothecary, simply saying, "Yes, this is very good!" The species accounts below are rooted in the research of Stephanie Mitchem,[3] Peter Burchard,[4] Dr. George Washington Carver, Herbert Covey,[5] Katrina Hazzard Donald,[6] Crystal Aneira,[7] Jerome Handler and JoAnn Jacoby,[8] and Wanda Fontenot[9] on African American herbal medicine and supplemented with Western perspectives by Andrew Chevallier,[10] Steven Foster and James Duke,[11] Jessica Houdret,[12] and Steve Brill and Evelyn Dean.[13]

Asafoetida (*Ferula assa-foetida*)

Victoria Adams recalls of the plantation, "We dipped asafetida in turpentine and hung it 'round our necks to keep off disease."[14] It was one of the herbs most commonly used by enslaved Africans. The fumes helped alleviate respiratory ailments. Asafoetida contains disulphids and foetidin. A tea of the herb

Disclaimer

The information about the medicinal uses of plants is for educational purposes only and not meant to substitute for consultation with a licensed physician for treating illness. The author and publisher assume no responsibility for problems arising from the reader's misidentification or use of medicinal plants. Please check with your doctor before using any herbal medicine if you are pregnant, breastfeeding, or have any pre-existing medical condition.

Known as the king of herbs, basil (*Ocimum basilicum*) has profound medicinal and spiritual properties. Photo by Neshima Vitale-Penniman.

helps with gas, bloating, indigestion, and constipation by altering the microflora in the gut. Inhaling or ingesting the herb aids with influenza, congestion, asthma, cough, and high blood pressure. It has a persistent aroma, similar to garlic, and is used as a flavoring in sauces. In spiritual medicine a bath of asafoetida is used to rid a person of negative habits. Known as Devil's Dung in Louisiana Voodoo, it is used to ward off negativity and keep away the police.

Ashwagandha (*Withania somnifera*)

Ashwagandha has been used for thousands of years in East and North Africa, India, and the Middle East as a tonic, aphrodisiac, narcotic, diuretic, anthelmintic, astringent, thermogenic, and stimulant. In Africa people use Ashwagandha to alleviate fevers and inflammatory conditions. In Yemen the dried leaves are ground into a paste that is used for treating burns and wounds, as well as for sunscreen. For external healing the berries and leaves have been applied to tumors, tubercular glands, carbuncles, and ulcers.

Basil (*Ocimum basilicum* and *O. gratissimum*)

Known as the king of herbs, basil has profound medicinal and spiritual properties. As a tea it is taken to reduce anxiety, remove phlegm from the bronchial tubes, alleviate fever, reduce cold and flu symptoms, expel kidney stones, reduce stress, cure headaches, and cure sores of the mouth. The leaves can also be chewed or gargled to relieve a sore throat. Applied topically, basil relieves itches and swelling from insect bites and may also cure leucoderma. Dried basil leaves make an effective tooth cleansing powder that removes bad breath and strengthens gums. A few drops of basil juice applied to the eyes helps with night vision and eye fatigue. In Santeria and Vodou basil is a common herb in baths for purification and good fortune. Basil is also burned to remove negative spirits from the home. Some of the important heritage varieties of basil include African Blue Basil, West African Basil or adefetue (*O. viride*), Tulsi basil (*Ocimum tenuiflorum*), and African Spice Basil.

Black Cohosh (*Cimicifuga racemosa*)

This native plant contains triterpene glycosides, isoflavones, isoferulic acid, and salicylic acid concentrated in the roots. A decoction or tincture of the root is used for gynecological problems, earning the name squaw root among the Penobscot people. Black cohosh reduces menstrual pain, menopausal symptoms, and depression. It also alleviates arthritis inflammation, high blood pressure, and asthma. In

Marigold (*Tagetes erecta*) is considered a bridge plant to the ancestors and has a central role in Day of the Dead celebrations. Photo by Neshima Vitale-Penniman.

spiritual medicine a bath of black cohosh is useful for releasing addiction, toxic relationships, and fear. **Caution:** *Do not use during pregnancy.*

Calendula (*Calendula officinalis* and *Tagetes erecta*)

Calendula is a powerful medicine for the healing of skin. Calendula oil and salve can be used to heal wounds, reduce inflammation, soothe burns, heal acne and rashes, kill fungal infection, and soothe diaper rash, cradle cap, and nipples sore from breastfeeding. Calendula tea reduces inflammation of the digestive system, relieving conditions such as gastritis, peptic ulcers, and colitis. In spiritual medicine, a calendula bath is used to bring respect and admiration. Calendula and marigold are considered bridge plants to the ancestors and have a central role in Day of the Dead celebrations. In Voodoo inhalation of the herb brings good fortune.

Chamomile (*Chamomilla recutita* and *Matricaria recutita*)

This sweetly aromatic herb contains proazulenes, faresine, spiroether, flavonoids, bitter glycosides, and coumarins. Chamomile tea is suitable for young

In spiritual medicine chamomile (*Chamomilla recutita*) tea is taken to clear up mental confusion, to prevent nightmares, and to attract love. Photo by Neshima Vitale-Penniman.

children and relieves colic, indigestion, bloating, Crohn's disease, irritable bowel syndrome, hiatus hernia, and peptic ulcers. It has an antispasmodic effect, relaxing aching muscles and relieving menstrual pain. The steam can be inhaled to relieve allergies. Externally, it can be applied to the eyes to relieve strain and to the skin to reduce itchiness. In spiritual medicine the tea is taken to clear up mental confusion, to prevent nightmares, and to attract love. The flowers are scattered about the home to remove negativity.

Echinacea (*Echinacea angustifolia* and *E. purpurea*)

Echinacea is a popular immunostimulant that contains isobutylamides, caffeic esters, humulene, echinolone, and betaine. A tincture or decoction of the root is taken to cure chronic infections, colds, flu, skin disorders, and respiratory infections. As a gargle, it combats throat infections. It inhibits the ability of viruses to enter and take over cells, and has antibacterial and antifungal properties. In spiritual medicine echinacea flowers are used as offerings to river spirits. The flowers can also be worn around the neck to bring strength in challenging times.

Epazote (*Dysphania ambrosioides*)

West African epazote, 'Megbezorli,' is a milder form of the well-known Mexican herb. It originates in the Volta region of West Africa and is used to mask strong cooking odors. A paste of macerated epazote and water is taken medicinally to treat anxiety and shortness of breath. In spiritual medicine the permission of the spirits is sought to allow the plant to intervene in nightmares.

Echinacea (*Echinacea purpurea*) is a popular immunostimulant. Photo by Neshima Vitale-Penniman.

Horehound (*Marrubium vulgare*)

Horehound is a much-overlooked herb that was central to health care on the plantation. It contains the diterpenes marrubiin and marrubenol, flavonoids, alkaloids, and volatile oils. It is a strong expectorant that is used for chest problems, usually taken as a syrup made with sugar or honey. A decoction of the herb can alleviate the symptoms of asthma, bronchitis, and tuberculosis. The herb also acts to normalize heart rhythm and improve eczema and shingles. In spiritual medicine, a tea of horehound is taken to increase concentration and mental clarity. Ash of horehound attracts healing energies. Horehound candy is offered to guests to bring blessings.

Hyssop (*Hyssopus officinalis*)

The beautiful flowering tops of hyssop are rich in terpenes, camphor, pinocamphone, hyssopin, and resin. It is a strong expectorant that encourages the production of a more liquid mucus, while stimulating coughing. This clears thick and congested phlegm after an infection has passed. Hyssop is also a sedative and lessens the symptoms of asthma. In spiritual medicine hyssop tea is used for cleansing and purification of the body and home. In the Book of Exodus in the Torah, lamb's blood was applied to the doorposts of the Israelites' homes using hyssop, and it was also used by priests in the Temple of Solomon for purification rites. **Caution:** *Hyssop essential oil can induce epileptic seizures.*

Lavender (*Lavandula angustifolia*)

Lavender is a sweet, fragrant herb rich in 40 volatile oils, flavonoids, and coumarins. A tea or tincture of lavender has a soothing and calming effect that relieves insomnia, irritability, headaches, and depression. It also relieves gas, bloating, and indigestion. The essential oil is strongly antiseptic and can be applied to heal wounds and sores. Massaging a few drops of essential oil on the temples eases headaches. A bath of lavender calms the nervous system and promotes sleep. In spiritual medicine lavender is used to attract love. A bath of lavender purifies energy and promotes psychic dreams. The ashes of lavender are scattered to bring peace and harmony to the home.

Lemon Balm (*Melissa officinalis*)

Lemon balm is a tender, aromatic herb that contains citral, caryophyllene oxide, linalool, citronellal, flavonoids, triterpenes, and polyphenols. A tea of lemon balm is taken to lift the spirits, as well as to relieve anxiety, depression, headaches, restlessness, irritability, indigestion, nausea, and bloating. Lemon balm can be chewed to relieve a toothache and cold sores. It also has an antithyroid effect, useful for people with an overactive thyroid. A salve of lemon balm can be massaged into the skin to relieve pain. In spiritual medicine lemon balm tea is taken to attract love and dispel melancholy. A satchel of lemon balm under the pillow promotes sleep.

Nzinzingrolo (*Solenostemon monostachyus*)

This African native plant is known as *magero* in Guinea-Bissau, *jewubue* in Sierra Leone, *ka mai tonto* and *te-te-vua* in Liberia, and *nzinzingrolo* in Ivory Coast, and has numerous medicinal uses.[15] The leaf sap is sedative and is used to treat fever, cough, headache, convulsions, colic, and sterility. The plant is applied externally to address eyesight troubles, foot infections, and snakebite. An extract of the herb is highly effective against panic attacks and anxiety, addressing both the physiological and spiritual dimensions of the distress.

Pennyroyal (*Mentha pulegium*)

Similar to its cousins in the mint family, pennyroyal is an aromatic herb that aids digestion and soothes skin conditions. It is rich in pulegone, menthol, and terpenoids. A tea of pennyroyal stimulates digestive juices, relieves gas, and kills intestinal worms. It is a good remedy for headaches and reduces fever and congestions. An infusion of pennyroyal can be applied externally for the treatment of itchiness, eczema, and gout. In spiritual medicine, pennyroyal is carried to protect the traveler. A bath of pennyroyal for a recently deceased person aids in passage to the afterlife. The herb brings peace to humans and nature, averting quarreling and calming stormy seas. In Louisiana Voodoo pennyroyal is a guardian plant used to protect people from enemies. **Caution:** *Essential oil of pennyroyal is highly toxic.*

Peppermint (*Mentha × piperita*)

The mint family (Lamiaceae) has over 6,000 species, of which several are native to Africa including *Mentha longifolia* (wild mint) and *M. aquatica* (wild water mint).[16] Mint contains menthol, methone, luteolin, menthoside, phenolic acids, and triterpenes. An infusion of mint increases the flow of digestive juices, reduces cramps and gas, soothes the lining and muscles of the colon, and relieves constipation. As Dr. George Washington Carver wrote in his herbal medicine catalog, "This plant is familiar to almost everyone as a specific for weak stomachs, diarrhea and as a stimulant." Applied to the skin, peppermint relieves pain. It can be inhaled to relieve headaches and improve respiratory function. In spiritual medicine mint is taken to revive hope and restore energy. A mint tea supports healthy grieving.

Rue (*Ruta graveolens*)

Rue is a powerfully aromatic herb rich in volatile oils, flavonoids, furanocoumarins, and rutin. It is useful for strengthening the inner lining of blood vessels and reducing blood pressure. It stimulates the muscles of the uterus to promote menstrual blood flow. Rue tea has also been used to treat mental illness, epilepsy, vertigo, colic, intestinal worms, poisoning, multiple sclerosis, Bell's palsy, and eye problems. An infusion used as eyewash brings relief to strained and tired eyes, and improves eyesight. Recalling times on the plantation, elder Sam Rawls shared, "When anybody got sick, the old folks made hot teas from herbs that they got out of the woods. One was a bitter herb called rue. They give it to the children, and to the grown ups, too."[17] In spiritual medicine rue is known as the mother of herbs and has sacred utility in Christianity, Islam, Santeria, and indigenous European traditions. A tea of rue is used to consecrate sacred objects, to wash the eyes to develop "second sight," and for purification and protection. **Caution:** *Rue is poisonous when taken in excess. Never consume during pregnancy.*

Sage (*Salvia officinalis*)

Sage is a powerfully curative plant containing thujone, diterpene bitters, flavonoids, phenolic acids, and tannins. Its antiseptic and astringent qualities make it useful as a gargle for sore throats and a rinse for canker sores and irritated gums. Taken as tea, sage cures mild diarrhea, calms the nervous system, dries up milk flow, and stimulates digestion. It also reduces hot flashes during menopause and helps the body adapt to hormonal change. The dried leaves are smoked to treat asthma. In spiritual medicine sage

Sage (*Salvia officinalis*) is a sacred, purifying plant in both Indigenous and Black traditions. Photo by Neshima Vitale-Penniman.

is burned to cleanse and purify spaces and taken as tea to enhance wisdom. Burned during funeral rituals, sage supports healthy grieving. A sachet of sage beneath the pillow is said to drive away nightmares. **Caution:** *Do not take therapeutic doses during pregnancy or while breastfeeding.*

Sorrel (*Hibiscus sabdariffa* or *H. acetosella*)

African rosemallow, false roselle, and sorrel are some of the names for this Angola-native plant, which is rich in plant acids, including citric acid, malic acid, tartaric acid, and allo-hydroxycitric acid lactone. The flower tea of sorrel has been shown to reduce cholesterol and blood pressure in clinical studies. It is also useful for treating loss of appetite, respiratory infections, inflammation, stomach irritation, poor circulation, and constipation. In spiritual medicine sorrel is an aphrodisiac, used to attract love and incite passion. The flower can also be floated in water to aid divination rituals.

Spilanthes or African Power Cress (*Spilanthes filicaulis*)

Native to West Africa, this plant is known to help make one's speech persuasive, smooth, and well received. Local leaders chew on the yellow button flowers before important meetings. African power cress contains spilanthol, a mild analgesic that produces a tingling sensation on the tongue. In Brazil the plant is known as *Wolikpekpe*. The flowering herb sometimes known as toothache plant (*Acmella oleracea*) also contains spilanthol and is a temperate substitute for this tropical species.

Thyme (*Thymus vulgaris*)

Thyme is central to Haitian cooking and powerfully medicinal, containing thymol, methyl chavicol, cineole, borneol, flavonoids, and tannins. Harvest the fresh aerial parts in late summer or early fall. A tea or syrup of thyme can be taken as an expectorant; to relieve the symptoms of colds, flu, asthma, and bronchitis; and to counter the effects of aging. Externally, the essential oil can be used on the skin to relieve insect bites, athlete's foot, infections, muscle spasms, and parasites. A tincture can be used to treat vaginal yeast infections. The fresh leaves may be chewed to relieve sore throats. In spiritual medicine an herbal bath is used after winter to restore vibrancy. The herb bundles are hung around the home to attract love and financial prosperity.

Tobacco (*Nicotiana tabacum*)

While smoking tobacco cigarettes is highly correlated with cancer and lung disease, organic tobacco without additives has been shown to heal ulcerative colitis, sarcoidosis, endometrial cancer, uterine fibroids, and breast cancer among women who carry the BRCA gene. In spiritual medicine tobacco smoke is blown over the fields before planting, over lovers before sex, between parties in peace negotiations, and into warrior's faces before battle. The smoke is believed to carry blessings, protection, and purification. In Lukumi tobacco is offered to the orisa to bring blessings.

Wormwood (*Artemisia absinthium* and *A. afra*)

Wormwood is an aromatic bitter native to Southern and Eastern Africa that contains sesquiterpene lactones, thujone, azulenes, flavonoids, phenolic acids, and lignans. Wormwood tea and tincture increase stomach acid and bile production, enhance the absorption of nutrients, clear up bronchial disease, relieve anemia, gas, constipation, and bloating, and restore vitality after a long illness. The tincture is moderately effective for eliminating worms. Applied topically, wormwood is a good insect repellent and insecticide, and can heal skin infections. **Caution:** *Use only in small doses and for no more than four consecutive weeks.*

Species Accounts of Wildcrafted Plant Allies

In her brilliant book *Braiding Sweetgrass: Indigenous Wisdom, Scientific Knowledge, and the Teachings of Plants*, Potawatomi scientist and healer Robin Wall

UPLIFT
Harriet Tubman

Harriet Tubman was a master herbalist and wildcrafter who used her knowledge of plants to heal Black and white soldiers in the Union army during the Civil War and to keep her passengers safe on the Underground Railroad.[19] She famously cured a soldier who was near death from dysentery by digging up some water lilies and cranesbill to make him a medicinal infusion. Tubman used paregorics to quiet babies on the journey north, as well as other herbs that were taught to her by her grandmother.[20]

Enslaved Africans kept their herbal traditions alive at great personal risk. Because enslavers were fearful of being poisoned by the people they enslaved, they forbade the practice of herbal medicine. African herbalists knew how to use plants to cause maladies that mimicked common diseases, like dysentery, making the transgression easy to conceal. By the mid-18th century both Virginia and South Carolina made it a capital offense for enslaved people to teach or learn about herbal medicine and prohibited us from working in apothecaries. Further, European Americans borrowed from African medicinal knowledge, then erased the stories of the originators over time. For example, in the 1863 edition of *Resources of the Southern Fields and Forests*, the entry on boneset (*Eupatorium perfoliatum*) includes, "This plant is extensively employed among the negroes on the plantations in South Carolina as a tonic and diaphoretic on colds and fevers, and in the typhoid pneumonia so prevalent among them."[21] By the time the Peterson Field Guide was published in 1990, boneset was merely described as a "common home remedy of 19th-century America, extensively employed by American Indians and early settlers."[22] The Black herbalists were erased. Similarly, Cesar (b. 1682) was an accomplished Black herbalist who developed a cure for enslavers who believed they were poisoned, using plantain and horehound as the active

Kimmerer summarizes the sacred law of wild foraging for plants.

Know the ways of the ones who take care of you, so that you may take care of them. Introduce yourself. Be accountable as the one who comes asking for life.
Ask permission before taking. Abide by the answer.
Never take the first. Never take the last.
Take only what you need.
Take only that which is given.
Never take more than half. Leave some for others.
Harvest in a way that minimizes harm.
Use it respectfully. Never waste what you have taken.
Share.
Give thanks for what you have been given.
Give a gift, in reciprocity for what you have taken.
Sustain the ones who sustain you and the earth will last forever.[18]

This is the law. The rest is commentary.

When wildcrafting, ecological conservation and personal safety come first. While Robin Wall Kimmerer explains that we must take less than half, our practice is to take no more than one-third of any plant population. Never harvest any plant that is rare or endangered, and avoid harvesting in protected ingredients. In a rare act, the South Carolina legislature of 1749 awarded Cesar his freedom and an annual stipend in exchange for his recipe and commitment to continue developing medicine. They published his cure-all poison remedy in the May 1750 issue of the *South Carolina Gazette*. Yet in 1887 the book *American Medicinal Plants* simply said of plantain, "Prominent folk cancer remedy in Latin America. Used widely in folk medicine throughout the world. Confirmed antimicrobial; stimulates healing process," again omitting the contribution of Black herbalists.[23]

Enslaved herbalists had intimate knowledge of hundreds of native and naturalized plants, including snakeroot, mayapple, red pepper, boneset, pine needles, comfrey, pokeweed, sassafras, goldenseal, belladonna, lobelia, sage, henna, rhubarb, bloodroot, wild cherry, jimsonweed, peppermint, saffron, pleurisy, horehound, elecampane, skunk cabbage, spikenard root, Alexandria senna, catnip, High John root, pennyroyal, and red oak bark. Black herbalists understood the physical healing properties of these plants as well as their spiritual dimension. In addition to preparing teas and poultices, these healers carried on the string- and knot-tying traditions of West and Central Africa. Strings of leather or vines were infused with plant essence, then tied to various parts of the body, including the neck, ankle, wrist, and waist, to bring spiritual power and strength. This practice originated in Kongo, Gabon, Ghana, and elsewhere on the mother continent and persisted in an unbroken chain through slavery and into northern Black urban communities after emancipation.[24]

While many herbalists in our communities were forgotten by history, at least one enslaved herbalist, Willie Elfe, published his own prescription book. Further, the legacy of Dr. James Still (b. 1812), likely America's most famous 19th-century root doctor, known to New Jersey locals as "Dr. James of Pine Barrens," lives on. His contemporary "Dr. Buzzard," Stepheney Robinson of St. Helena Island, South Carolina, was the best-known conjurer of the time and is still lifted up by practitioners today.

wilderness areas. Our wild nonhuman cousins rely on these plants for their survival.

Developing a relationship with plants takes time. Many plants are easiest to identify when they are not edible, so you may need to identify where they are growing in one year and return to harvest them the following year. It can be helpful to learn some basic botany terms, so that you know the difference between a compound and a simple leaf, and between a regular and an irregular flower, for example. It is essential to positively identify a species, so bring along a few references and cross-check to be sure you don't have a poisonous look-alike in front of you. The wildcrafter's pack needs to be stocked with snacks, water, plant identification guides, offerings of tobacco or corn for the earth, a magnifying lens, and an emergency whistle. Pack plastic bags to gather the herbs and rigid containers for berries. The colors beige, green, and white repel ticks and biting insects and are the best colors to wear for wildcrafting. Take care to only harvest in areas free of pesticide spray, as many public parks, road edges, railroads, farms, and homeowners spray their properties with poison. Even when you are relatively certain of the cleanliness of the property, rinse your harvest before consuming.

Here are some plant allies we harvest at Soul Fire Farm that also have a rich history in the Black community:

Black Cherry (*Prunus serotina*)

The berries are high in vitamin C and can be eaten raw, dried, jellied, or juiced. Collect the inner bark of the tree in the fall, when the amygdalin level is highest. The tea is highly effective at suppressing coughs and also works as a sedative, decongestant, expectorant, disinfectant, fever reducer, and gargle for sore throats. Enslaved Africans additionally simmered the bark as a remedy for malaria. A poultice of the bark can be applied to wounds and burns as an external disinfectant and astringent. In spiritual medicine the tea is taken to promote long life. **Caution:** *Partially wilted cherry leaves produce cyanide, which can cause breathing difficulty, spasms, coma, and death in humans and livestock.*[25]

Boneset (*Eupatorium perfoliatum*)

The leaves and flowers of boneset contain sesquiterpene lactones, polysaccharides, flavonoids, diterpenes, and sterols, the last of which are immunostimulants. An infusion of boneset is used to treat colds, fever, arthritis, and "break-bone fever" (dengue fever). In the 18th century enslaved Africans commonly used the plant as a fever reducer.[26] The plant stimulates resistance to viral and bacterial infections and stimulates sweating. In spiritual medicine dried boneset leaves or boneset tea can be used to wash the body and remove negative energy and illness. When using dried leaves for spiritual medicine, burn them outdoors after rubbing over the skin. **Caution:** *There are concerns about pyrrolizidine alkaloid toxicity with boneset. Avoid the internal use of boneset during pregnancy and nursing, in children under 12, and in those with liver disease. Others should limit internal dosage to once per week.*

Burdock (*Articum* spp.)

Collect burdock root at any time during the first year of its biennial cycle. The root has a mild, nutty, sweet flavor and can be used in sauces, soups, and sautés. It is delicious prepared with sesame oil, tamari sauce, and ginger. In the second year, collect the flower stalk (cardone) for its tender celerylike core that tastes like artichoke heart. The cardone can be boiled or fried. Burdock root contains vitamins B_1, B_6, B_{12}, C, E, biotin, potassium, sulfur, silica, and manganese. It provides inulin, which helps regulate the metabolism. A tea or tincture of the root is used for liver dysfunction, urinary tract disorders, weight loss, immune support, colds, eczema, psoriasis, and acne. A poultice of burdock leaf or seeds cleanses the skin and heals bruises.[27] In spiritual medicine burdock is worn as a protective amulet, infused to make a purifying wash for the home, or made into an oil to rub on the genitals to restore potency.

Common Dandelion (*Taraxacum officinale*)

Collect dandelion leaves in the early spring and late fall after the frost, when they are least bitter. As Dr. George Washington Carver affirmed, "Never a spring came that we didn't have our wild greens. They were a part of our regular diet . . . They did indeed have distinct medicinal value. Our medicines before we learned how to make so many artificial products came from plants largely."[28] The greens and young crown are delicious in salads, sautéed with alliums, or steamed. The yellow portion of the dandelion flower can be picked or battered and fried. The leaves contain more beta-carotene than carrots, as well as vitamins C, B_1, B_2, B_5, B_6, B_{12}, C, E, P, and D, biotin, inositol, potassium, phosphorus, magnesium, and zinc. The taproot is also edible in soup. The root contains inulin, a prebiotic fiber, as well as the detoxifying chemical tarazin, which cleanses and reduces inflammation in the liver and gallbladder. The roots and leaf tea also act on the kidneys as a gentle diuretic, without leaching potassium the way pharmaceutical diuretics do. Dandelion's diuretic properties are what give it the Haitian Kreyol name *pisanli* or "urinate in bed." Dandelion tea is recommended to combat stress, sluggishness, diabetes, obesity, and indigestion. The leaf's white, milky sap removes pimples, warts, and sores. In spiritual medicine the root tea is taken before divination to strengthen communication with ancestors and spirits.

Burdock (*Articum* spp.) has a delicious root with a mild, nutty, sweet flavor. Consuming burdock helps regulate blood sugar. Photo by Neshima Vitale-Penniman.

Jimsonweed (*Datura stramonium*)

All parts of the plant contain atropine, scopolamine, and other alkaloids used to treat eye ailments, Parkinson's disease, cancer, rheumatism, and vertigo. Smoking the leaves may have antispasmodic effects in the treatment of asthma. Clergy in Ethiopia consume the seeds of datura to "open the mind" to be more receptive to learning and creative thinking. The Algonquin, Navajo, Cherokee, and other Indigenous peoples use the seeds to commune with deities through visions. Known as the zombie cucumber in Haiti, *D. stramonium* is used in the medicine for zombification. **Caution:** *Jimsonweed is violently toxic at amounts just above the medicinal dose and has resulted in many fatalities.*

Mugwort (*Artemisia vulgaris*)

The tea of mugwort relaxes the nervous system, promotes digestion, and relieves the symptoms of colds, bronchitis, fever, sciatica, kidney disorders, and anxiety. A bath alleviates menstrual cramps and tones the uterus. Chinese acupuncturists burn incense over mugwort placed on the skin to alleviate arthritis and joint pain, a process called moxibustion. In spiritual medicine the tea is used to enhance communication with the ancestors. Mugwort is also placed under the pillow to promote vivid dreams.

Mullein (*Verbascum thapsus*)

Mullein tea provides vitamins B_2, B_5, B_{12}, D, choline, hesperidin, PABA, sulfur, magnesium, and saponins. It is an expectorant and one of the safest, most effective cough remedies. The tea is anti-inflammatory, antibiotic, demulcent, and astringent, and inhibits tuberculosis bacillus. Dried mullein can be smoked with coltsfoot to relieve asthma. A compress of mullein relieves the symptoms of hemorrhoids and arthritis. An oil extract of the flowers can cure ear infections and relieve migraines. As Dr. George Washington Carver affirmed, "I wish to say that mullein is one of the oldest of our medicinal plants and is a noted remedy for all kinds of coughs and colds, rheumatic troubles, stopping of blood, asthmatic affections, and all manner of things that human ills are heir to. It is of unusual value along that line, one of the best known of household remedies. The flowers are especially valuable in aggravated cases of earache."[29] In spiritual medicine mullein tea is taken for prophetic dreams and astral travel. An incense of mullein assists with communicating with the ancestors.

Pine (*Pinus* spp.)

The seeds of pine trees are rich in protein, fat, and calories and have a sweet and buttery flavor. Roasting the dry cones facilitates seed removal. Chop and infuse pine needles to make a tea rich in vitamin C and beta-carotene. Pine resin contains a natural antibiotic, so the inner bark can be used as a poultice or bandage for wounds. Pine bark tincture is an effective cough syrup. Pine sap can be mixed with oil to make a salve to heal skin irritation. Turpentine and pitch tar are strong resinous constituents of the pine tree, extracted by distillation. Enslaved Africans used turpentine topically to treat toothaches and joint pain, and inhaled it to relieve congestion associated with bronchitis. In spiritual medicine pine is burned to purify the home and attract joy and peace. The tea is taken to restore youthfulness and vitality in old age. **Caution:** *Two western species of pine,* Pinus ponderosa *and* P. taeda, *are poisonous, so only use easterly pines for tea. Turpentine is now known to be carcinogenic; it is not recommended to ingest or inhale it.*

Plantain (*Plantago* spp.)

"I feel called to plantain because it grows in places it wasn't invited, places that have been trampled down, and it naturalized to those places, just like African American culture. I pray with this herb to remind me of the things that are left when all else is crushed and blown away," shared Black farmer Chris Bolden-Newsome. Plantain leaves provide beta-carotene, calcium, and a fiber called mucilage that reduces cholesterol, helping to prevent heart diseases. Plantain also contains monoterpene alkaloids, glycosides, triterpenes, linoleic acid, and tannins. An infusion of the leaves or ground seeds treats diarrhea, sore throats, gastritis, fevers, inflammation, and kidney disorders. It can be used as an internal wash to cure yeast infections. Plantain leaves can be chewed or mashed and applied externally to quickly remedy the pain of insect bites, stings, and poison ivy rash. The leaves can also be made into a salve for skin irritation. The seeds are rich in fiber and can be used as a gentle laxative. Chewing a plantain leaf also freshens the breath. Dr. George Washington Carver instructed us to eat the leaves directly: "Take the tender leaves of plantain and cook the stem, along with it the seed on, and they are so good for the system. I am so glad that you are keeping up with the program of the old time way of living. I can tell that you are getting stronger, just keep it up."[30] In spiritual medicine a footbath of plantain is used to remove weariness, and a bath of the head to cure headaches. An amulet of the herb is worn around the neck of children for healing and protection.

Red Clover (*Trifolium pratense*)

Flowers in full color can be dried for a delicious tea or ground into flour. Clover flower infusion contains beta-carotene, vitamins C, B_1, B_2, B_3, B_5, B_6, B_9, and B_{12}, biotin, choline, inositol, and bioflavonoids, as well as the minerals magnesium, manganese, zinc, copper, and selenium. The tea is detoxifying and anti-inflammatory, and relieves the symptoms of asthma, cough, and bronchitis. Clover compresses and tea have also been used as a cancer treatment. In spiritual medicine, the flowers are added to a bath to bring financial prosperity and sprinkled around the home to remove negativity. **Caution:** *Do not eat the decaying leaves of clover. They contain dicoumarin, which stops the blood from clotting.*

Sassafras (*Sassafras albidum*)

"Sassafras root tea was a popular seasonal blood cleanser [among enslaved Africans] believed to 'search de blood' for what was wrong and go to work on it."[31] The leaves, twigs, bark, and roots can be harvested to make a tasty reddish brown herb tea. Simmer rather than steep the plant material. Add seltzer water and sweetener to make root beer. The dried inner bark can be used in place of cinnamon. The dried leaves are called gumbo filé, added to southern soups to flavor and thicken. Sassafras root decoction is a blood purifier and diaphoretic useful in the treatment of colds, fever, arthritis, gout, high blood pressure, measles, kidney problems, stomachaches, and eczema. Compresses relieve external infections, burns, and poison ivy rash. In spiritual medicine sassafras tea is used as an aphrodisiac and love potion and as a body wash to cleanse spiritual energy. It contains safrole, the active ingredient in MDMA (ecstasy), which promotes empathy and sensory stimulation. In African American conjuring traditions, sassafras root chips are stored in the wallet or under the carpet to attract prosperity.

Winged Sumac (*Rhus copallinum*) and Staghorn Sumac (*R. typhina*)

Harvest the berry clusters in late summer or early autumn. The crushed berries can be mixed with cold water and strained to make a pink lemonade rich in vitamin C. A more concentrated infusion can substitute for lemon juice in salad dressings and recipes. A tea made from the berries or decoction of the inner bark can treat colds, fevers, diarrhea, asthma, urinary infections, sore throats, gum infections, and cold sores. A compress made from the leaves stops bleeding and reduces swelling. The dried berries can be mixed with tobacco for smoking. **Caution:** *Avoid poison sumac (*Toxicodendron vernix*), notable for its drooping clusters of white berries. The edible species have upright flowers.*

Yarrow (*Achillea millefolium* and *A. lanulosa*)

Yarrow contains over 120 active medicinal compounds including cyanidin, achilleine, azulene, and salicylic acid. A tea of yarrow relieves heavy menstruation, reduces blood pressure, and helps you sweat out fevers. A compress of yarrow stops bleeding and reduces bruising. In spiritual medicine yarrow tea is used to increase focus during divination and to increase abilities to see things from another point of view. A bundle of yarrow is hung over a couple's bed to ensure lasting love.

Yarrow (*Achillea millefolium*) is used to increase focus during divination and to increase abilities to see things from another point of view. Photo by Neshima Vitale-Penniman.

Growing an Herb Garden

Perennial herbs require human generosity to thrive. Most need to be divided every few years to stay healthy and avoid overcrowding. The result of dividing is several identical daughter plants that can be shared with others in the community. Almost all of the herbs in our garden came from our friends and mentors. We spread the word that we were planting a new herb garden several years back, and generous offers of mint, pennyroyal, lady's mantle, marshmallow, and dozens of other herbs poured in.

The easiest way to get started in herbal cultivation is to volunteer to help a knowledgeable neighbor divide their perennials in early spring and ask to take a few sections home to your land. Divide plants in cool weather when they are at their peak of health and beauty, not waiting until they become overcrowded or stressed. Start by carefully digging up the entire mother plant, then separate the root ball into four or more sections. Many root types simply pull apart, but tubers and taproots need to be sliced apart with a sharp knife. Replant a section 20 to 25 percent of the size of the original plant back in place and transplant the other portions in your garden or offer them to friends. Provide the plants with ample organic matter and water to help them settle back into the soil. While most herbs need to be divided every several years, including echinacea, bee balm, yarrow, lady's mantle, wormwood, and rhubarb, a few species do not tolerate division, including lavender and sage.

Similar to dividing, certain herbs propagate by stem cuttings. Herbs like rosemary, southernwood, rue, and hyssop can grow from cuttings. Clip off sturdy,

Harriet's Apothecary, an intergenerational healing village led by Black cis women, queer, and trans healers, holds a "healing village" at Soul Fire Farm. Photo by Neshima Vitale-Penniman.

UPLIFT

Black Herbalists Today

One of the most sacred days of the year at Soul Fire Farm is when Harriet's Apothecary, a collective of Black healers convened by Harriet Tubman and Adaku Utah, hold a healing village on the land. While I offer baths for emotional and spiritual healing, other members share herbal medicine, energy work, and sound healing to the community. Harriet's Apothecary is one example of Black people continuing the legacy of sacred plant allyship that our ancestors catalyzed.

Mentorship is an essential component of the practice of plant medicine. We are blessed that we have the unbroken knowledge in our own communities to educate one another in this art. Here is an incomplete list of Black and Indigenous herbalists offering classes:

Ancestral Apothecary School, California
Blue Otter School of Herbal Medicine, California
Centro Ashé (Ayo Ngozi), Washington, DC
Eagletree Herbs (Daphne Singingtree), Oregon
Farm School NYC & Sustainable Flatbush (Sheryll Durrant), New York
Femme Science Herbs (Meghan Elizabeth), California
Gold Water Alchemy (Hayden), Florida
Harriet's Apothecary (Adaku Utah), New York
Hattie Carthan Farm (Yonnette Fleming), New York
Herbal Tea House (Meadow Queen), New Jersey
Indigenous Remedies (Dr. Turtle), Texas
Medicine Woman Healing (Gogo Ateyo Nkanyezi), Georgia
mindbodygreen (Sokhna Heathyre Mabin), Michigan
Natural Choices Botanica School of Herbalism & Holistic Health (Angelique Moss Greer), Tennessee
Osain Yoruba Herbal Medicine, Nile River Medicine
Queen Afua Wellness Center, New York
Queering Herbalism, Herbal Freedom School (Toi Scott)
Roots of Resistance (Sade Musa)
Rootwork Herbals (Amanda David), New York
Sacred Roots Wellness (Tanya Henderson), California
Sacred Vibes Apothecary (Karen Rose), New York
Sitting Bull College (Linda Black Elk), North Dakota
Soul Flower Farm (Maya Blow), California
Southeast Wise Women (Kifu Faruq), North Carolina
The Stinging Nettle (Abi Huff), California
Sweet Rose of Sharon Afro American Herbalists, Texas
Third Root Community Health Center (Julia Bennett), New York
Well of Indigenous Wisdom School of Herbal Studies and Traditional Thought (Olatokunboh Obasi), Pennsylvania

nonflowering, 4-inch (10 cm) stems with lots of leaves and remove the leaves from the bottom half. Place the cuttings into a pot of soil and keep moist while the cuttings develop new roots. Plants like mints, soapwort, wild bergamot, and horseradish can propagate from a tiny, 1-inch (2.5 cm) piece of root as long as the root has a single bud. Plant these roots in a tray of soil and water them as the daughter plants emerge. Shrubby herbs like rosemary, bay, sage, and thyme can also be propagated by layering. Trim the lower leaves from a side stem, bend it over, and staple it to the ground. Roots will grow from the stem into the soil. The daughter plant can then be divided and replanted.

Growing perennial herbs from seed is also rewarding and relatively simple. Growing herbs from seed is very similar to propagating vegetable crops. In early spring plant your herb seeds into 1020 flats in a warm greenhouse or under growing lights, then transplant outdoors in late spring. Some herb seeds such as catnip, marjoram, thyme, chamomile, and lemon balm require light to germinate, so are scattered on the surface of the soil rather than covered. Of course, this makes the seed more susceptible to drying out, so a plastic cover over the flat is recommended. Other herbs are planted in the soil at the standard two times the depth of their length. After the danger of frost has passed, you can also plant herbs that prefer direct seeding directly into the soil, including calendula, basil, holy basil, feverfew, German chamomile, borage, garden sage, skullcap, echinacea, boneset, and anise hyssop. There are a few plants that do not produce seed, such as French tarragon and golden sage, and hybrid cultivars that do not breed true, including all mints, most lavenders, and ornamental thyme. These sterile plants and hybrids can only be propagated through division or cuttings.

Certain herbs require a vernalization period (cold season) in order to germinate. To reproduce these conditions artificially, put the seeds in a plastic bag with moist sand and leave in the refrigerator for two weeks to two months before sowing, depending on the herb. This process is called stratification and is required for boneset, ginseng, blue vervain, butterfly weed, blue cohosh, bloodroot, goldenseal, trillium,

Rosemary can be propagated from vegetative cuttings. Each stem from the mother plant will grow roots and become a new individual. Photo by Neshima Vitale-Penniman.

wild yam, wild ginger, false unicorn root, culver's root, mullein, skullcap, wormwood, and echinacea, among others.

Some hard-coated seeds will germinate more readily if first rubbed with fine sandpaper to break up the outer coating and allow moisture to germinate. Rub the seeds between two pieces of sandpaper until you see a little bit of the endosperm, which is usually a lighter color than the seed coat. This process, called scarification, is recommended for astragalus, wild indigo, hollyhock, licorice, marshmallow, passionflower, red root, and rue.

In areas where perennials will be sown, it is even more crucial to weed thoroughly as compared with annual cropping areas. A cover of black plastic mulch for six-plus weeks of warm weather in advance of planting will take care of even the most persistent weeds and create a sterile seed bed. Loosen the soil with a fork, add organic matter, and transplant the young seedlings to the same depth as they were in the pots. Herbs are easygoing plants and will tolerate low maintenance. Provide ample mulch and water at the time of planting. Once established, most herbs do not require watering unless there is a drought. A few times per season, you will need to weed and trim the plants. A mulch of comfrey leaves in the herb garden provides potassium and micronutrients.

Store certain seeds in moist sand in the fridge for a few weeks to simulate winter and promote germination. Mix the seeds, water, and sand in a bowl before transferring the mixture to a sealed plastic bag.

Herbal Preparations

Before harvesting or preparing herbal medicines, it is important to give thanks and pay homage to the spirit of the leaves. Many practitioners also leave a small offering of tobacco, flower petals, or cornmeal at the base of the plants before harvesting. Here are Yoruba prayers of thanksgiving for medicinal herbs and water.

Iba se Ori ewe.	I pay homage to the Spirit of the Leaves.
Ire alafia,	The good fortune of peace,
Ire'lera,	The good fortune of a stable home,
Ire ori're.	The good fortune of wisdom.
Ewe, mo dupe, ase.	Leaves, I thank you, so be it.

Iba se omi tutu.	I pay homage to the Spirit of cool water.
Ire alafia,	The good fortune of peace,
Ire'lera,	The good fortune of a stable home,
Ire ori're,	The good fortune of wisdom,
Ire ori tutu.	The good fortune of calmness.

Mo dupe gbogbo ire, omi tutu, ase.	Thank you for the many blessings of water, May it be so.

Harvesting herbs is a continuous process rather than a onetime annual event. Most herbs grow strongly enough to offer plenty of regrowth for repeat picking. Harvest in the morning of a sunny day to maximize the essential oil content of the plants. Wait until any dew has evaporated. Use sharp scissors to harvest rather than pulling at the plant with bare hands, as this prevents disease and damage. Leaves are best picked before they come into flower. Flowers should be cut soon after they open and not left to drop their petals. Seeds are harvested once ripe, but

UPLIFT

Sacred Herbalism of Cuba and Haiti

In the African traditions of Lukumi and Vodou, illness is primarily seen as the disruption of the connection between the person and Nature. Soul force, or ase, can be found within the plants, imbuing herbal medicine with its power. Each December 24–25 in Haiti, the community gathers to make the herbal medicines that will restore the life force of the people in the year

In Haiti the *houngan* (traditional priest) makes offerings to the lwa as part of a medicine-making ceremony, called *piley fèy*.

before they drop and disperse. Roots and rhizomes are best collected during their dormant period in autumn or winter. Leave at least 25 percent of the root to allow the plant to regenerate. Herbs should never be left in heaps waiting to be processed, as deterioration sets in quickly. Wipe, rather than wash, foliage clean. Wash the roots and rhizomes in cold water and cut into pieces before processing.

Air-drying herbs is the simplest method of preservation. Harvest bunches of leafy herbs and strip the leaves off the bottom inch of the stem for banding. Hang the bunches by the bands in a dry, well-ventilated place out of sunlight. In dusty and urban areas, it is best to cover the bunches with a paper bag to keep the herbs clean. Herbs can also be laid out on trays or screens for drying. Aim to complete the drying in four to seven days, so that volatile oils are not lost. Once leaves and flowers are crackly-dry, they can be stripped off the stem by hand and stored in paper bags or glass containers in a dark, dry place. The

to come. Called pile fèy, this ceremony involves offerings for the lwa, singing, drumming, and pounding medicinal leaves in a large mortar to the beat of the songs. During the day the dry medicine is prepared, filling the space with spicy pungent dust that uplifts the power of the medicine makers. By night the liquid medicine is made by vigorously rubbing the leaves in water to sacred song.

Each orisa or lwa has a list of medicinal herbs and natural substances assigned to it and used to treat ailments related to particular areas of the body. For example, the Orisa Osun is said to govern the bloodstream and her sacred herbs are used to heal diabetes and anemia.[32] The preparation of medicines, called *ozain* and *omiero* in Lukumi, involves singing to specific orisa while their associated herbs are crushed and the juice extracted. The ceremony enhances the power of the *ewe* (medicinal leaves) and the resulting infusion can be used to cleanse and refresh sacred objects, bathe individuals during initiation, and provide healing for physical and spiritual complaints. For example, the plant *Tradescantia zebrina* (wandering Jew) is a central ingredient in certain omieros for its spiritual power, and also has active compounds that purify the kidneys, ease colitis, and encourage menstruation.

Similarly, Haitian Vodou practitioners make spiritual baths that have both curative and spiritual elements. Practitioners generally prefer wild plants over cultivated ones, because they are believed to have more ase. Gatherers of plants leave a coin or other offering in the place where the *fèy* (leaves) are collected. Herbalists have knowledge of hundreds of potent plants, including *Momordica charantia* and *Hamelia patens* for measles and smallpox, *Dysphania ambrosioides* and *Momordica charantia* for parasites, and *Vitex trifolia*, *Trichilia glabra*, *Alpinia speciosa*, and *Allophylus cominia* for spiritual cleansing. The baths are either immersive or applied head-to-toe and repeated for a number of days, usually three or seven, sacred numbers in the religion. Spiritual baths can be applied to purify the home as well, the cleansing mixtures rubbed on walls and floors, then "swept" out using branches or long stemmed flowers. A specific spiritual bath applied only to the head, called *Rogation* in Lukumi or *Lave Tet* in Haiti, is said to alleviate depression, mental confusion, high blood pressure, and temper, having an overall cooling effect.

ideal drying temperature for leaves and flowers is 80° to 99°F (27–37°C). Roots and tubers require higher temperatures to dry, 120 to 140°F (50–60°C). You can also cut them into small pieces and dry in the oven.

Preparations for both dried and fresh herbs include:

Infusion. *Infusion* is a fancy word for "tea." Combine 1 teaspoon of dry herb or 2 teaspoons of fresh herb per 1 cup (240 ml) of almost boiling water. Cover the mixture with a lid to prevent the volatile oils from escaping. Steep for 5 to 10 minutes, then strain out the herbs. Infusions can be taken hot or cold. A standard dose is 6 ounces (180 ml), three times per day.

Decoction. Roots, barks, and berries require more forceful treatment to extract their potent constituents. Add 1½ tablespoons of dried roots or 3 tablespoons of fresh roots to 3 cups (720 ml) of cold water. Simmer for 20 to 30 minutes until the liquid is reduced by one-third. Strain out the plant parts

Herbs hang to dry in the farm kitchen. Photo by Neshima Vitale-Penniman

and store the liquid covered in a cool place. A standard dose is 6 ounces (180 ml), three times per day.

Tincture. Tinctures are made by soaking the herb in vodka or rum to dissolve its active constituents. Combine 1 part chopped herb to 5 parts alcohol in a sterile glass jar, ensuring that the herb is covered. Shake well for one to two minutes then store in a cool, dark place for 10 to 14 days, shaking the jar every day or two. Strain out the herbs using a colander or wine press, then store the tincture in clean dark glass bottles. A standard dose is 1 teaspoon, two or three times a day, diluted in water or juice. Note that tinctures can also be made with glycerine for those who are avoiding alcohol.

Oils. Infusing an herb in oil dissolves its fat soluble constituents. To make a hot infused oil, combine 9 ounces (250 g) of dried or 18 ounces (500 g) of fresh herb with 3 cups (720 ml) of olive or sunflower oil. Stir the chopped herb and oil together in a glass bowl and double boil over a saucepan of water for two to three hours or in a slow cooker for two to three hours. Do not put the oil directly over the heat or it will burn—trust me, I've made that mistake! Allow the mixture to cool, and strain it through a cheesecloth. Similarly, cold infused oils are made by covering fresh or dried herbs with olive oil in a clear glass jar. Place the jar on a sunny windowsill for two to six weeks, then

An oil infusion of calendula and chickweed makes a curative base for salve.

strain through cheesecloth. Store the infused oil in dark glass bottles. Oils can be rubbed on the skin three-plus times per day to relieve pain and promote healing.

Salve. Combining infused oils with solid fats creates a salve that can be applied to the lips or skin for healing and pain relief. Use a double boiler to melt the infused oil together with beeswax in a 4:1 oil:beeswax ratio. Alternatively, use shea butter or coconut oil instead of the beeswax at a 1:1 shea butter:infused oil ratio. Once melted, you can add vitamin E or rosemary oil to increase shelf life and prevent rancidity as well as essential oils for their aromatic gifts. Pour the mixture into small jars with lids and set aside to cool. Store in a dark place. Salves can be rubbed on the skin three-plus times per day.

Poultice. A poultice is a mixture of fresh or dried herbs applied directly to the skin to ease pain and draw impurities from wounds. Add just enough water to a pan to cover the bottom, then simmer the herbs for two minutes. Rub oil on the affected area to prevent sticking, then apply the hot herb to the skin. Bandage the herb securely in place using gauze and leave for two to three hours. Repeat as often as necessary.

Steam inhalation. Inhaling steam is a powerful way to clear respiratory ailments. Make an infusion of 1 ounce (28 g) of the herb in 4 cups (1 L) of water. Pour the infusion into a bowl, cover the head and bowl with a towel, close the eyes, and inhale the steam for 10 to 15 minutes or until the preparation cools.

Baths. A warm or cold infusion of herbs can be used to bathe the skin for medicinal and spiritual purposes. Offer prayers and rub the mixture into the skin from head to toe. Spiritual baths are more potent if the herb is not strained. An herbal bath can also be made by rubbing and crushing herbs in cold water. If bathing the eyes, add a pinch of salt to the infusion to prevent leaching of minerals from the eye.

CHAPTER ELEVEN

Urban Farming

I grew up around a lot of violence. I lost a lot of friends. When there were drive-by shootings, I would get low to the ground and the smell of the earth meant I was safe.

—SOUL FIRE FARM ALUM, 2015

The Great Migration that carried 6 million of our people out of the rural Southeast of the US and into the urban North, Midwest, and West is often miscategorized as a voluntary exodus to exploit new economic opportunities. The truth is that our people were catalyzed to move because of segregation, land theft, racial terrorism, and lynching. Over 4,000 Black people were lynched in the South between 1882 and 1968. At the same time, the United States Department of Agriculture (USDA) was colluding with white supremacist groups to maintain sharecropping by denying independent Black farmers the loans and relief to which we were entitled by law, leading to the loss of millions of acres of Black-owned farmland. By the end of the Great Migration in 1970, 80 percent of African Americans lived in cities.[1] There was a trade-off for leaving the South; as playwright August Wilson wrote, "We were a land based agrarian people from Africa. We were uprooted from Africa and we spent 200 years developing our culture as black Americans and then we left the South. We uprooted ourselves and attempted to transplant this culture to the pavements of the industrialized North. It was a transplant that didn't take. I think if we had stayed in the South we would have been a stronger people and because the connection between the South of the 20s, 30s, and 40s has been broken, it's very difficult to understand who we are."[2]

In the North, Black people confronted racism in other forms: discrimination in housing, employment, schools, and policing. The National Housing Act of 1934 institutionalized pre-existing housing discrimination. The Federal Housing Administration (FHA) created "residential security maps" that ranked neighborhoods from A to D, listing areas from the most to least desirable for lending. The D neighborhoods were predominantly Black communities and were outlined in red, labeled too risky for mortgage support. These maps were used by public and private lenders to deny mortgages to Black people. Further, the FHA's Manual of 1936 advocated deed restrictions to "prevent the infiltration of inharmonious racial groups" and to "prohibit the occupancy of properties except by the race for which they are intended." These recommendations were in the same section that advised how to prevent "nuisances like pig pens." Redlining led to lower property values, abandonment, vacancy, and decline in Black neighborhoods. When the GI Bill was enacted during World War II, veterans who wanted to buy homes in their own redlined neighborhoods were denied the zero-interest mortgages to which they were entitled. Consequently, white military families moved to the suburbs while Black families had to turn to predatory lenders or rent from slumlords.[3]

In the 1960s and 1970s amid steep urban decline, arson, and blight in our urban neighborhoods, courageous Black and Latinx farmers revived their agricultural traditions by establishing community gardens. Neighbors transformed trash-strewn lots into urban oases with the support of their churches and neighborhood associations. Urban farmers of color removed rubble, planted trees, installed vegetable beds, and built structures for community gatherings. For instance, we pay homage to Hattie Carthan, Black environmental activist, who coordinated the planting of over 1,500 trees in the Bedford-Stuyvesant neighborhood of New York City in the late '60s and '70s. We also pay homage to Black growers John and Elizabeth Crews, who catalyzed subsistence farming in Detroit. We pay homage to Rufus and Demalda Newsome, elder leaders in the urban farming movement in Tulsa, and to all the other Black visionaries who helped us find our way home to land.

There are now an estimated 18,000 urban community gardens in the US, predominantly in neighborhoods once redlined. Today an overlay GIS map of New York City's redlined districts and community gardens elegantly correspond.[4] As Black and Brown people worked together to beautify their neighborhoods, city governments and white opportunists began to take interest. For example, in 1998 Mayor Rudolph Giuliani attempted to put over 700 community gardens up for sale and was only stopped because of powerful grassroots resistance through the courts and civil disobedience. Under the guise of color-blindness, public and private sectors have facilitated gentrification, promoting a vision of low-income communities as the "urban frontier," encouraging young, middle-class white people to act as urban "pioneers" and "homesteaders" by populating these communities building by building, block by block.[5] These white "pioneers" have co-opted urban farming in many locations, attracting grants, media attention, and public influence denied to the Black founders of the movement.

For most of our community, the city is now home and we need to find our "liberation on land" amid the pavement. This chapter elucidates specific strategies for accessing land, rooftops, and sunny corners in urban spaces. It offers ideas for managing farming challenges specific to urban areas, including small growing area, contaminated water, insecure land tenure, and community support. This chapter uplifts a few of the current-day Black urban farmers in the US as well as the Afro-Cuban farming technologies that helped Cuba become almost entirely self-sufficient in fruit and vegetable production using urban spaces.

Hattie Carthan was a Black environmental movement pioneer. Her legacy is stewarded by Farmer Yon in Bed-Stuy, New York. Photo courtesy of Yonnette Fleming.

Laws and Land Access

For a few years our illegal chickens flew under the radar of the city. Albany had an ordinance that disallowed chickens and other livestock under the belief that they were "incompatible with urban life." There was an exception for educational nonprofits, but no flexibility for residents trying to raise food for survival. We collaborated with three other neighbors with adjoining backyards to take down our fences and build a collaborative food landscape. All combined, there were two chicken coops, a goat pen, vegetable gardens, and mulberry trees. Our young children could roam far without ever having to cross a street. Unfortunately, code enforcement caught on and evicted the chickens. Many neighbors took the fight to city hall. After nine months of organizing, activists convinced the city council to allow backyard chickens in 2011. Sadly, the mayor vetoed the vote and the chickens could not return.

In many cities these nuisance ordinances were created to specifically target immigrants and people of color who were more likely to be preserving agrarian practices. Whether you decide to abide by the law, skirt the law, or challenge the law, it is helpful to understand what restrictions apply to urban agriculture in your area. Before you invest substantial resources into your plot of earth, investigate the following questions:

A small flock of hens is compatible with urban life.

UPLIFT

D-Town Farm

In 2006, 40 people came together to form the Detroit Black Community Food Security Network (DBCFSN).[6] Some of the key contributors included Baba Malik Yakini, Nefer Ra Barber, Kwamena Mensah, Kadiri Sennefer Ra, Babatunde Bandele, Aisha Ellis, Iythiyel Elqanah, Shakara Typer, Karanji Kaduma, and Tiffany Harvey. In just a decade DBCFSN has catalyzed a food policy council, cooperative grocery store, youth education program, and a 7-acre urban farm. For most of their history, the project had no outside funding, relying exclusively on sweat equity. DBCFSN now has 70 members who make decisions about the organization democratically.

DBCFSN took leadership in reclaiming some of the more than 90,000 vacant lots in Detroit for food production. They approached the city council in 2006 asking for 2 acres of land to start D-Town Farm. The city granted them a license agreement for those acres and later added 5 additional acres to the deal. DBCFSN arranged legal representation at no cost with the University of Michigan Economic Clinic to help negotiate the license. As Baba Malik explained, "We started out initially with two acres and we currently have seven acres—so it's a pretty large project. We're trying to stay focused and not have various locations around the city. Frankly, we don't have the capacity to manage that. Even managing the seven acres we have now is challenging! We consider ourselves to be a model. Rather than trying to start gardens all around the city, what we're doing is creating a learning institution where people who are interested in doing this work can come and learn various techniques and strategies that they can take back to their neighborhoods. So we see ourselves as a catalyst."

D-Town Farm hires local residents to grow the food as well as engage collective work ethic to get tasks done on the farm. D-Town Farm sells their produce at farmers markets, restaurants, and grocery stores. One of their goals is to have locally grown options available at the stores where people normally shop. "Rumor has it that D-Town has the sweetest collard greens in Detroit," offered Baba Malik. They grow other heritage crops like black-eyed peas, sweet potatoes, okra, and watermelon. Their growing practices are "in line with how our ancestors farmed and how most human beings farmed for the last several thousand years."

D-Town Farm Staff. Photo by Malik Yakini.

Zoning. Does the zoning of the land permit agriculture? What "public nuisance" laws exist in your city? Are there "right to farm" laws in your county?

Greenhouses and high tunnels. Is a building permit required for greenhouses? Is there a size limitation? Note that some states categorize greenhouses as permanent structures and high tunnels as temporary structures, influencing how they are taxed.

Animal housing. Is animal agriculture permitted? Is a building permit required for chicken coops and other animal housing? How far away must animals live from residential buildings or property lines?

Rooftop and wall gardens. What building code requirements exist for a rooftop garden? Do rooftop gardens require permits? Is a permit required to grow plants on the exterior wall of a building?

Food safety. Is it legal to sell processed food out of your home kitchen per a "cottage food law"? If not, determine whether a commercial kitchen is available to rent. Learn more about food safety law from the Urban Agricultural Legal Resource Library, a project of the Sustainable Economies Law Center.

Should you decide to advocate for changes in zoning or ordinances that are more favorable to urban agriculture, it helps to look at best practices in other cities. For example, Cleveland, Ohio, has an Urban Garden District Zoning ordinance that gives community gardens additional protection from being sold and converted to other uses. Seattle, Richmond, Portland, Oakland, Minneapolis, Milwaukee, Chicago, and Baltimore also have useful policy templates for promoting urban agriculture.[7] In advocating for policy change, try to get your hands on any community food assessments that can provide data to support the need to uplift urban agriculture. Engage with the local food policy council to find allies for your cause.

Lack of accessible and affordable land can be one of the greatest constraints to urban farming. Empty lots, utility rights-of-way, private backyards, parks, institutional land (schools, hospitals, churches, prisons, universities, senior homes), and rooftops are all examples of vacant land that might be reclaimed for agricultural use. When you encounter vacant land with agricultural potential, take note of the address of the adjacent parcels and take that information to the local tax assessor or department of finance. Ask to see the tax map and property records to determine the parcel number, and research the site's ownership history. You can then contact the most recent owner to discuss license, lease, or sale of the property. Alternatively, many cities have passed legislation to allow citizen-controlled land banks and land trusts to manage vacant land. If your city has a land bank, approach them directly to determine what properties are available, their land-use history, zoning designation, and any public programs that provide incentives for their purchase. The national nonprofit Trust for Public Land also stewards urban lots that can be used for agriculture.

If you are unable to buy property outright, you will need to enter into a lease or other contractual agreement to guarantee access to the land. The Urban Agricultural Legal Resource Library outlines important elements of land-use agreements for both public and private land and sample land-use agreements. Your land-use agreement should include the following provisions, at minimum:

Provisions of Land-Use Agreement

Land. Specifications of size and location.

Rent. Cost to tenant.

Use of land. Specification of permitted uses and prohibited uses (sales, tree removal, fires, and the like).

Term. Duration of lease, options for lease renewal, and expected tenure of project on land.

Building and improvements. Clarification of building types prohibited and permitted (carports, storage, temporary shelters, and so on) and improvements (fencing, garden beds, landscaping).

Right of entry. For example, restrictions to farm employees, contract workers, volunteers.

Hours of use. Days and times of activities, clarification of overnight stay.

Noise. Expected decibels of noise pollution created.

Animals. Use of animals and restrictions thereof.

Expected traffic. Estimated number of trips to the site and number of people expected on a plot at any given time.

Growing practices. Farmers' use of tools/machinery and use of pesticides, fertilizer, fungicides, and so on. (On the city's end, this could be a selection criterion—for example, projects growing organically could rank higher than projects proposing to use these chemicals.)

Environmental impacts. Management of runoff and water pollution.

Water usage. Agreement on source, use, and payment.

Routine maintenance. Specifies responsibilities of landowner and farmer in maintenance of plot's appearance and preventing hazards.

Subleasing policy. Permitted/prohibited and where liability for subtenant lies.

Garden produce. Clarification of ownership of produce from the land and whether sales are permitted.

Compost. Agreement on use and location of compost pile and perhaps use of landowner's acceptable yard and kitchen wastes.

Payment. Type and amount of payment; can be monetary or in-kind through share of crops.

Liability. Two-way release of liability; each party gives indemnity to the other over specific scenarios and legal responsibilities for their respective uses of the land.

Clean Soil, Clean Water

The ¼-acre lot at the end of Oread Street in the Main South neighborhood of Worcester, Massachusetts, was partially covered in asphalt and entirely covered in garbage. There were high levels of soil lead, no secure land tenure, no infrastructure, and no access to water. Where others may have seen despair, the mostly Afro-Latinx residents on the block saw a future youth farm. We supported neighbors to negotiate with the grumpy absentee landlord and get a license agreement, then convinced the city to pick up the trash and asphalt for free. Of course, we first had to remove the trash and asphalt. With borrowed crowbars, pickaxes, and wheelbarrows, our intergenerational team set to work clearing the lot over a series of weekends. We installed rainwater catchment on a neighbor's shed as our initial irrigation and implemented best practices to keep gardeners and consumers safe from contaminated soil. Over 15 years later we are proud to say that the Oread Street Farm is still in operation, managed by local youth growers.

Clean Soil

A legacy of environmental racism has contaminated many of the sacred soils in our neighborhoods with arsenic, cadmium, copper, lead, nickel, selenium, and petroleum pollution. From the demolition of buildings containing toxic materials, to atmospheric deposit of air pollution, to industrial effluent, the land in our communities bears a disproportionate burden of harmful toxins. This legacy, however, should not make us fear the earth. According to Malik Yakini, only 13 percent of soil samples in garden lots across Detroit have heavy metals above the federal guidelines, despite public perception that all are contaminated. Further, it is possible to grow in moderately contaminated areas so long as you follow certain safety precautions.

A thorough discussion of the possibilities for remediating contaminated soil using specific plants and phosphate-rich compost is provided in chapter 4, "Restoring Degraded Land." Even for those without the time and resources to remove contaminants through remediation, there are simple precautions you can take if your heavy metal levels are within the range presented in table 11.1. It is imperative that you have the soil-tested for heavy metals at a professional soil testing laboratory. Your local agricultural extension office can provide you with a list of certified laboratories. At Soul Fire Farm we mail our soil samples to the University of Massachusetts, Amherst for analysis.

Provided that your soil contaminant levels are below those in the "unsafe for any gardening or

Teens from Worcester's Youth Grow urban farming program visit Soul Fire Farm. Photo by Jonah Vitale-Wolff.

Table 11.1. Unsafe Levels of Heavy Metals in Soil

Heavy Metal	Typical Levels for Noncontaminated Soils	Take Precautions, Unsafe for Leafy or Root Vegetables	Unsafe for Any Gardening or Contact
Arsenic	3–12 ppm*	>50 ppm	>200 ppm
Cadmium	0.1–1.0 ppm	>10 ppm	>50 ppm
Copper	1–50 ppm	>200 ppm	>500 ppm
Lead	10–70 ppm	>500 ppm	>1,000 ppm
Nickel	0.5–50 ppm	>200 ppm	>500 ppm
Selenium	0.1–3.9 ppm	>50 ppm	>200 ppm
Zinc	9–125 ppm	>200 ppm	>500 ppm

Source: Hannah Koski, *Guide to Urban Farming in New York State* (Ithaca, NY: Cornell Small Farms Program, 2012), http://www.ruaf.org/sites/default/files/GuidetoUrbanFarminginNYS_Revised2.12.13-2jpbu08.pdf.

* parts per million

contact" column, follow the subsequent best practices to minimize risk.

- Plant crops away from building foundations, painted structures, and heavily traveled roads.
- Plant vegetable crops in raised beds containing uncontaminated soil and compost.
- Remove obvious contaminants such as scrap metal and construction materials.
- Use mulch and cover crops to minimize exposure to contaminated dust, and to maintain high levels of organic matter.
- Because concentrations of heavy metals are highest in roots and leaves, avoid planting and

UPLIFT

Karen Washington, Garden of Happiness

Karen Washington, founder of the Garden of Happiness in New York City, as well as Black Urban Growers, is a movement elder who exemplifies the teaching "We protect what we love." Karen learned to love and cherish wholesome food and agriculture as a child. She comes from a long legacy of talented cooks, and her father was the first Black produce manager for a major supermarket chain. He was also a fisherman and brought home "tons" of fresh seafood and vegetables for the family. Mama Karen shared, "We lived in the projects and every Saturday mother would be knocking on doors to share the extra fish. We had it wrapped in newspaper in the freezer ready to give to everyone in the building." As a child, she watched the "farm report" on television before the cartoons and tasted milk fresh from a cow's udder on a school field trip, while the other children squealed.

Mama Karen moved to the Bronx in 1985 as a single parent of two children and had a big backyard for the first time. She called upon her lifetime love of farming and taught herself to grow food from library books. She said, "It was the tomato that changed my life. I used to hate them because they were pale and did not taste good. But that first tomato was so good, I decided right then that I wanted to grow everything."

In 1988 she founded the Garden of Happiness on a vacant lot right across the street from her home. At the time there were a lot of empty lots in the Bronx; many considered it a "war zone," with buildings literally burning down everywhere you turned. Mama Karen got support from Bronx Green Up to start that landmark garden 30 years ago. The garden was centered on community hope and survival in the midst of a cocaine and heroin epidemic. The gardeners removed the garbage and made the lot into a beautiful, safe space, a beacon that encouraged people not to flee from the community. Mama Karen explained, "Anytime Black and Brown people do something to be self-reliant, the system does something to buck it, dismiss it, or co-opt it. 'Look at those

eating leafy or root vegetable crops in soils with heavy metals above typical levels.
- Lime, compost, or amend soil to keep pH close to neutral, or even slightly alkaline, and ensure adequate draining to reduce the mobility and availability of lead and heavy metals.
- Work in the garden only when soil is moist or damp to avoid inhaling particles.
- Wear gloves, long sleeves, and pants while gardening to prevent skin exposure.
- Wash hands after gardening.
- Wash all vegetables thoroughly.
- Remove gardening shoes and garments before entering the home, and wash gardening clothes separately from other clothing.

Clean Water

In order to save money, the Emergency Manager of the city of Flint, Michigan, poisoned its residents. In 2014 Flint switched its water supply from Lake Huron to the notoriously filthy Flint River, resulting in several instances of fecal coliform bacteria contamination. The city responded by adding more chlorine to the water, monkeys, look at what they live in, they live in filth,' said the white police officers. I heard it with my own ears."

Then Mayor Giuliani came in with the developers, enticed by what the gardeners had done to improve the neighborhood and eager to develop these "empty lots." They auctioned off our gardens and we fought back. "My garden was not going to be on that auction block," Mama Karen pledged. She garnered the support of the NY Botanical Garden, Green Thumb, and Green Guerillas and got the garden transferred to the protection of the Parks Department. Another 294 gardens followed suit, saving many of their properties.

The fight is far from over. Not only are gardens under threat from gentrification, but the urban garden movement is being "swindled by a high-tech vision." Individuals and corporations from outside the community are working to commercialize gardening and negate the original vision of home provisioning, safe spaces, and cultural maintenance for immigrants. "We have been here from the beginning. We tested the soil, put in raised beds, and have grown food sustainably and efficiently. Community gardeners are *farmers*. We are strong and we have done so much with nothing. We are the ones who can feed the world. We know how to work and what to do. If you give us the capital, land, and opportunities, we can move mountains."

Karen Washington is founder of the Garden of Happiness in the Bronx and the national organization Black Urban Growers. Photo courtesy of Karen Washington.

causing a buildup of cancerous trihalomethanes. They failed to add an anticorrosive agent to the water, so iron and lead leached from pipes into the water supply, making children sick. In one home the tap water had 13,200 parts per billion (ppb) of lead. The EPA limit is 15 ppb. After persistent organizing by residents, the city declared a state of emergency and several lawsuits ensued.

In our own rural community, several towns recently had to install water filtration plants because private wells were contaminated with carcinogenic perfluorooctanoic acid (PFOA) that had leaked from the Taconic Factory and Saint-Gobain Performance Plastics. Fortunately, Soul Fire Farm's well was not contaminated, but many people lost their health and their homes to the disaster.

Clean water is life. In urban and low-income communities, we cannot assume that society will provide clean water for us. We need to be proactive about testing our water and seeking clean sources. To find a water-quality testing lab in your area, call the EPA's Safe Drinking Water Hotline at 800-426-4791 or visit www.epa.gov/safewater/labs.

Table 11.2. Recommended Residential Water-Quality Testing

Analysis	Recommended Level	Concerns
Coliform bacteria	0 colonies	Indicator of disease-causing contaminants
Lead	0.015 mg/L	Brain, nerve, kidney damage
Nitrate	10 mg/L as N	Methemoglobinemia (blue baby syndrome)
Turbidity	5 NTU	Carries contaminants, interferes with disinfection

Source: "Individual Water Supply Wells—Fact Sheet #3: Recommended Residential Water Quality Testing," New York State Department of Health, https://www.health.ny.gov/environmental/water/drinking/regulations/fact_sheets/docs/fs3_water_quality.pdf.

A simple rain barrel can be constructed from a 55-gallon (210 L) barrel. Photo by Scott Kellogg.

For those without access to a roof, you can construct a rain barrel with an inverted satellite dish to collect the water. Photo by Scott Kellogg.

Even when we know that the public water supply is clean, it may be prohibitively expensive to access. D-Town Farm in Detroit received a $1,500 water and sewer bill from the city for one month of summer usage. Of course, the farm was not using the sewer system at all, but rather diverting water from city management by maintaining a large permeable surface and a rainwater retention pond. While they work to reform policies, they are also taking action by developing their rainwater catchment systems to reduce the water they need to purchase from the city.

Rainwater harvesting saves money for farmers and has positive environmental impacts. Capturing rainwater reduces pressure on the sewage system, which maintains the health of local water bodies. While rainwater is not clean enough to drink or wash produce, it is suitable for irrigation. If you have a roof with gutters, a simple rain barrel can be made using an HDPE 55-gallon barrel that you can pick up from a car wash, from a food distributor, or on Craigslist for under $10. Select an opaque barrel that has not been used to store any toxic chemicals and rinse it thoroughly. Drill a hole on the side of the barrel toward the top and install a hose adapter for overflow. Drill another hole on the side of the barrel toward the bottom and install a hose bib to release the water. Set up the rain barrel on cinder blocks under the downspout of the gutter system and use a downspout adapter to direct the flow into the barrel. Multiple rain barrels can be attached to one another via the hose adapter.

Growing in Small Spaces

Like the exemplary urban farmers in Cuba who have sustained themselves for over 25 years on tiny city lots, we often have paltry spaces in which to grow the food that will nourish our community. It is possible to coax abundance from cramped places using a few strategies.

Sprouts

Our team of Afro-Caribbean farmers, artists, and healers were making our annual solidarity delegation to work with partners and family in Haiti on sustainable farming projects. We arrived in Leogane well after dark, exhausted from hours of hot dust on the bumpy road from Port-au-Prince, and eager for sleep. My roommate and womb sister, Naima, removed a glass jar with a perforated lid from her bag. She filled it with a handful of mung beans, rinsed the seeds with water, and set the jar on our bathroom sink to grow sprouts. "I always carry my greens with me," she announced casually while I looked on with awe.

Sprouts are nutrient-dense, containing vitamins C, A, and K, fiber, manganese, riboflavin, copper, protein, thiamin, niacin, vitamin B_6, pantothenic acid, iron, magnesium, phosphorus, potassium, and live enzymes. In 1997 John Hopkins University researchers showed that broccoli sprouts contained major amounts of the substance glucoraphanin, a precursor to the natural antioxidant and cancer-inhibiting detoxifying isothiocyanate called sulforaphane (SGS).[8]

Sprouts are simple to raise and require negligible space and light. The only equipment you need is a quart-sized glass canning jar, a sprouting lid or sprouting screen, clean water, and sprout seeds. Lentil, alfalfa, broccoli, buckwheat, clover, cress, fenugreek, mung bean, onion, pea shoot, radish, and sunflower are among our favorites. You can buy sprout seed from a health food store or order it online, though we find that it saves money to simply buy seeds marketed as grains or legumes. We purchase organic lentil, buckwheat, and sunflower seeds directly from the bulk bins at the health food store and find that they germinate effectively. Place 3 tablespoons of sprout seeds into the bottom of the jar and affix the lid. Fill the jar with clean water and let the seeks soak for 6 to 10 hours: less time for small seeds, more time for large ones. Rinse and drain the sprouts after soaking, allowing the jar to rest at a 45-degree angle in a bowl or rack in between rinsing. Rinse two or three times per day. The sprouts will grow to fill out the jar and then turn green, all in just four to six days. Store dry sprouts in a plastic bag in the refrigerator. They keep for four to five days.

We distribute sprouts in our Ujamaa Farm Share. To get the quantity and quality we need, we grow them in trays rather than jars. We use sterilized 1020 seedling flats with plastic dome lids. One plastic dome lid is inverted and used as the drainage tray. We place six metal rings from canning jars evenly spaced in the tray to hold up the 1020 flat. Next the 1020 flat is placed on the metal rings. Inside the flat, we evenly distribute about 2 cups (480 ml) of soaked

UPLIFT
Cuban Urban Agriculture

After the food crisis of 1989, when Cuba was cut off from trade with the rest of the world and residents experienced a loss of one-third of their daily calories, urban residents took action. Cubans created 8,000 *parcelas* (small lot gardens) and 1,000 small livestock enterprises in Havana alone. They implemented closed-loop, agroecological techniques and relied on compost and recycled organic matter for soil nutrition. The government caught on to the value of urban farming and created policies to support and expand the practice. The state offered training programs, set up subsidized agricultural stores, and created three compost production sites, seven artisanal pesticide labs, and 40 urban veterinary clinics in Havana, and set up a nationwide network of urban farming infrastructure. People drew on traditional farming techniques to creatively farm in small spaces. Although the urban agriculture movement came to incorporate new research on agroecological production, it was largely informed by knowledge that had been passed down over the generations, including from African ancestors.

Cuba became the global model for urban agriculture, sustaining food security for 25 years in an oil-scarce environment. Currently more than 86,000 acres of land are being used for urban agriculture in Havana. One rooftop farmer in the El Cerro neighborhood of Havana raises six chickens, two turkeys, 40 guinea pigs, and more than 100 rabbits on the roof, plus maintains systems for vegetables, compost, and rainwater collection.[9] Urban and other small-plot farms have played a major role in increasing vegetable consumption in Cuba, and in allowing the country to domestically produce almost all of the fruits and vegetables that its people consume, without relying on imported inputs.

Organoponico Vivero Alamar in Havana. Photo by Elizabeth Henderson.

Sprouts can be grown in jars or trays on the countertop.

sprout seed. Another plastic dome lid goes on top to hold in the moisture. Over the dome lid, we place a clean hand towel to keep out the light until the last day. Restricting light makes the sprouts grow taller. We remove the towel for the last 24 hours so that the green color develops. Similarly to the jar method, rinse and drain the sprouts two or three times per day and harvest in five to six days.

Most seeds contain natural toxins to protect against herbivory. Precautions are necessary to ensure that you do not consume these toxins in your sprouts.

- Legumes contain lectins, a natural toxin. Certain types of legumes have less of this toxin and are preferable for sprouting, including lentil, sugar pea, adzuki, urad, and garbanzo. Do not sprout peas in the Lathyrogen family.
- Flaxseeds contain cyanogenic glycosides—natural toxins that decompose into hydrogen cyanide when crushed. Medical professionals recommend that you not consume either raw or sprouted flaxseed in large quantities.
- Sunflower seeds have a high content of the heavy metal cadmium, which can accumulate in the body. Medical professionals recommend eating sunflower seeds in small quantities.
- Buckwheat seeds contain the natural toxin fagopyrin on their hard outer shell. Some people are sensitive to this substance and may experience skin irritation if they consume large quantities of buckwheat sprouts.
- Alfalfa contains the natural plant toxin L-Canavanine. To neutralize this substance, grow the sprouts until they produce green leaves. The small concentration of this toxin in the sprouts should not harm a healthy individual.

Microgreens

Like sprouts, microgreens are packed with nutrients and beneficial enzymes. The difference is that microgreens are grown for three to four weeks in a thin layer of soil and only the stems and leaves are consumed, not the roots. They are also less likely to suffer contamination as compared with sprouts. Our favorite microgreens are lettuce, kale, spinach, radish, beet, watercress, mustard greens, cabbage, sunflower, buckwheat, and culinary herbs.

Grow microgreens in a south-facing window or under a grow light. Place an inch of organic potting soil in the bottom of a 1020 flat or other shallow tray. Evenly scatter about 1 ounce (28 g) of small seeds or 2 ounces (57 g) of larger seeds over the soil. Cover the seeds with a thin layer of soil and use a spray bottle to water the seeds. Mist twice per day. Some farmers like to use a plastic blackout dome over the greens to encourage them to grow taller before turning green. Such a dome is optional. Harvest the greens with shears when they are about 2 inches tall, rinse, drain, and enjoy!

Containers and Rooftops

In urban spaces we may only be able to access a fire escape, front stoop, rooftop, or tiny yard for growing our crops. In these cases container gardening can be a space-efficient way to provide for ourselves. Containers can be made of plastic, clay, ceramic, or wood so long as they have drainage holes at the bottom. Avoid lumber treated with chemicals. Locate your container garden in a spot that gets at least six hours of direct sunlight per day, such as a south-facing patio.

Vegetables that do best in containers have a compact growth pattern and can tolerate the limited volume of soil. Most vegetables require at least 8 inches of soil for their roots to develop properly; the deeper and more voluminous the container, the higher the yields. Table 11.3 lists crops that grow well in containers and their space requirements. You can fill your containers with a mixture of compost and garden soil, or purchase organic potting soil. Container gardens require fertilization since the roots are limited in their ability to scavenge for nutrients naturally. A biweekly dose of compost tea or seaweed emulsion will help your plants thrive. For crops with heavy fruits, like tomatoes, eggplants, and pepper, provide supportive staking.

The frequent watering required by container gardening is one of the barriers that discourages some farmers. Containers need to be watered daily, saturating the medium such that water drains out of the holes at the bottom. If daily watering is not possible for you, consider constructing a container garden with reservoirs. The farmers at the Rooftop Garden Project in Quebec published free designs for several types of reservoir gardens that can be constructed with scavenged materials. My favorite simply uses two 5-gallon (20 L) buckets stacked with a spacer

Table 11.3. Container Sizes for Vegetables

Crop	Minimum Container Size (diameter inside top), in inches	Minimum Container Size (volume of soil), in gallons	Number of Plants per Container
Basil	6 (15 cm)	½ (2 L)	1
Beets	10 (25 cm)	2 (8 L)	12
Cabbage	10 (25 cm)	2 (8 L)	1
Carrots	10 (25 cm)	2 (8 L)	12
Cilantro	6 (15 cm)	½ (2 L)	1
Cucumber	10 (25 cm)	2 (8 L)	2
Eggplant	10 (25 cm)	2 (8 L)	1
Green beans	8 (20 cm)	1 (4 L)	2–3
Leaf lettuce	8 (20 cm)	1 (4 L)	5
Parsley	6 (15 cm)	½ (2 L)	1
Pepper	10 (25 cm)	2 (8 L)	1
Radishes	10 (25 cm)	2 (8 L)	16
Spinach	8 (20 cm)	1 (4 L)	10
Swiss chard	8 (20 cm)	1 (4 L)	1
Tomato, cherry	10 (25 cm)	2 (8 L)	1
Tomato, standard	13 (33 cm)	4 (16 L)	1

Source: Richard Jauron, *Container Vegetable Gardening* (Ames: Iowa State University Extension and Outreach, 2013).

Sajo Jefferson tends rooftop crops at Top Leaf Farms in Berkeley, California. Photo by Rucha Chitnis.

in between. The bottom bucket is filled with water through a pipe. The top bucket is perforated and filled with the soil and the plants. Through capillary action, the water is drawn up into the soil medium. This setup allows the farmer to water once every few days rather than on a daily basis, and results in a healthier crop.[10]

Intensive Growing

In urban spaces, every square foot of growing space is precious. When we managed community gardens in Worcester, I remember clandestine deals made between growers to maximize the number of plots they could access: "You sign up your sister for a plot and then I will grow the food and tend it, but don't tell nobody." Personally, I have done my fair share of negotiating, squatting, and fabrication to find a place for my seeds to germinate. Land is our right and fundamental to our survival. For that reason, we need to make best use of the space we have.

Youth at Gardening the Community, Springfield, Massachusetts, prepare an area for biointensive growing. Photo by Ibrahim Ali.

Table 11.4. Plant Spacing for Square-Foot Gardening

Crop	Number of Plants Per 1-Foot (30 cm) Square
Basil	4
Beans	8–9
Beets	9
Broccoli	4 plants per 9 squares
Cabbage	4 plants per 9 squares
Cauliflower	4 plants per 9 squares
Carrots	16
Celery	1
Cilantro	9
Collard greens	1
Corn	3
Cucumbers	2
Dill	1
Eggplant	1
Garlic	9
Kale	1
Leeks	9
Lettuce, head	1
Mustard greens	16
Okra	1
Onions	9
Parsley	1
Parsnips	16
Peas	8
Peppers	1
Potatoes	1
Radishes	16
Spinach	9
Sweet potatoes	1
Swiss chard	4
Tomatoes	4 squares per plant with stake or trellis
Turnips	9
Watermelon, cantaloupe	2 squares per plant with trellis
Winter squash	2 squares per plant with trellis
Zucchini and summer squash	4 squares per plant with cage

Source: Mel Bartholomew, *All New Square Foot Gardening II: The Revolutionary Way to Grow More in Less Space* (Brentwood, TN: Cool Springs Press, 2013).

Intensive growing maximizes productivity. Pile on the compost and inoculants to magnify the health of the soil so that it can support more plants per unit area. Rather than plant in rows, organize the garden into a grid comprised of 1-foot (30 cm) squares. Within each square, evenly space plants according to their requirements, as shown in table 11.4. Stake or cage any crop that has the capacity to grow vertically. Pole beans, winter squash, cucumbers, tomatoes, and peas are all encouraged to grow up rather than sprawl, further maximizing the use of space. As soon as one crop is harvested, remove it immediately, add an inch of compost, and plant the next succession.

Mushrooms

As Haitian people we cherish the color, taste, and aroma that the *djon-djon* mushroom imparts on rice. In Nigeria we make soups and stews with plump mushrooms, called *ero* or *elo* in Igbo, or *olu* in Yoruba. Harvesting and cooking with fungi predates recorded history in Africa.

Cultivating mushrooms requires very little space, light, or maintenance. Mushrooms are low in cholesterol, high in vitamin D and protein, and said to boost the immune system, regulate diabetes, and combat cancer. All it takes to grow mushrooms are logs, fungal spores, water, and a drill bit.

At Soul Fire Farm we grow several strains of shiitake mushrooms (*Lentinula edodes*), and look forward to trying oyster, lion's mane, box elder, and nameko mushrooms in the future. We purchase spawn from Field & Forest Products. Growing mushrooms on logs, in contrast with sawdust, imbues them with higher nutrient content. Shiitakes prefer oak logs, but we are not blessed with oak on our land, so we use sugar maple. Hophornbeam, red maple, ironwood, alder, poplar, and yellow birch also work well for growing shiitakes. In early spring, cut 4-foot (1.25 m) logs from fresh, living trees. Then, let the logs rest for a week or two while the antifungal coumarins dissipate. Drill 5/16-inch (8 mm) holes, 1 inch deep into the logs at 10-inch (25 cm) intervals down the length of the log. Spin the log to provide a gap of 2 inches (5 cm) between the rows, and stagger the holes from one row to the next until the entire log surface is drilled. Insert the plugs into the log and seal the holes with paraffin wax to prevent other fungal species from colonizing the log. Additionally, apply a layer of wax to either end of the log to retain moisture. Stack the logs in a crisscross pattern under the shade of trees or a building. Keep the logs moist by either using a sprinkler or soaking them in a tank for 24 hours every month. Some growers pound their logs with a heavy bar or drop them forcefully to encourage fruiting. Fruits emerge after four to six months and continue producing for several years.

Vermicomposting

Vermicomposting puts red wiggler or red earthworms to work eating your food scraps. Vermicomposting is an ideal strategy for the urban dweller who does not have access to an outdoor compost pile. The rich castings produced by the worms are suitable to fertilize container gardens and houseplants. Source a long, wide plastic bin with a lid, sized to about 1 cubic foot (0.03 cubic meter) per person in your household. Use a hammer and nail to poke ample holes into the lid. Add shredded newsprint, food scraps, and approximately 1 pound (0.5) of worms to the bin. Maintain a ratio of 70:30 paper (brown matter) to food scraps (green matter) by adding food to your bin as available, at least every couple of weeks. Stir the contents of the bin gently once every one to two weeks to help with aeration. Always keep a layer of shredded paper on top of the pile to discourage odor and insects. Keep the contents of the bin as moist as a damp sponge and in a location between 55° and 75°F (13–24°C). After 1 month you can harvest your first castings. Begin feeding the worms only on one side of the bin to encourage them to migrate over. Then remove the castings from the side without the food and pick out any straggling worms. The compost can be applied immediately to your container gardens.

Soul Fire Farmers drill holes in maple logs to prepare them for fungal inoculation. Photo by Jonah Vitale-Wolff.

Red wiggler worms transform kitchen scraps into Black Gold. Photo by Elizabeth Vitale.

Community

Do not protect yourself by a fence, but rather by your friends.

—CZECH PROVERB

Urban agriculture is distinct from rural growing, not just in terms of access to space, but also in regard to proximity of people. Neighborhood gentrifiers often view the humans on the block as obstacles to the beautiful gardens they want to establish. They view community in terms of vandalism, theft, and trespassing. Gardens that are created by and for community members see humans as essential components of the urban ecology, integral to the health of the space.

There is no substitute for creating a garden space together with neighbors from the ground up. It is much more difficult to "outreach" to stakeholders after key decisions have been made. Ownership comes from authentic involvement from the beginning and the shared power that ensues. Once a garden is established, certain practices encourage ongoing community involvement:

Youth investigate a sunflower with Black urban farmer and author of *The Color of Food*, Natasha Bowens. Photo courtesy of Natasha Bowens.

- Transparent membership process and decision-making structures so new people can get involved
- Regular community volunteer days, tours, workshops, meals, and public celebrations
- Open gate, low fence, or no fence
- Multilingual signage indicating name and purpose of the garden, and contact information for getting involved
- Donations or low-cost sales of produce to neighbors
- Designated planters at the garden edge with signs indicating that people can help themselves
- Maintenance of a beautiful and organized space that inspires pride

We want to offer a shout-out to some urban farms and urban farming initiatives that were founded and led by Black people and exemplify community ownership. This list is certainly incomplete. Reach out in your city to connect with the courageous urban farmers who are leaders in the Black land sovereignty movement.

- D-Town Farm, Detroit, Michigan
- East New York Farms!, Brooklyn, New York
- Farm a Lot Program, Detroit, Michigan (founded by Black mayor Coleman Young, 1974)[11]
- Farms to Grow, Oakland, California
- Garden of Happiness, Bronx, New York
- Gardening the Community, Springfield, Massachusetts
- Growing Power, Milwaukee, Wisconsin (now stewarded by Green Veterans)
- Hattie Carthan Community Garden, Brooklyn, New York
- La Mott Community Garden, Cheltenham, Pennsylvania
- RID-ALL Green Partnership, Cleveland, Ohio
- Ron Finley Project, Los Angeles, California
- Sankofa Farm at Bartram's Garden, Philadelphia, Pennsylvania
- Soilful City, Washington, DC
- Truly Living Well, Atlanta, Georgia
- Urban Growers Collective, Chicago, Illinois

CHAPTER TWELVE

Cooking and Preserving

I need to hear the bumping of pestles making percussion with sunrise and twilight. I need the scratch of grating tubers and the grinding of spices on stone. I need the sonic world of the ancestors, lullabies said while babies are fed, bawdy songs as the land is smoothed for planting. I need to understand the sound of the wind in the rice and the complexities of the yam mounds intercropped to save space. I need the rustle of the oil palm fronds so I can hear the generations speak.

—MICHAEL W. TWITTY, *The Cooking Gene: A Journey Through African American Culinary History in the Old South*

The two-hour round trip to pick up 80-year-old Mama Isola from her modest apartment in a senior housing complex was rousing and challenging. We were on our way to the farm to co-facilitate a workshop on canning and food preserving. Mama Isola, the canning expert with decades of life experience, would teach the content. I, the young organizer, would convene the people and ensure that all necessary supplies and equipment were in place. As we drove, Mama Isola told me about life as a young girl in Mississippi. Her family raised all of the vegetables and meat they needed to survive and only went to the store for "flour and sugar." She also explained to me that I needed to learn how to cook properly so that my husband would never leave me. "Men love through their stomachs," she explained, waving her tattered copy of *The Way to a Man's Heart* for emphasis. I lovingly countered, "My man needs to learn how to cook so I don't leave him!" She just shook her head.

Mama Isola offered the 20 aspiring young homesteaders plenty of insider tips on canning during her afternoon in our farm kitchen. "Put a metal spoon in with the hot beets so your jar doesn't crack . . . Scald the tomatoes with boiling water and the skins come right off . . . Did you wash your hands properly, young lady, or do you want to poison your husband with nasty germs?" Learning from elders involves more effort and complexity than looking up procedures online. We decided to embrace the rich texture of intergenerational transference and were blessed with a day of learning, laughter, and heritage.

Too often we are told that healthy cooking and food preservation is a "white people thing," when in fact, the unhealthy aspects of our cuisine resulted from the deprivations instituted under slavery. In this chapter we uplift traditional and contemporary African Diasporic diets, recipes, and food preservation methods. We also offer strategies for stocking up and eating healthy on a limited budget.

Mother Isola teaches a canning workshop at Soul Fire Farm. Photo by Jonah Vitale-Wolff.

African Food Pyramid

In addition to mass incarceration, one of the most insidious and pervasive forms of state violence against our people is the flooding of our communities with foods that kill us. In fact, Black people are 10 times more likely to die from poor diets than from all forms of physical violence combined. From the corner store, to the public school lunchroom, to the prisons, our federal government subsidizes the processed foods that undermine the health and future of our community. The USDA invests $130 billion annually into industrial agriculture and commodity foods, such as wheat, soy, milk, and dairy, and comparatively little into "specialty crops" like vegetables.[1] White neighborhoods have an average of four times as many supermarkets as predominantly black communities. Fast-food chains and junk-food corporations disproportionately target their advertising to children of color, resulting in an epidemic of childhood obesity and diabetes.[2] In our Black communities nearly 40 percent of the children are overweight or obese, a higher percentage than for other ethnic groups. African Americans are also two times as likely to have diabetes as whites, and 29 percent more likely to die prematurely of all causes than Americans as a whole. And perhaps most insidious, in this wealthy nation, one in three black children and one in four Latino children go to bed hungry at night. Clearly, the current food system does not have the best interests of our community in mind. We believe that the term *food desert* is too passive to describe the inequity in today's food system. Our mentor, Black farmer-activist Karen Washington, taught us to recognize

America for what it is, a deliberate "food apartheid" where certain populations live in food opulence and others cannot meet their basic survival needs.

Traditional African diets are inherently healthy and sustainable, based in leafy greens, vegetables, fruits, tubers, and legumes. Communities that maintain our traditional diets have much lower rates of heart disease, high blood pressure, stroke, diabetes, cancer, asthma, glaucoma, kidney disease, low-birthweight babies, obesity, and depression. When our ancestors, survivors of the Middle Passage, first reached plantation America, they attempted to continue their traditional diets, growing and preparing boiled yams, eddoes (or taros), okra, callaloo, and plantain, which they seasoned generously with cayenne pepper and salt. Under enslavement and colonization, their diets began to shift and their health began to deteriorate. On meager rations of corn and the offcuts of pork, and with little time to grow vegetables, our enslaved ancestors suffered from deficiencies in protein, thiamine, niacin, calcium, magnesium, and vitamin D. As a result of poor diet and living conditions, half of the infant children of enslaved mothers died before their first birthday, twice the mortality rate of white children. Children who survived often suffered from night blindness, abdominal swellings, swollen muscles, bowed legs, skin lesions, and convulsions from chronic undernourishment.

A litany of recent scientific studies has verified the value of traditional African diets for long-term health. One recent study involving 4,543 participants showed that rural Ghanaians ate more roots, tubers, plantains, and fermented corn products as compared with Ghanaians living in Europe, who ate more rice, pasta, meat, sweets, dairy, and oils. The rural Ghanaians had a lower body mass index (BMI) than their European counterparts.[3] Similarly, studies show that colon cancer impacts a greater proportion of African Americans than rural Africans. Researchers put 20 middle-aged African Americans on a traditional African heritage diet (averaging 55 grams of fiber daily and 16 percent calories from fat, with foods like mangoes, bean soup, and fish) and 20 middle-aged rural South Africans on a typical American diet (averaging 12 grams of fiber daily and 52 percent calories from fat, with foods like pancakes, burgers, fries, and meat loaf). After two weeks on the African heritage diet, African Americans increased the diversity of their healthy gut bacteria, increased levels of butyrate, an anticancer chemical, and reduced inflammation of their colons. On the other hand, the Africans on the American diet produced more bile acid and decreased the diversity of healthy gut bacteria.[4]

It is imperative that we decolonize our diets and reclaim African traditional foodways. Disease and early death are not part of our ancestral heritage. From the Black American South to the Caribbean, from South America to West Africa, there are commonalities in the traditional diet. We base our meals on leafy greens, vegetables, and tubers enlivened with ample herbs, spices, and sauces. We consume fresh fruits often and only make decadent desserts for special occasions. We use fish and meat in small quantities, usually to flavor a stew, and only eat meat in quantity on celebration days. Dairy is rare in our cuisine, and when we do consume it, we first

UPLIFT

Traditional African Diet

According to a 1730s enslaver, the diets of the Dahomey people made them strong and able to resist: "Those from the Gold Coast, who are accustomed to Freedom and inhabit a dry Champain Country and feed on nutritious and solid aliments, such as Flesh, Fish, Bread of Indian Corn, are healthy and robust; little subject to Mortality; very hardy and turbulent, as well as much disposed to rise on the White People . . ."[5] May the reclamation of our traditional diet fortify us for resistance against tyranny in present time.

ferment the milk into yogurt or buttermilk. Most important, we cook and eat together. In Kroboland, Ghana, whenever a person is eating and someone passes by, the person says, "*Ba eh no*" (come and eat). Our tables are healing tables, fellowship tables, and living history tables. We offer our gratitude to the Oldways Preservation Trust for creating the African Heritage Diet Pyramid to uplift our healthy, traditional ways of eating (see table 12.1).

Recipes

The teens are often reluctant to touch one another when we ask them to join hands around the center island in the kitchen to set intentions before we cook lunch together. We let them know that it is fine to touch elbows instead, so long as they focus their attention on putting love into the food we are about to bring forth. Each young chef shares an intention in the presence of the modest kitchen altar: "I want the food to taste good . . . I want to use the big knife . . ." The learners relax a little bit when they see Goya Adobo and hot sauce on the shelf, feeling more assured that we will not be creating some bland white folks food. We break into teams for chopping, grating, and mixing, and soon the kitchen is filled with the rich smells of hot chili and corn bread and the crisp sounds of

Table 12.1. Common Foods of African Heritage by Oldways Preservation Trust

Frequency	Ingredients
Base every meal on these foods	**Leafy greens:** Beet greens, cabbage, callaloo, chard, collard greens, dandelion greens, kale, mustard greens, spinach, turnip greens, wild foraged spring greens.
	Vegetables. Asparagus, beets, brussels sprouts, broccoli, carrots, cauliflower, eggplant, garlic, green beans, jicama, lettuce, long bean, okra, onions, peppers, pumpkin, radish, scallion, squashes, zucchini.
	Fruits. Avocados, bananas, blackberries, blueberries, cherries, dates, dewberry, figs, grapefruit, guava, horned melon, lemons, limes, mangoes, oranges, papaya, peaches, pineapple, plums, pomegranates, oranges, tamarind, tomatoes, watermelon.
	Grains. Amaranth, barley, couscous, fonio, kamut, maize, millet, rice, sorghum, tef, wild rice.
	Tubers. Breadfruit, cassava, plantains, potatoes, sweet potatoes, taro, yams, yucca.
	Beans. Black-eyed peas, broad beans, butter beans, chickpeas, cowpeas, kidney beans, lentils, lima beans, pigeon peas.
	Nuts and seeds. Benne seeds, Brazil nuts, cashews, coconuts, dika nuts, ground nuts, peanuts, pecans, pumpkin seeds, sunflower seeds.
	Herbs, spices, and sauces. Annatto, arrowroot, bay leaf, cinnamon, cilantro, cloves, coconut milk, coriander, dill, ginger, mustard, nutmeg, oregano, paprika, parsley, peppers, sage, sesame, vinegar.
Use sparingly in meals	**Oils.** Coconut oil, olive oil, palm oil, sesame oil, shea butter.
2–3 times per week	**Fish and seafood.** Bream, catfish, cod, crappie, crayfish, dried fish, mackerel, mussels oysters, perch prawns, mackerel, rainbow trout, sardines, shrimp, tuna.
Small portions 2–3 times per week	**Poultry, eggs, and other meats.** Beef, chicken, eggs, lamb, turkey.
	Dairy. Buttermilk, yogurt.
Special occasions	**Sweets.** Cakes, custards, cobblers, pies made with fruits and nuts.

Source: "African Heritage Diet," Oldways, https://oldwayspt.org/traditional-diets/african-heritage-diet.

Young people prepare a meal together at Soul Fire Farm. Photo by Capers Rumph.

Soul Fire Farm participants enjoy a 100 percent African heritage diet during the program, based on leafy greens, vegetables, and tubers enlivened with ample herbs, spices, and sauces.

UPLIFT
Chef Kabui and Other Radical Black Chefs of Today

Chef Njathi Kabui has dedicated his life to decolonizing food by preparing and teaching others how to prepare what he calls "Afro-futuristic diasporic cuisine." He was born in rural Kenya to an agricultural family that never traveled outside of the region. To those who believe that cuisine and ideas should be static, he challenges, "How unfair would it be for the first person in the family to leave the continent, the first to get an undergraduate degree, the first to get a graduate degree, the first to be published, the first to be a chef—for that person to then cook the same as his grandmother who never left home?" At the same time, he believes that farming must carry on ancestral legacy, advising, "Every one of us has a family history in growing and preparing food. So, find it. Connect with it, with that ancestor. Carry it on."

Kabui now lives in an intentional community in North Carolina where he hosts decolonization dinners, manages a busy touring schedule, and grows African vegetables like moringa, spider plant, ngai ngai (*Hibiscus sabdariffa*), amaranth, and black nightshade. An example of a dish that honors both legacy and innovation is Chef Kabui's "kuriama." Its base is millet, the first ever domesticated grain and an African native. The boiled millet is topped with a sauce of celery, beets, tomatoes, garlic, mushrooms, rosemary, basil, and "Kenya delight" spice mix containing cloves, fennel seeds, coriander, cumin, and cayenne. In Kenya beets are only fed to the cows, and people generally dislike fennel and celery. Chef Kabui has found ways to adapt the traditional millet dish to incorporate locally available spices and vegetables in his new home.

Chef Kabui is one of the innumerable radical Black chefs paving the way for our people to decolonize and re-indigenize our diets. We also want to shout out Tennessee-born Afro-vegan chef Bryant Terry, who remixes our beloved recipes from the Caribbean, West Africa, and the American South with nutrient-dense plant ingredients, each accompanied by a sweet musical track to play while you cook. We also honor Chef Michael Twitty, the spokesperson for the movement for "culinary justice," whose research and culinary creations honor the vast number of Black cooks who created the Creole cuisines of the Atlantic world, and whose names are often forgotten. We also want to shout out some up-and-coming Afro-Latinx chefs, Gabriela Alvarez of Liberation Cuisine and Merelis Catalina Ortiz and Ysanet Batista of Woke Foods. Both women-led collectives are using Caribbean ancestral foods as tools for community healing.

salad-in-the-making. Every single person loves the food they created together. In fact, all 1,768 young people who have cooked a meal at Soul Fire Farm have enjoyed it without exception. One participant explained, "I thought lettuce was nasty, but this lettuce is actually good."

In this section, we share the most popular heritage recipes that we prepare at Soul Fire Farm during our youth programs and Black Latinx Farmers Immersion. All of our recipes are based on locally available, whole, healthy, plant ingredients. All of the recipes have a sacred connection to our ancestry as Black people.

Soup Joumou

Soup Joumou is the soup of Independence, the soup of remembrance, and the soup that celebrates the New Year. The soul-warming dish commemorates January 1, 1804, the date of Haiti's liberation from France. The soup was once a delicacy reserved for white enslavers but forbidden to the enslaved people who cooked it. After Independence, Haitians took to eating it to celebrate the world's first and only successful revolution of enslaved people resulting in an independent nation.

Active time: 30 minutes | **Total time:** 1 hour, 15 minutes | **Yield:** 6–8 servings

1 pound (0.5 kg) Kabocha squash or Caribbean pumpkin, peeled and chopped
Oil (canola, safflower, or sunflower)
4 cloves garlic, crushed
1 celery stalk, chopped
1 large onion, chopped
2 potatoes, chopped
½ pound (0.25 kg) cabbage, chopped
1 turnip, diced
2 carrots, chopped
2 leeks or scallions, chopped
1 can (12 ounces/360 ml) whole coconut milk
8 cups (2 L) water
1 cup sweet corn, fresh or canned
1 tablespoon chopped parsley
1 whole Scotch bonnet pepper or other spicy pepper
1 tablespoon lime juice
2 whole cloves
Salt
Pepper
Thyme
Splash of sweetener (optional)
8 ounces (226 g) pasta (optional)

Coat the squash/pumpkin in oil and roast until golden brown and tender. Simultaneously, in a separate pan, roast the remaining vegetables (except the corn, parsley, and hot pepper) in oil and a bit of salt until golden and tender. Blend the cooked squash with the coconut milk in a blender or food processor, then add this to the water in a medium stockpot and bring to a low boil. Stir in the roasted vegetables and the corn, parsley, and hot pepper. Add spices to taste. Cook for 15 to 20 minutes to blend the flavors. If you are using pasta, add it when there are 10 remaining minutes of cook time.

Note: The squash and hot peppers are essential ingredients. All other ingredients can be replaced with similar vegetables that are locally available.

Soup Joumou cooks over coals during the New Year festivities in Leogane, Haiti.

Groundnut Stew

Mafe, or groundnut stew, is common throughout West and Central Africa. This traditional stew can include meat, vegetables, or seafood, and it is always based on a savory sauce made from peanut butter and tomatoes.

Active time: 20 minutes |
Total time: 45 minutes | **Yield:** 6–8 servings

1 medium onion, chopped
6 cloves garlic, diced
1 tablespoon minced ginger
1 teaspoon cumin
1½ teaspoons coriander
½ teaspoon black pepper
Salt
3 tablespoons coconut or safflower oil
3 cups (720 ml) vegetable stock
2 medium sweet potatoes, cubed
3 cups (1.25 kg) chopped fresh tomatoes (or 1½ cups/360 ml tomato sauce)
1 cup (250 g) peanut butter, preferably unsweetened
Optional: carrot, celery, coconut milk, and white beans

Sauté the onion, garlic, ginger, and spices in the oil until slightly softened. Add the broth, sweet potatoes, and tomatoes. Bring to a boil, then reduce the heat to medium and simmer, uncovered, for 20 to 25 minutes, until the potatoes are soft. Whisk in the peanut butter until smooth. Then transfer the soup to a blender in batches and purée. Return to the pot and warm over medium heat. Serve with rice. Possible toppings include toasted coconut, boiled eggs, chopped cilantro, or fresh diced onion and tomato.

Akara Pwa Ak Sos

Akara is a Nigerian black-eyed pea fritter that made its way across the Atlantic Ocean to become *akara pwa* in Haiti and *acaraje* in Brazil. In addition to being a popular breakfast and snack food, akara is used as an offering for the orisa and ancestors. It is one of the most anticipated dishes at Soul Fire Farm's Black Latinx Farmers Immersion.

Active time: 35 minutes |
Total time: 8 hours, 35 minutes | **Yield:** 8 servings

3 cups (480 g) black eyed peas, soaked overnight
1 large onion
1 hot pepper
Salt and pepper
Safflower, sunflower, or canola oil for frying (using palm oil is traditional)

Soak the black-eyed peas in 10 cups (2.5 L) of water overnight. Drain the peas and remove the outer skins by rubbing them together in the palms of your hands. Alternatively, remove the skins by pulsing briefly in a food processor, transferring to a bowl of water, then rubbing the skins off the peas and pouring out the floating skins. Put the peas back into the food processor or blender with the chopped onion and pepper, and purée into a fine paste. Usually the water from the onion is enough to achieve the desired consistency, but add a little water if needed. Add salt and pepper to the batter to taste. Heat the oil in a skillet on high. You may add a large slice of onion to flavor the oil while you fry. Once the onion turns brown, replace it with a fresh piece. Spoon the batter into golf-ball-sized balls and cook in the oil, flipping to achieve a golden-brown color on both sides. Serve with Haitian Sos, or any sauce made of tomato, onions, and peppers.

Epis

Similar to Puerto Rican sofrito, epis is a Haitian seasoning base that can be used in any soup, stew, or sauce. We keep it in the refrigerator at all times. Epis can even make a fried egg exciting.

Active time: 10 minutes |
Total time: 10 minutes | **Yield:** 2½ cups (600 ml)

1 medium onion
2 sweet peppers (green, yellow, or red)

Akara is a black-eyed pea fritter found in many variations across the African Diaspora.

Epis is a Haitian seasoning base that can be used in any soup, stew, or sauce.

6 scallions
6 garlic cloves
1 cup (25 g) parsley leaves
6 basil leaves
Juice of 1 lime
½ cup (120 ml) olive oil
Optional: cilantro, thyme, clove, hot pepper

Coarsely chop the vegetables, then blend all ingredients in a food processor until smooth. Epis keeps fresh in the refrigerator for 5 days and can be frozen up to 1 month.

Hoppin' John

Hoppin' John is a traditional Black southern dish eaten on the New Year for good luck. Place a penny under your plate of Hoppin' John to amplify good fortunes. It is served with collard greens, representing paper money, and corn bread, representing gold.

Active time: 30 minutes |
Total time: 8 hours, 30 minutes | **Yield:** 6–8 servings

¾ cup (120 g) black-eyed peas
2 cups (400 g) brown rice
1 medium onion, diced
½ teaspoon paprika
½ teaspoon chili powder
¼ teaspoon dried thyme
½ teaspoon dried oregano
Cayenne, to taste
Optional: garlic, bay leaf, thyme, parsley
2 tablespoons olive oil

Slice your collards super thin for a tender result.

4 cups (960 ml) vegetable broth
Salt
1 cup diced fresh tomatoes (455 g)

Soak the black-eyed peas overnight and drain. Sauté the dry brown rice, onion, and spices in the oil until aromatic. Add the broth, salt, and black-eyed peas to the sauté. Bring to a boil, then reduce the heat to low and simmer for 30 minutes. Add the tomatoes and simmer for an additional 10 minutes or until the rice is tender. Note that cooked red beans can substitute for the soaked black-eyed peas for a more Caribbean flavor.

Collard Greens and Cabbage

Dark green vegetables are the foundation of the African Heritage Diet Pyramid. While collard greens and cabbage are popular in the Black American South, this recipe can use amaranth, sweet potato greens, turnip greens, dandelion greens, or any other leafy green vegetable. The quick cooking method in this recipe retains the vitamins in the vegetables while offering a melt-in-your-mouth experience.

Active time: 25 minutes |
Total time: 25 minutes | **Yield:** 8 servings

1 pound (0.5 kg) green or savoy cabbage, thinly sliced
1 pound (0.5 kg) collard greens, destemmed and thinly sliced
8 cloves garlic, minced
3 tablespoons olive oil
1 cup (240 ml) vegetable stock
Salt

Put a large pot of water on to boil. In batches, blanch the thinly sliced cabbage and collard greens by dropping them in the water for 30 to 45 seconds. The greens should turn a deeper color but retain their firmness. Remove immediately and run under cold water. Sauté the garlic in the oil for 1 minute. Add the blanched greens and the vegetable stock and cook for another few minutes. Keep the batches small to ensure that the overall cook time is under 5 minutes. Add salt to taste.

Variation: To attain a tender result without blanching, slice the collard greens "Brazilian-style"—super angelhair thin—and sauté in a pan with a small amount of water.

Gluten Free Skillet Corn Bread

Corn bread is a classic Southern dish that originated as hoe cakes cooked on hoe blades or griddles outdoors. Other variants include baked corn pones, skillet-baked Johnny cakes, ash cakes, and corn bread kush. Corn bread is traditionally served with molasses.

Active time: 15 minutes |
Total time: 55 minutes | **Yield:** 12 servings

3 cups (720 ml) coconut milk
4 eggs
¾ cup (180 ml) safflower, sunflower, or canola oil
3½ cups (450 g) cornmeal
1¼ cups (160 g) brown rice flour
1¼ cups (160 g) arrowroot starch
(or whole wheat flour)
4 teaspoons baking powder
2 teaspoons baking soda
2 teaspoons salt
¾ cup (150 g) sugar

Combine the wet ingredients—coconut milk, eggs, oil—in a mixing bowl. Combine the dry ingredients—cornmeal, flour, starch, baking powder, baking soda, salt, sugar—in a separate bowl. Pour the dry mixture into the wet mixture and mix well. Bake in a greased cast-iron skillet at 375°F (190°C) for 30 to 45 minutes or until a toothpick inserted into the center comes out clean.

Mayi Moulin ak Fèy

Mayi Moulin ak Fèy, Haitian grits with spinach, is one variation on the breakfast porridge that is the staple food to start the day across the Diaspora. The rural farmers of Haiti say that mayi moulin is what makes them strong. Breakfast porridge can be made with cornmeal, millet, amaranth, rice, oats, or any other grain. Porridge can be prepared savory or sweet.

Active time: 12 minutes | **Total time:** 12 minutes | **Yield:** 3–4 servings

½ medium onion or 3 scallions, finely chopped
1 clove garlic, minced
2 large tomatoes, diced
1 cup (30 g) spinach, finely chopped
2 tablespoons coconut oil or butter
2 cups (480 ml) water
1 hot pepper (optional)

Corn bread is an easy-to-make African American staple dish.

1 cup (140 g) coarse cornmeal (or other grain)
2 teaspoons salt
Salt and pepper
1 avocado (optional)

Sauté the onion, garlic, tomatoes, and spinach in the oil until the tomatoes dissolve. Add the water and bring the mixture to a boil. Add the hot pepper (if using), but do not crush it. Then add the cornmeal to the boiling water, stirring constantly to avoid sticking and clumps. Continue to stir intermittently until the cornmeal is cooked, about 8 minutes in total. Add salt and pepper to taste. Remove the hot pepper. Serve with fresh avocado, if you like.

Variation: To make a sweet version, replace the onion and vegetables with chopped fruits and nuts, such as apples, raisins, dates, peaches, almonds, walnuts, and/or sunflower seeds. Add an additional cup of water or milk to replace the tomato. Add cinnamon, nutmeg, ground ginger, and sweetener to taste.

Sweet or Savory Roasted Roots

From candied yams to baked potatoes, roasted roots are staples of the Black Southern kitchen. A savory version and a low-sugar sweet version are provided. We always keep some baked roots on hand for quick meals. They work well as salad toppers and sides.

Active time: 12 minutes | **Total time:** 45 minutes | **Yield:** 6–8 servings

Sweet Version

3 medium sweet potatoes, peeled and sliced into half-moons
1 cup (240 ml) apple cider or 1 cup (240 ml) water with 4 tablespoons honey
2 teaspoons cinnamon
1 teaspoon nutmeg
1 teaspoon cloves
1 teaspoon salt
⅓ cup (80 ml) coconut oil or ⅓ cup (75 g) butter

Savory Version

4 cups (570 g) chopped root vegetables (sweet potato, carrot, potato, turnip, and/or parsnip)
⅓ cup (80 ml) coconut oil or safflower oil
2 teaspoons salt
1 teaspoon black pepper
3 teaspoon curry powder

Toss all ingredients in a bowl to thoroughly coat the roots. Spread the mixture in a single layer on a baking pan and bake at 375°F (190°C) for 30 to 45 minutes or until the roots are tender and golden. For the sweet version, keep the baking pan covered for the first 25 minutes of the cook time.

Cabbage Salad

What kind of Southern Black family reunion does not involve some coleslaw? This light, healthy, versatile recipe works perfectly for that cookout and as an everyday side.

Active time: 15 minutes | **Total time:** 15 minutes | **Yield:** 6–8 servings

½ head green or red cabbage, very thinly sliced (ideally with a mandolin)
3 carrots, sliced with a mandoline
1 medium red onion, sliced with a mandoline
½ cup (120 ml) apple cider vinegar
1 tablespoon olive oil
2 teaspoons lemon juice
1 teaspoon salt
2 teaspoons chopped dill or parsley

Toss all ingredients in a large bowl and ensure that the dressing thoroughly coats the vegetables. Cabbage salad keeps in the refrigerator for 3 days.

Peanut version: Add sesame oil, carrots, and peanuts.
Mint version: Add chopped mint and thinly sliced radishes.
Caribbean version: Add mango and crushed cashews.
Fruity version: Add chopped apples and dried cranberries.

Fruit Cobbler

Grandmommy's peach cobblers were the highlight of holiday meals. While the decadent cakes and pies of the Black South make our hearts swoon, the fruit-based cobblers are most adaptable to a healthy lifestyle and should be our go-to dessert.

Active time: 15 minutes | **Total time:** 50 minutes | **Yield:** 6 servings

8 peaches, sliced, or 6 apples, peeled, cored, and cut into wedges
1 tablespoon lemon juice
⅓ cup (80 ml) apple cider
2 teaspoons cinnamon
1 teaspoon nutmeg
1 teaspoon ground ginger
½ teaspoon salt
½ cup (50 g) oats
1½ cups (170 g) raw nuts or seeds, chopped (pecans, cashews, almonds, peanuts, sunflower seeds)
2 tablespoons honey or maple syrup
4 tablespoons butter or coconut oil
Yogurt (optional)

For the fruit filling, toss together the fruit, lemon juice, apple cider, half of the spices, and the salt. In a separate bowl combine the oats, chopped nuts, the remainder of the spices, the sweetener, and the oil. Spread the fruit into a shallow baking pan. Sprinkle the topping evenly over the fruit. Bake the cobbler at 350°F (180°C) until the fruit is bubbling and the topping is golden, 35 to 45 minutes. Serve with yogurt.

Variation: Cook the topping separately on a baking sheet at 325°F (170°C) until crispy but not burned. Use as a crumble on fresh fruit.

Food Preservation

Fall evenings at Soul Fire Farm are about stocking up for the winter. As Fannie Lou Hamer said, "When you've got 400 quarts of greens and gumbo soup canned for the winter, nobody can push you around or tell you what to say or do."[6] Our kitchen is filled with piles of red hot chili peppers and garlic, filling the air with pungent fire and getting ready to become hot sauce. On the rug are bowls of maize and pods of beans waiting to be casually shelled while we chat after dinner. The cars serve as makeshift drying ovens, filled with racks of herbs and greens that infuse the seats with the aroma of their airborne oils. Sauerkraut bubbles out of its jars on the counter, freezer bags are laden with berries, and tomato sauce boils down in the Crock-Pot. Every year we try to beat our record for how many days we can hold out between last harvest and first trip to the supermarket for vegetables. So far we have managed to make it to January before the craving for the flavors outside of our region overtakes us. Having a full larder must be part of our collective survival strategy. We are beholden to those who feed us and would much prefer to be beholden to ourselves and one another rather than the industrial food system. In this section we share our tried-and-true food preservation strategies.

Preserving in Soil and Ash

Hardy vegetables can be stored for months in the soil and retain their freshness and vitamin content. Candidates for soil storage include root vegetables like potatoes, carrots, parsnips, turnips, beets, and sweet potatoes, as well as cabbage, brussels sprouts, and celery. Dig a hole 8 inches (20 cm) deep and wide enough to accommodate the amount of root vegetables to be preserved. At the bottom of the hole spread sand and wood ashes to keep out slugs. Place wire-mesh hardware cloth at the bottom and around the edges of the hole to stop rodents. Remove damaged or rotten parts of the vegetables before stacking them in the hole. Place a small bundle of twigs upright in the center of the vegetables to encourage ventilation. Pile the vegetables to a height of 8 to 12 inches (20 to 30 cm) above the level of the ground, then cover the mound with 8 inches of straw. Use the soil you dug out of the hole

UPLIFT
Food Preservation of West Africa

Jean Nduwimana, a farmer in Eastern Burundi, sifts the ash from his fire three or four times to remove residues and debris. Then he places the ash into a paper carton and stacks his tomato harvest in neat rows amid the ash. With this technique, Mr. Nduwimana keeps his tomatoes fresh for five to six months.[7] He is part of a long legacy of African farmers who developed innovative methods to preserve the harvest.

In Ghana farmers immerse fresh cassava root in 195°F (90°C) water for three to five minutes to slow microbial action and extend the storage life of the crop by two weeks. Alternatively, they bury the roots in moist soil to extend their freshness. When an even more stable food product is desired, these farmers prepare *gari* by roasting grated and fermented cassava in large pans over open fires. They peel, wash, and grate the roots, then pack the pulp into jute bags. These bags are stacked on wooden racks for three to four days and pressed to remove the starchy juice. The pulp is dried in the sun and roasted in pans. The farmers also make *kotonte* by slicing cassava roots and setting them out to dry in the sun. Similarly, hot peppers and okra are blanched in hot water and then dried on mats in the sun.

Ghanaian farmers preserve fish through smoking, drying, fermenting, and salting. For example, whole mackerel, cassava fish, and seabream can be arranged in layers separated by sticks in smoking ovens. The smoke is generated by burning wood chips, sugarcane chaff, and coconut husks at low temperatures for a full day. To preserve fish by salting, farmers place crude salt in the gut cavity of the fish then arrange the fish in wooden barrels with more salt sprinkled on each layer. After two to three days, the fish are spread out to dry in the sun. To make "stink fish" to flavor stews, the fish is left to ferment for 9 to 48 hours before salting.

In many cases the food preservation methods increase the nutritional content of the product. After smoking the fish, 100 percent increases occurred in the riboflavin, thiamin, and niacin content (compared with a dry-weight basis). While unprocessed cassava contains toxic cyanogenic glucosides, processing the cassava into gari hydrolizes these compounds.[8]

to cover the straw, further encouraging airtightness and insulation. The last step is to cover the air shaft made by the twigs with a piece of plastic or metal to keep out rain. You can remove vegetables from the pit throughout the winter, so long as you reconstruct the insulative barrier each time. Some farmers like to dig a separate pit for each vegetable. Personally, I like to layer vegetables such that I can open the pit once every two weeks and have a ready assortment of the vegetables I am likely to use.

Drying

If you have access to a hot radiator, woodstove, electric drying, or ambient climate that is hot and dry, preservation by food drying is a suitable method for you. One of our favorite dried foods to make is vegetable bouillon powder. We use celery, carrots, garlic, onion, basil, peppers, and tomato, though almost any vegetable or herb can be included. Finely chop or pulse your desired herbs and vegetables and place them in the dehydrator for one to two days until completely desiccated. In lieu of a dehydrator, you can use an oven at 140°F (60°C), or spread the ingredients on a screen hung over a hot radiator or woodstove. The resulting powder is shelf-stable and makes an excellent base for soups and stews.

Fruits also dry very well, retaining their nutritional value and flavor. Larger fruits, like apples, peaches, and strawberries, should be thinly sliced and soaked in water with lemon juice before drying. No processing is necessary for smaller berries. Spread the fruits in a single layer in the dehydrator and dry for one to two days or until your desired texture is achieved. Chewy dried fruits retain some moisture content and have a shorter shelf life than crispy dried fruits.

Dehydrated vegetables make an excellent bouillon powder for winter soups and stews.

Fermentation

The miracle of lactic acid fermentation is that it is the only food preservation method that actually increases the nutritional value of the food. The vegetables are grated or chopped, combined with non-iodized sea salt, and packed into a tight space with no air. Lactic acid bacteria thrive in the saline environment and multiply, producing lactic acid. The resulting acidity prevents undesirable, pathogenic bacteria from colonizing the food.

Foods made with fermented grains are indigenous to Africa, including *kenkey*, *kunuzaki*, *injera*, *ogi*, and many more.[9] To make *banku*, a sourdough dumpling from Ghana, place whole maize and salt in a pot and cover with water. Allow it to set for two to three days, then drain the water and blend the maize in a food processor. A thick, pasty dough will form. Roll the dough into fist-sized balls and put them into a pot of boiling water. Cook for 20 minutes and serve with stew. A similar method can be used for any grain or soft legume. When using rice, lentils, or sorghum, the resulting dough is thinner and should be cooked on a hot skillet like a flatbread rather than boiled like a dumpling. Fermented dough keeps in the refrigerator for up to one month.

Fermenting vegetables is also an effective way to preserve the harvest and provide fresh, raw nutrition in the winter. The best vegetables for lactic acid fermentation are cabbage, turnips, radishes, carrots, cucumbers, and green beans. Cabbage is the easiest vegetable for beginners. Combine the sliced cabbage with non-iodized sea salt at a ratio of 1 pound (0.5 kg) of vegetables to 1 teaspoon of salt. Use your hands to massage the salt in the cabbage. Let it sit and brine while you sterilize the jars in boiling water. Note that a standard quart-sized canning jar holds about 2 pounds (1 kg) of vegetable. Pack the brined cabbage tightly into the canning jar, pressing out air as you go, so that the cabbage fills the jar up to the bottom of the rim. Then pour the liquid brine over the cabbage to completely fill the jar. Place the lid on very loosely. Arrange the filled jars on a pan and place them at

The author grates cabbage on a mandoline as an initial step in fermentation. Photo by Emet Vitale-Penniman.

room temperature for three days. They will bubble and release liquid as a result of the hardworking bacteria completing the fermentation. Top off each jar with a brine of 1 teaspoon of salt per 4 cups (960 ml) of water. Then place the lids on tightly and transfer the jars to a cool basement or refrigerator. The same method can be used for other vegetables. To jazz up your ferment, try adding garlic, dill, mustard seeds, caraway seeds, juniper berries, or other spices.

Preserving in Vinegar

While vegetables know how to collaborate with bacteria to make their own acid, some people prefer the taste and convenience of pickling vegetables directly in vinegar. *Pikliz* is a ubiquitous Haitian side, a spicy and sour raw vegetable pickle. In many Haitian homes a jar of pikliz resides on the counter as a permanent fixture. To make pikliz, thinly chop cabbage, carrots, and onions and pack them into a clean glass jar. Add enough distilled vinegar to just cover the mixture. Then add spices to taste: thyme, whole cloves, lime juice, salt, and hot peppers. Cover and shake the mixture and allow it to sit at room temperature for at least three days before consuming. Use a clean spoon each time you remove some pikliz to jazz up your meal. Experiment with other vegetables to make pikliz, especially cucumber, sweet pepper, fennel, radish, turnip, cauliflower, green peas, green beans, cooked beets, or boiled eggs.

Canning

Canning produces shelf-stable vegetables, fruits, soups, and even meats, but requires some specialized equipment. Use brand-new canning jars and lids, as used jars are likely to be weaker and crack. Obtain a large canning pot, canning rack, and jar tongs. While almost any food can be canned, acidic foods are much easier. Our staple canning recipes are tomato sauce and hot sauce.

To make "Soul On Fire Hot Sauce" combine 2 pounds (1 kg) of hot peppers with seeds, 20 cloves of garlic, 1 tablespoon of salt, 9 tablespoons of vinegar, and 9 tablespoons of sugar in the food processor. Cook the mixture over medium heat for about five to seven minutes, stirring frequently. Pour the hot sauce into sterilized canning jars and wipe the rims with a clean towel before putting the tops on. Boil the jars in the hot-water bath for 20 minutes. Remove the jars and place them on a towel. As they cool, the lids will seal, making a *pop* sound and becoming concave. Sealed jars are shelf-stable and can store up to one year.

Similarly, you can make tomato sauce by cooking down tomatoes, onions, peppers, summer squash, basil, and any other vegetables and herbs you have available. Once the sauce loses much of its water and has the desired consistency, add 2 tablespoons of lemon juice or vinegar per quart or liter to ensure shelf stability. Pour the sauce into canning jars and boil in the hot-water bath for 40 minutes.

The author cans tomato sauce under the guidance of elder Mama Isola. Photo by Jonah Vitale-Wolff.

Freezing

Freezing fruits and vegetables retains much of the flavor and nutrition of the original crop. Fruits, including berries and cherry tomatoes, require no special processing to freeze. Simply spread them out on a tray in a single layer until frozen. Then transfer the fruits into a freezer bag, carefully pressing out as much air as possible. When freezing vegetables, the first step is blanching, which stops the enzyme actions that cause loss of flavor, color, vitamin content, and texture. Blanching also removes bacteria from the surface of the vegetables. Submerge your chopped vegetables in a pot of rapidly boiling water for one to two minutes for tender vegetables like peas and greens or three to four minutes for chunky vegetables like broccoli and cauliflower. Then remove the vegetables and immediately transfer them to ice water. Drain the vegetables and freeze them on a tray in a single layer, then transfer them to a freezer bag. In our opinion, the tastiest frozen vegetables are corn, peas, edamame, and dark leafy greens. Note that you can also freeze, rather than can, sauces, salsas, and pestos.

Our freezer is well stocked with a variety of pestos made from herbs, greens, and garlic. Photo by Neshima Vitale-Penniman.

No Money, No Time

Even for those of us who can navigate the bureaucracy and be approved for government food benefits such as WIC, SNAP, and subsidized school meals, the average benefit is still meager at about $4 per person per day.[10] Given that commodity foods are subsidized by the USDA, the least expensive options in the grocery store are packed with refined starches and sugar. Systematic shifts are imperative to uproot food apartheid. As we do that organizing work, we can also take immediate steps toward our personal food sovereignty. In this section we explore strategies for making a dollar and an hour stretch in service of our physical health.

Sunday Soup. Allocating a few hours on a Sunday to make food for the week is an effective time and money saver. Make a big pot of Soup Joumou, groundnut stew, curry chickpeas, or vegetable soup plus some rice to store for the next several days. Also on Sundays, soak and boil beans for later use. You can store drained, cooked beans in the freezer indefinitely. While those pots are boiling

My Mother, Reverend Adele Smith-Penniman

Growing up, our family developed an intimate knowledge of the challenges presented by mental illness, substance addiction, unemployment, and violence. My two siblings and I spent our summers and school vacations with our brilliant and beautiful mother, who was often forced by circumstance to receive us in halfway houses or apartments in the projects. Looking back through adult eyes, I can see what a stretch it must have been to care for three children with no reliable income. Yet never, not once, did we go hungry. Our mother made it her number one priority to ensure that there was food on the table and protected us from any worry about scarcity. We offer a low bow of gratitude to Reverend Adele Smith-Penniman and all of the parents whose ingenuity and sacrifice make it possible for their children to be nourished.

Stock up on dry staple foods as a low-cost foundation for your meals. Photo by Neshima Vitale-Penniman.

on top of the stove, you might as well use the oven to bake some root vegetables as salad toppers or glaze some sunflower seeds to sprinkle on fruit for breakfast. Make extra food and freeze it so you can use it on the days when you really don't have time to do more than heat up your portion.

Power ingredients. There are a few high-calorie-per-penny food staples that are also nutrient-dense. Stock up on these ingredients and use them as the basis for meals. Brown rice, dry beans and peas, oatmeal, sweet potatoes, eggs, and peanut butter are at the top of the list. The following is a complete shopping list for the budget- and health-conscious household. You are more likely to keep costs down if you go to the store with a shopping list and stick to it.

Fruits and vegetables. Apples, bananas, carrots, spinach, potatoes, sweet potatoes, onions, garlic, raisins, and in-season fruits and vegetables.

Oils and spices. Sea salt, vinegar, olive oil, safflower or sunflower oil, baking powder, molasses, tamari, lemon juice, cinnamon, nutmeg, ginger, cumin, curry, black pepper, chili powder, cayenne pepper, paprika, thyme, clove, and oregano.

These "truffles" made from nuts and dried fruit are an energy-dense, healthy snack.

Grains and flours. Rolled oats, brown rice, whole wheat pasta, whole wheat flour or rice flour, cornmeal, popcorn kernels, corn tortillas, and other whole, unprocessed grains and flours.

Dry legumes and seeds. Black-eyed peas, red lentils, green split peas, garbanzo beans, red beans, black beans, roasted peanuts, coconut flakes, sunflower seeds, sesame seeds, pumpkin seeds, and other whole, unprocessed legumes.

Canned foods. Crushed tomatoes, peanut butter, and coconut milk.

Dairy. Milk or almond milk, cheddar cheese, and eggs.

Purchase sparingly. Almonds, walnuts, honey, dates, and coconut oil.

Make processed foods at home. Yogurt, granola, epis, pesto, hummus, salsa, pikliz, salad dressing, cooked beans, and baked goods.

Drink water. All your body needs to stay hydrated is water. Most bottled drinks are packed with sugar and contain no valuable nutrients. To liven up your water, add a splash of lemon or lime juice or a sprig of mint. Alternatively, make tea with fresh herbs from your garden and store it cold in the refrigerator. Our favorite is sorrel tea, mulled with cinnamon and oranges and brightened with a touch of honey.

Carry snacks. We tend to spend money on unhealthy packaged or prepared foods when hunger catches us off guard. Eat before you leave home and certainly before you go grocery shopping. To prevent impulse buys that are hard on your wallet and your health, try carrying these tasty nut truffles on the go. They are packed with protein, vitamins, minerals, and healthy fats.

Nut truffles. Choose your desired combination of nuts, seeds, and dried fruits from the options below. The ratio of nuts to fruit should be 2:1. Blend the nuts and spices in the food processor until floury, then add the fruits and any sweetener. Blend until very fine then add just a dash of water (if needed) so the mixture balls up. Roll into balls or press and cut into squares and coat with grated coconut, crushed nuts, or cocoa powder. Serve immediately or chill.

- **Apple pie.** Walnuts, dried apples, raisins, honey, cinnamon, nutmeg, ginger.
- **Cardamom delight.** Almonds, dates, cardamom, cinnamon, nutmeg, ginger.
- **Peanuts and chocolate.** Peanuts, dates, raisins, oats, cocoa powder, honey.
- **Fruit pie.** Dried cherries or blueberries, dates, almonds.
- **Carrot cake.** Almonds, walnuts, dates, raisins, grated carrot.

- **Gingersnap.** Pecans, almonds, ginger, dates, cloves, cinnamon.
- **Sunflower butter.** Sunflower seeds, almonds, dates.
- **Cashew butter.** Cashews, dates, cocoa powder.

10 minutes or less. Even with the best of intentions, there are times when we simply cannot make "Sunday Soup" or freeze meals ahead of time. When you only have 10 minutes to get some calories into your body, here are some quick, healthy, and affordable options:

Apple toast. Toast bread and top with peanut butter, honey, and sliced apples.

Everything salad. Toss salad greens or spinach with whatever you have on hand—chopped apples, boiled egg, dried fruits, sliced nuts or seeds, crumbled cheese—and splash on some lemon juice and olive oil.

Quick veggie grits. Boil 2 cups (480 ml) of water and add ½ cup of any vegetables—for example, chopped onion, spinach, tomato, broccoli, or carrot. Stir in ⅔ cup (95 g) of cornmeal and salt to taste. Continue to stir over medium heat for six to eight minutes, until all the water is absorbed.

Hot taco. Scramble or fry an egg with onions, garlic, and greens. Fold the egg into a warm tortilla and top with salsa, tomato, salad greens, and/or cheese. You can also skip the egg, simply melting cheese over the tortilla and topping the result with a mound of salad and some black beans.

CHAPTER THIRTEEN

Youth on Land

It is easier to build strong children than to repair broken men.

—Frederick Douglass

Dijour arrived on the farm last summer with 13 of his Albany peers, full of excitement and trepidation. We asked him to find an object from the natural environment that represented how he felt that day and bring it to a circle of sunny benches for an introductory conversation. Dijour hesitated for words during those initial minutes, explaining later that he does not usually talk in front of groups and was significantly outside of his comfort zone on the unpaved and unpredictable ground. We toured the farm and the young people quickly saw that their gleaming sneakers would be ruined in short order, so the courageous idea to go barefoot quickly spread. Amid giggles, warm mud oozed between toes and worms found their way into hands. The spoken content of the tour was nearly drowned out in deference to the more compelling tactile experience of land connection. The rest of the day was filled with hands-on-the-land practical farming experience, cooperative preparation of a plant-based meal, spontaneous dance cypher, and creative live commercials serving as an antidote to corporate media's promotion of food that kills.

At our day's end gratitude circle, Dijour found his voice. He shared an experience so profound that the truth of it alone justifies the immense grit required to maintain Soul Fire Farm's youth food justice programs. Djour explained that when he was very young his grandmother had shown him how to garden and to gently hold insects. She died long ago and he had forgotten these lessons. When he removed his shoes on the tour and let the mud reach his feet, the memory of her and the memory of the land literally traveled from the earth, through his soles, and to his heart. He arrived "home." He went on to complete his summer film program, where he created a documentary honoring the memories of his ancestors.

No land-based project is complete without the integration and empowerment of young people. As soon as we gain knowledge for ourselves, it's incumbent upon us to share it with the next generation. This chapter offers strategies to honor and inspire young people with authentic farm experiences. We offer our favorite farming and food justice curriculum examples, refined from over a decade of work with youth targeted by state violence. A list of farming programs that honor the dignity of young people is included.

Why Youth on Land?

Human beings have evolved for thousands of years in natural environments, having direct contact with the fullness of ecological cycles, and witnessing plants and animals grow, reproduce, die, and cycle back into new life. For only a few generations, we have lived in cities, encased in human-created artifacts and bombarded with artificial stimulation. Black Americans are even more likely to live in cities than the population at large and less likely to live in neighborhoods with parks or even green landscaping.[1] In the entire history of our existence on the planet, we have never had so little contact with the earth.

A plethora of scientific studies have confirmed what we can already intuit: We need nature not just for the material sustenance she provides, but for our physiological, emotional, and spiritual well-being. Spending time in nature, particularly free play on uneven ground, reduces stress, social anxiety, depression, disease, and impulsivity, and increases concentration, creativity, conflict management skills, agility, balance, academic performance, eyesight, and life satisfaction. Hospital patients with a view of nature through their window recover more quickly from surgery and have fewer complications and less pain. School children with ADD who are given a chance to play outside before class show greater focus, academic achievement, and ability to delay gratification. Adults who take walks in nature, as compared with walking in the mall, experience lower blood pressure and cholesterol. We now understand that experiences in nature stimulate the parasympathetic nervous system, associated with the restoration of physical energy.[2]

We may also consider that so-called nature-deficit disorder has a metaphysical component.[3] For example, trees in a forest form a superorganism that utilizes

No land-based project is complete without the integration, empowerment, and leadership of young people. Photo by Neshima Vitale-Penniman.

UPLIFT

Brother Yusuf Burgess

Brother Yusuf Burgess (1950–2014) grew up in the Marcy Projects section of Brooklyn, and would walk himself over to Prospect Park to collect tadpoles in a paper cup and gather acorns. Of that time, he remembered, "I often reflect back to my early childhood in Prospect Park, when my world was fresh and new and beautiful, full of wonder and excitement. I know now that there was an innate part of me that was drawn to nature." As an adult, Brother Yusuf served a tour in Vietnam, endured incarceration, and struggled with substance abuse. A counselor helped him remember his powerful relationship with nature and "prescribed" kayaking as therapy to help him readjust to civilian life. After that, it was rare to see the Burgess Ford without at least one boat strapped to the roof.

Brother Yusuf decided to share the healing power of nature with the young people in his community of Albany, New York, saying, "Many of today's children are growing up in busy cities without nearby parks or 'special places' to experience the beautiful and awe-inspiring. They stand to lose a very important part of what it is to be human." He started the Youth Ed-Venture & Nature Network to take urban youth on trips to the wilds of the Adirondacks, Catskills, and even once to Yosemite National Park. He coordinated the Department of Environmental Conservation's Capital District Campership Diversity Program and worked with young people at Green Tech High Charter School and the Albany Boys and Girls Club. He also frequently brought youth out to Soul Fire Farm, where I witnessed firsthand his unequivocal dedication and love for our youth and his reverence for the Earth. After his death, 90 children who loved him came together to build a community garden dedicated to his memory.[4]

Brother Yusuf (in blessed memory) brought the Youth Ed-Venture Network out to Soul Fire Farm.

the underground network of fungal hyphae to share sugars, minerals, and warnings of pest outbreaks. Trees give away their food to one another, caring for members of the same species as well as members of distant families. When we spend time observing, touching, and learning in the forest, we receive these lessons about how to live in a spirit of cooperation and mutual aid. In many African traditional religions, we believe that the ancestors and certain orisa dwell under the earth and under the water. We experience their wisdom and love through physical contact with the earth, enhanced by ceremony, song, and prayer. Our children need the earth, just as did generations past. As stewards of the future, it is incumbent upon us to repair the broken threads in the fabric that weaves together our people and the land.

Best Practices in Youth Programming

Soul Fire Farm's Youth Food Justice program attempts to address the dual challenges of increasing access to the outdoors and access to healthy food. With young people at the center, we tend the earth, cook for one another, explore the forest, tell stories, climb the trapeze, and sit with the novel discomfort of dirt in the creases of our hands. Perhaps more important than the tangible skills transfer is our effort to create an environment that challenges internal self-degradation. Of all the threats facing Black and Brown youth, we are most concerned about the pervasive perspective that "there is no future worth investing in—it does not make sense to care for our bodies or for the land because incarceration and violent death are our destiny." While there is no prescription for creating a healthy and healing space on land, some of the principles that guide our youth programming are listed here.

Youth Programming Principles at Soul Fire Farm

- We treat you as part of our family, welcome in our home and at our meal table.
- We honor your Black, Latinx, and Indigenous ancestors and heroes with storytelling.

We need contact with nature for our physical, emotional, and spiritual health. Photo by Neshima Vitale-Penniman.

- We incorporate art, music, rhythm, and creative expression into your learning experiences.
- We form a circle often, and each of you is given a chance to raise your voice.
- We trust you by telling the whole truth, even when it is hard.
- We trust you by sharing real "adult" skills, tools, and complex concepts.
- We honor your dignity by shielding you from outside "observers" who want to document and consume your image, appropriating your story for their own gain.
- We make space for you to meet in identity caucus groups to share experiences.
- We design our curriculum and activities around the topics you want to learn and the themes that are most important to you.
- We engage you in thoughtful critique of media, capitalism, mass incarceration, white supremacy, and other institutions that oppress youth.
- We provide mentors and leaders with overlapping demographics and life experiences to you.
- We offer you tangible take-aways that you apply back home, so the experience on land is not just a novelty.
- We strive to eliminate the oppressions facing youth and, in doing so, ultimately render our organization obsolete. This mission is more important than institutional self-preservation.

UPLIFT

Rooted in Community

Rooted in Community (RIC) is a national organization that brings together the efforts of over 100 youth empowerment and food justice formations across the nation. RIC holds member organizations accountable for uplifting youth dignity and leadership, engaging in courageous conversations about oppression and privilege, and honoring ancestral ways of embodying sustainability, respect, honor, and beauty. RIC convenes a national youth summit, regional gatherings for youth leaders, and a winter training for adult allies of youth leaders. The youth of RIC wrote a "Youth Food Bill of Rights" in 2013, listing their demands for a sustainable and just food system, including the "right to culturally affirming food." We offer a shout-out to Rooted in Community's member organizations, including the Food Project DIRT crew, School of Unity & Liberation (SOUL), and East New York Farms.

One of the most powerful recent developments in our youth program is the collaboration we formed with Mission Accomplished Transition Services and the Albany County District Attorney's Office to interrupt a pathway in the school-to-prison pipeline. A good friend and farm-share member who worked as counsel for the DA approached us with concern about the way the criminal punishment system traps black youth into a downward spiral of increasingly harsh sentencing. Boys in their early teens are being arrested for loitering and petty theft, then assigned a public defender who encourages them to "cop a plea" and settle for a lighter sentence regardless of their guilt or innocence.* Once they have a record, these children are more likely to be targeted by law enforcement, and it becomes disconcertingly easy to predict which middle school students will be in

* *Cop a plea* is slang for a plea bargain, in which an accused defendant in a criminal case agrees to plead guilty or no contest to a crime in return for a promise of a recommendation of leniency in sentencing to be made by the prosecutor to the judge and/or an agreement by the prosecutor to drop some of the charges.

Youth Food Bill of Rights

by Rooted in Community

1. **We have the right to culturally affirming food.** We demand the preservation, protection and reconstruction of traditional farming, cultural history and significance of food and agriculture. We demand that indigenous peoples have the right to establish their own autonomous food systems should they choose.
2. **We have the right to sustainable food.** We demand an end to the mistreatment of animals and the environment, that is caused by our current food system.
3. **We have the right to nutritional education.** We demand government funding to educate and inform youth and parents about nutrition. Education on things such as seasonal eating, organic farming, sustainability, and diet-related illness should be provided so that people can make better informed decisions. We recommend that schools recognize youth-led fitness programs as tools for success.
4. **We have the right to healthy food at school.** We the youth demand more healthy food choices in our schools, and in schools all over the world. We want vending machines out of schools unless they have healthy choices. We need healthier school lunches that are implemented by schools with the ingredients decided on by the Youth. We demand composting in schools and in our neighborhoods.
5. **We have the right to genetic diversity and GMO-free food.** We the youth, call for the Labeling of Genetically Modified seeds, plants, and produce. We demand a policy from the government that labels all GMOs.
6. **We have the right to poison-free food.** We the youth absolutely don't want any chemical pesticides in our food!
7. **We have the right to beverages and foods that don't harm us.** We the Youth demand a ban on High-Fructose Corn Syrup and other additives, and preservatives that are a detriment to our communities' health. This must be implemented by our government, and governments around the world.
8. **We have the right to local food.** We demand food to be grown and consumed by region to cut the use of fossil fuels and reduce the globalization of our food system.

prison by the time they turn 18. Together with our lawyer friend, we decided to introduce legislation for a restorative justice project that we termed Project Growth. Instead of receiving punitive sentences, court-adjudicated youth would be able to complete a 50-hour training at Soul Fire Farm and other community organizations and earn money for the time they invest. The youth were obligated to first use the funds to pay off any restitution owed to those impacted by their crimes and then use the remainder as discretionary income. Our proposal was approved for a three-year period, and we mentored 15 young people through the program.

Very early on in the piloting phase, the DA threatened to cancel the program after an incident where one of the youth stole an iPhone from an adult working on our farm. While the phone was found and returned, the DA was trapped in

9. **We have the right to fair food.** We the youth demand that everyone working in the food system must be treated with respect, treated fairly, and earn, at the minimum, a just living wage. For all those who are working in the food system we demand a model like the Domestic Fair Trade Association to be implemented.
10. **We have the right to good food subsidies.** We demand an end to the subsidy of cash crops, including corn and soybeans. Rather than our tax dollars going to subsidies for industrial farming, we demand financial support for small organic farmers.
11. **We have the right to organic food and organic farmers.** We demand a restructuring of the process of being certified organic and fair trade to improve the thoroughness and accessibility of these programs.
12. **We have the right to cultivate unused land.** We demand that a policy be enacted allowing for unused land to be made available for communities to farm and garden organically and sustainably.
13. **We have the right to save our seed.** We believe farmers and all people should have the freedom to save their seed. Any law that prevents this should be reversed; no law shall ever be made to prevent seed saving.
14. **We have the right to an ozone layer.** We the youth demand a 20% decrease of industrial farms every 5 years, to decrease the high levels of greenhouse gas emissions associated with industrial farming.
15. **We have the right to support our farmers through direct market transactions.** We demand that the number of farmers markets be increased every year until there are more farmers markets than corporate super markets.
16. **We have the right to convenient food that is healthy.** We want healthy options in corner stores while empowering the community to make better food choices. We demand more jobs for youth to work with our communities to make this happen and help them control their food systems.
17. **We have the right to leadership education.** We the youth demand that there be more school assemblies to inform and empower more youth with the knowledge of food justice. The continuation of the movement for Food Justice, Food Sovereignty and cultivation of future leaders is necessary for feeding our youth, our nation, and our world.[5]

a "no-second-chances" mentality and wanted all of the youth in the program to be dismissed and returned to the traditional punishment system. We convinced them to give us a few days to demonstrate our techniques for accountability and justice. They reluctantly agreed. Unfortunately for our tenuous bargain, Sam,* the young person who stole the phone, did not show up to the program the next day. Our son, Emet, then nine years old, had formed a close bond with Sam and was devastated at his absence. The two had planned to make bows and arrows and play a game in the forest. Emet contacted Sam in tears and convinced him to return the next day. To the surprise of the DA staff, Sam did. We engaged him right away with "rock therapy," a practice I learned at Many Hands Organic Farm, in Massachusetts. We gathered

* Name altered to protect privacy.

stones from the fields and threw them forcefully into the forest, while yelling and cursing loudly about the things that frustrated us—for example, "#%&@ my probation officer!" After letting off some steam, Sam apologized for his theft, signed a behavior contract, and completed the program with excellent focus and no further breaches of trust.

One thing we must remember is that healing is an extended process during which patience is in order. Just as the Emancipation Proclamation did not erase the effects of 500 years of enslavement, so a few days of a restorative justice program cannot erase the emotional and spiritual impacts of inherited and lived trauma on the youth. All of our young people deserve to enjoy a sense of purpose, the acquisition of real skills, and the opportunity to contribute to the betterment of the community.

Youth in Soul Fire Farm's Project Growth are diverted from the criminal injustice system and given the choice to complete a training program on the land.

Youth Food Justice Curriculum

In collaboration with the thousands of youth who have blessed Soul Fire Farm with their footsteps, we have built a curriculum that connects learners with the land and advances their personal food sovereignty. Of course, the central "curriculum" of our youth programs is direct experience with the land herself. We spend most of our time together farming, cooking, chasing the dog, Rowe, and tasting edible wild plants. After eating the burrito lunch that we prepare together, we settle in for a few structured activities to overlay a conceptual understanding on our felt experiences from earlier in the day. The activities detailed in the rest of this chapter are a handful of our favorites. They were designed with and for teens, but are easily adaptable to younger and older audiences.

Move Your Butt

Overview: Participants learn the three components of food justice through an interactive, kinesthetic game.

Time: 15–20 minutes.

Materials: Benches or chairs in a circle.

Activity

1. Participants sit in a tight circle. The facilitator says "Move your butt if . . ." followed by a statement. If that's true for a participant, they need to find a new seat across the circle. This is noncompetitive in that there are enough seats for everyone. The purpose is to notice how many people move.
2. **Round 1:** Say, "Move your butt if . . ."
   a. You know someone who has diabetes.
   b. You know someone who has high blood pressure or heart disease.
   c. You know someone who struggles with obesity.
3. Ask participants, "What do these three things have in common?" Let participants guess. Then

share that diabetes, heart disease, and obesity are all diet-related illness. The first component of food justice is food security. Currently Black, Latinx, and Indigenous people are more likely to suffer from diet-related illness and hunger because of food apartheid. Access to affordable, healthy, culturally appropriate food is a basic human right.

4. **Round 2:** Say, "Move your butt if . . ."
   a. You value your cell phone.
   b. You value your friends.
   c. You value clean water.
5. Ask participants, "What do these three things have in common?" Allow them to guess, letting them know that this one is trickier. Then share, "They all come from the earth." Explain how each comes from the earth: Minerals and oil are mined for phones, natural filtration cleans our water, and humans are made of the food we consume. Ask, "How should we treat something that is the source of all we value?" The second component of food justice is about caring for the earth and leaving something for the next generation. Our ancestors knew how to farm sustainably; many of the techniques that are now called organic—like raised beds, intercropping, and natural pest management—were created by Black and Indigenous people.
6. **Round 3:** "Move your butt if . . ." Remind participants that this last round is very personal and to remember to treat each person with dignity and respect throughout the process.
   a. Your ancestors were stolen from land or had land stolen from them.
   b. Your ancestors were enslaved or worked as sharecroppers.
   c. Your ancestors participated in stealing land from others or enslaving others.
7. Talk about the history of genocide and land theft perpetrated against Native Americans; the enslavement, sharecropping, and land theft perpetrated against African Americans; and the forced migration and poor working conditions on farms perpetuated against Mexican and Asian Americans. The food system was built on stolen land and exploited labor, and this continues today. Over 80 percent of our food is grown by Latinx people while only 2 percent of farm managers are Latinx. African American farmers control around 1 percent of the farmland. The

The Black Child's Pledge

by Shirley Williams, published in The Black Panther *newsletter, October 26, 1968*

I pledge allegiance to my Black people.

I pledge to develop my mind and body to the greatest extent possible.

I will learn all that I can in order to give my best to my people in their struggle for liberation.

I will keep myself physically fit, building a strong body free from drugs and other substances that weaken me and make me less capable of protecting myself, my family, and my Black brothers and sisters.

I will unselfishly share my knowledge and understanding with them in order to bring about change more quickly.

I will discipline myself to direct my energies thoughtfully and constructively rather than wasting them in idle hatred.

I will train myself never to hurt or allow others to harm my Black brothers and sisters for I recognize that we need every Black man, woman, and child to be physically, mentally and psychologically strong. These principles I pledge to practice daily and to teach them to others in order to unite my people.[6]

Youth gather for opening circle and orientation at Soul Fire Farm. Photo by Neshima Vitale-Penniman.

third component of food justice is equitable distribution of power and resources in the food system. People of color are working for our fair share of land, business ownership, representation in government, and wealth.

Media Does Not Have My Mind

Overview: Participants create short skits about healthy food and learn about the role of advertising in determining our diets.

Time: 35–50 minutes.

Materials: Projector, screen, audio, various props for skits, advertising facts.

Preparation: Select two or three popular fast-food commercials, print out the facts.

Activity

1. Hold a brief opening discussion around these questions: What are your favorite foods? Who decides what you love to eat? Take a poll, "How many of you think that *you* decide what you like to eat?" At this point do not reveal anything about the role of advertising in influencing our diets.
2. Watch two or three video clips of popular fast-food commercials. The activity works best if the commercials use a variety of strategies to engage the audience, for example celebrities, sex, fun, travel, popular music, "cool" teens, Black culture, and wealth. After each clip, ask the participants what strategies were used by the advertisers to get the attention of viewers.
3. Pass out the following facts about fast-food advertising. Ask participants to read the facts out loud. These facts come from Rudd Center for Food Policy & Obesity's publication, "Fast Food FACTS."

   a. Preschoolers view 1,023 fast food ads per year, an average of 2.8 per day.
   b. Over $4.6 billion is spent each year to advertise fast food, compared with $116 million to advertise fruits and vegetables. That's 40 times as much on fast food.

UPLIFT
Black Panther Party Oakland Community School

One of the demands of the Black Panther Party (BPP), founded in 1966 by Huey P. Newton and Bobby Seal, was, "We want education for our people that exposes the true nature of this decadent American society. We want an education that teaches us our true history and our role in the present day society . . . We believe in an educational system that will give to our people a knowledge of the self. If you do not have knowledge of yourself and your position in the society and in the world, then you will have little chance to know anything else."

The BPP ran over 60 survival programs, ranging from daily free breakfast for 10,000 children, to health clinics, to schools. The BPP's Oakland Community School opened in 1973, under the leadership of Erika Huggins and Donna Howell. With an enrollment of 160 students and a daunting waiting list, including the unborn, this school provided a culturally relevant education to its predominantly black pupils. The students studied traditional subjects—math, English, history, and science—but through a critical analytical lens. One alumna commented, "They taught us how to think, not what to think." The students ate three hot meals daily in school, practiced meditation before afternoon classes, and ran their own discipline committee. The "Justice Committee" was a group of peers who listened to the student's offense and issued them a consequence known as a "correction." For example, the young students on the committee told a peer that because she didn't do her homework, she would not get any free time. "You know we're giving you this correction because we love you," one of the young committee members told her.[7]

Perhaps most important, the staff at OCS were predominantly Black, and those from other races went through training to expose and correct their implicit bias. Studies show that white teachers demonstrate negative bias in their grading of the work of Black students and in their predictions of student graduation rates and life success.[8] In a phenomenon called the Pygmalion Effect, students internalize the expectations of their teachers and perform accordingly.

While OCS closed in 1982, its legacy lives on. Colin Kaepernick, inspired by the work of the Erika Huggins and BPP, started a "know your rights" summer camp for Oakland youth. The curriculum is based on the 10-point platform of the Black Panther Party.[9] The Malcolm X Grassroots Movement, with chapters in Atlanta, Jackson, Oakland, New York, Philadelphia, and DC, organizes Black youth and elders to know their history and "free the land." Black Youth Project (BYP100) also carries on BPP values by convening young Black activists to work for economic, social, political, and educational freedom.

c. African American children are targeted by junk-food advertisers, viewing 40 percent more calories daily in fast-food ads compared with white children.

d. African American children and teens in the US are more than twice as likely to see an advertisement for candy and soda on TV as their white counterparts.[10]

Young people use theater to explore the history of land loss and resistance in the Black community. Photo by Neshima Vitale-Penniman.

4. Come back to the initial question, "How many of you think *you* decide what you like to eat?" Encourage participants to consider external influences on their dietary choices.
5. Ask the participants to break into small groups that each create their own health-food commercial in skit form. Give time to prepare, offer props, and encourage theatrics, singing, rhythm, and stunts. If participants need a little inspiration, show them the 2016 "Grow Food" music video, published on YouTube by the organization Appetite for Change. Challenge participants to outdo the ads we viewed on the screen. Perform skits to one another with lots of snaps and claps.

Collage Biographies

Overview: Participants develop a deeper understanding of people who have led the movements for food sovereignty by creating collages of these historical and contemporary leaders.

Time: 40–60 minutes.

Materials: Glue, scissors, magazines, cardstock paper, devices with internet access.

Preparation: Research and print out information on selected leaders.

Activity

1. Ask participants to discuss the following quote by Howard Thurman: "Don't ask what the world needs. Ask what makes you come alive, and go do it. Because what the world needs are people who have come alive." Have participants share moments when they were doing work that made them feel alive. In this activity they will be honoring leaders in the movement for food sovereignty who work at the intersection of what the world needs and what makes them feel alive.
2. Read off the list of names from which participants can select a leader to investigate. Each participant will create a visual collage about that person. Images and text should be combined on

half-page cardstock and shared out to the group. The objective is to understand the strategies and victories of the farm activists who came before us. Below is a list of notable leaders, though any activists can be highlighted.

a. Asha Carter, BYP100 member, presidential appointee in the Obama administration at the US Environmental Protection Agency
b. Ben Burkett, president of the Mississippi Association of Cooperatives
c. Booker T. Whatley, agricultural professor at Tuskegee University and pioneer of CSAs and pick-your-own farms
d. Cesar Chavez, co-founder of the National Farm Workers Association
e. Charo Minas Rojas, Proceso de Comunidades Negras—PCN (Afro-Colombian solidarity network)
f. Cynthia Hayes, founder of SAAFON (Southeastern African American Farmers' Organic Network)
g. Dolores Huerta, co-founder of the National Farm Workers Association
h. Dominique Hazzard, BYP100 organizer, outreach specialist at DC Greens
i. Fannie Lou Hamer, founder of Freedom Farm
j. Francia Márquez, coordinator of Mobilization of Women for Care for Life and Ancestral Territories
k. Gail Taylor, director of Three Part Harmony Farm
l. George Washington Carver, professor at Tuskegee Institute and regenerative farming pioneer
m. Karen Washington, founder of Black Urban Growers and Rise & Root Farm
n. Malik Yakini, founder of the Detroit Black Community Food Security Network
o. Nelson Carrasquillo, general coordinator of CATA Farmworker Support Committee
p. Owusu Bandele, professor of sustainable agriculture at Southern University Agricultural Center
q. Ralph Paige, 30-year executive director of the Federation of Southern Cooperatives
r. Rosalinda Guillen, founder of Community to Community Development in Bellingham, Washington
s. Shirley Sherrod, founder of the New Communities Land Trust and plaintiff in the *Pigford v. Glickman* lawsuit
t. Tanya Fields, founding director of BLK ProjeK
u. Tavia Benjamin, Earthseed and Black/Land Project
v. Wallie and Juanita Nelson, tax resistors and market farmers in Deerfield, Massachusetts
w. Wangari Maathai, founder of the Green Belt Movement and the 2004 Nobel Peace Prize Laureate
x. Will Allen, founder of Growing Power urban farm
y. Zachari J. Curtis, owner and operator of DC's Good Sense Farm & Apiary

3. Allow each participant to have a turn sharing their collage card. Encourage them to state the person's name, their contribution to food sovereignty, and one specific detail that inspired the artist making the collage. Circle back to the quote at the beginning. Ask, "In what way have these leaders 'come alive' in their work? What type of justice work can you do that would make you also 'come alive'?"

Land Loss and Resistance

Overview: Participants create skits to uncover the historical factors that led to the decline of Black farm ownership from 14 percent in the 1920s to less than 1 percent today.
Time: 30–45 minutes.
Materials: Props, handouts.
Preparation: Print historical summaries.

Activity

1. Ask participants to consider the following quotes, discussing why Ralph Paige and Malcolm X place such strong value in the land. Ask participants whether they agree with this perspective on land.

a. "Land is the only real wealth in this country, and if we don't have any of it, we'll be out of the picture." —Ralph Paige, Federation of Southern Cooperatives
b. "Revolution is based on land. Land is the basis of all independence. Land is the basis of freedom, justice, and equality" —Malcolm X

2. Break the participants up into groups of three to five. Each group will receive a paragraph describing certain events in the history of land loss and resistance. They will act out their scene in front of the other groups. Encourage groups to think creatively about their skit—for instance, instead of a play with dialogue, they could also make a freeze-frame human sculpture, a poem or song, or a human library where each person gives a personal narrative. *Note: Especially if there are white people among the participants, reinforce that it is* never *okay to use racial slurs or act out violence, even in a skit. Generally, white actors should not assume an "oppressor" role over black actors.*
3. The following historical vignettes are summaries of the information contained in the PBS special *Homecoming: The Story of African-American Farmers*.

a. **Group 1:** The wealth of this nation was built on the backs of enslaved Africans. The Emancipation Proclamation of 1863 and the ratification of the 13th Amendment at the end of the Civil War in 1866 gave 4 million enslaved African Americans their legal freedom. The federal government provided a limited number of opportunities for Blacks to acquire land. In 1865 General William T. Sherman's Field Order #15 deeded "40 acres and a mule" over to Black families on the South Carolina and Florida coasts. President Andrew Johnson reversed the policy, and most never received their allotments. Most Southern Black farmers worked as sharecroppers or tenant farmers on white-owned plantations, a system very similar to slavery. Despite this, African Americans worked very hard and saved money to purchase land. Black people purchased 120,738 farms by 1890. By 1910 Black farmers had accumulated 218,972 farms and nearly 15 million acres, 14 percent of the nation's farmland.
b. **Group 2:** As African Americans acquired land, resentment from southern whites mounted. After 1877, and the election of Republican president Rutherford B. Hayes, the South quickly replaced Reconstruction laws with new ones that restricted the rights of Blacks. White secret societies began forming to address the "Negro problem." In 1881 the first "Jim Crow" law was born when Tennessee required racial segregation in railroad cars. By 1896 the *Plessy v. Ferguson* case put the federal stamp of approval on Jim Crow, excluding blacks from public transport and facilities, jobs, juries, and neighborhoods throughout the South. The laws helped spur racist hysteria, lynching, rioting, and the rise of the Ku Klux Klan. In the two-year period 1900–01, 214 lynchings were reported. While many Black people moved North to escape the violence, other African Americans organized with poor white farmers to form the Southern Tenant Farmers Union. It was the only interracial union of its time and used nonviolent protest to demand their fair share of government support during the Great Depression.
c. **Group 3:** A 1964 study exposed how the US Department of Agriculture (USDA) actively worked against the economic interests of Black farmers. The USDA's loan agencies, such as the Farmers Home Administration (FHA), denied Black farmers ownership and operating loans, disaster relief, and other aid. One practice was to deny credit to any Black farmer who assisted civil rights activists, joined the NAACP, registered to vote, or simply signed a petition. The study further revealed that there had never been an African American elected to a county agricultural committee—a structure established by the USDA. Currently, less than

1 percent of US farmland is owned by African Americans. In 1999 farmers sued the US government for discrimination and won over $1 billion for 13,300 farmers who lost their land. It is the largest civil rights settlement in US history. Another 70,000 farmers are still fighting to receive their compensation.

d. **Group 4:** Today African Americans and other communities of color suffer from high rates of diabetes, heart disease, obesity, and other diet-related illnesses. People of color are also more likely than whites to live in food deserts—neighborhoods that do not have access to affordable, culturally appropriate, healthy foods. One of the solutions is urban farming, but a huge challenge is that urban land is often owned by corporations or individuals with a profit motive. Urban farms and gardens can be taken over at any time. The Detroit Black Community Food Security Network, a collective of motivated African Americans from Detroit, did not give up even after losing two of their farms to development. They now have a long-term lease on a 7-acre site, and people from the community can grow vegetables and fruit for their families. DBCFSN also started a food coop to get more healthy food into their neighborhoods.

4. In conclusion, ask participants to summarize what factors led to the decline in Black landownership. What do they think we can do to reclaim landownership in our communities? If they get stuck, discuss the efforts of National Black Food & Justice Alliance, Federation of Southern Cooperatives Land Assistance Fund, Black Family Land Trust, and Movement for Black Lives.

Scavenger Hunt

Overview: Participants work together in small groups to find items around the farm. The objectives are to get moving, have fun, cooperate, observe the environment, and review what was learned on the farm earlier in the program.

Time: 20 minutes.

Materials: Printed lists, containers, healthy prize (optional).

Preparation: Print list.

Activity

1. Divide the group into teams of two to four. Challenge learners to collect as many items from the list as they can find in 15 minutes and put them in the container. Remind learners not to step on the beds, and which plants are okay to pick. You can offer one "lifeline" per team whereby they can ask the facilitator to help them find an object.
2. Adapt the following list to the items available on your site and the prior knowledge of the group.

   a. Edible "weed"
   b. Cover crop
   c. Animal that helps "till" the soil
   d. Compost
   e. Mulch
   f. Pest insect
   g. Insect that kills pest insects
   h. Mature seed
   i. Medicinal plant
   j. Chicken feather
   k. Water from a place where fish live

During a scavenger hunt, one of the young people encounters a snake and befriends it. Photo by Neshima Vitale-Penniman.

l. Leaf from a deciduous tree
m. Hair from an animal who keeps coyotes away
n. Leaf from a plant that originated in Africa
o. Leaf from a plant that originated on this continent

3. Invite participants to circle up after time is up. Call off the items one at a time and ask participants to hold it up if they found it. Discuss any misconceptions as you go along. Offer a prize to the team that collects the most items from the list.

Nature Is My Teacher

Overview: Participants hone skills of observation, gain comfort being alone in nature, and learn to listen carefully to nature.
Time: 30–45 minutes.
Materials: Cell phone basket, a patch of nature (for example, a night sky, sunrise, sunset, river, pond, ocean, forest, or farm).

Activity

1. Explain to the learners that there is great wisdom in stillness and silence. When we are quiet we can hear nature, our ancestors, and our own consciousness. We are rarely still anywhere, especially in nature. Give a specific example of something that nature teaches us by example, such as the trees sharing sugars through mycelial networks, demonstrating cooperation and generosity. Explain that we can learn a lot by following nature's example, and the first step is to pay attention.
2. Collect cell phones and other electronic devices. Guide learners to each choose their own spot in nature, where they cannot see or interact with anyone else. Tell them you will pick them up at the conclusion of the allotted time, and no one should move from their spot.
3. Ask learners to listen, look, breathe, and question while they are in their spots. *Alternative: Give learners a notebook and pen and ask them to sketch or write.*
4. Return to a circle at the end and ask each person to share what they learned. Give some examples of things that can be learned by watching nature—for instance, how the forest floor shows us the way compost works and how the opening tree buds tell us it's time to plant our seeds in the soil. Ask participants what feelings came up for them, how difficult or easy it was to be still in nature, and what they learned about themselves.

We Gonna Be All Right

Overview: Participants chant a positive message while passing stones in sync. This activity is rooted in a children's game from Ghana where players sing about traveling to Kumasi as follows: "O sami nami nami nami na. O sami nami na abele abele. Kumasi banti ma bele o. O sami nami na e e. Abele o."
Time: 10–15 minutes.
Materials: One small stone or palm nut for each participant.

Activity

1. Tell participants that you are going to challenge them with a game to see whether they have rhythm and know how to cooperate. This game comes from Ghana, and children as young as 4 and 5 years old can play it without making a mistake. It is simple. You just pass a stone around the circle to the beat. Ask whether they think they can do it.
2. Ask participants to stand, sit, or kneel in a tight circle so that they are very close to the person next to them. This will facilitate passing the stone.
3. First, teach the chant; the bolded type indicates the emphasis: "They **try** to cut us **down**, but **we** gonna be all **right**." Repeat it several times until everyone can say it in unison.
4. Next have the participants hold out their left hand. You will say the chant again, but this time use your right hand to clap the outstretched left hand of the person to your right, on the emphasized beat. On the interim beats, clap your own hand.

Words	Hand Clap
They	own
Try	neighbor
To **cut** us	own
Down	neighbor
But	own
we	neighbor
Gonna **be**	own
all **right**	neighbor

5. Finally, pass one stone to each participant. They should hold the stone in their left hand and pass it to the person on their right using the same pattern as was just practiced. If done right, everyone passes in unison and everyone continues to have one stone at a time. Much laughter should ensue and many mistakes will be made. Keep trying.
6. Ask the participants, "What does this activity teach us about unity?" and "What does this chant mean to you?"

Hope Tree

Overview: Participants write their wishes for health, justice, peace, and sovereignty on colorful ribbons to be caught and dispersed in the wind.

This activity is inspired by *drapo servis* (Haitian Vodou prayer flags) and the S.T.I.T.C.H.E.D. project of Climbing Poetree.

Time: 30 minutes.

Materials: A tree or collection of branches, colorful ribbons, a fine-point permanent marker for each person, contemplative music (optional).

Preparation: Cut the fabric or ribbons into lengths of approximately 12 inches and width of 1 to 2 inches. Give each participant a permanent fine-point marker and a ribbon.

Activity

1. Explain to participants that cultures around the world engage in the practice of inscribing things that are sacred to them on cloth and letting the wind interact with the fabric. From Tibet to Haiti prayer flags are central to spiritual practice. Each of us has wishes and intentions around our own health, justice, peace, and sovereignty, and for the community at large.
2. Let each participant select a piece of fabric and marker upon which to write their hopes, prayers, and intentions. Play contemplative music and encourage people not to talk to one another while they decide what to write.

Participants write their wishes for health, justice, peace, and sovereignty on colorful fabric to be caught and dispersed in the wind. Photo by Neshima Vitale-Penniman.

3. Come back to the group and invite participants to read aloud their prayer and then tie the ribbon to the "hope tree." After each person says their prayer, the group should respond, "We got your back!"

It Is Our Duty

Overview: Participants close out the program by making action commitments and pledging to support one another in those commitments. This activity was created by Amani Olugbala, Soul Fire Farm.
Time: about 10 minutes (30 seconds per person).
Materials: None.

Activity

1. Offer a few highlights of the time together and encourage participants to think about how they will take this powerful experience back into their everyday lives and back to their community. They may choose to make a dietary shift, get involved in a campaign, advocate for community gardens, or connect with an elder to learn more about food histories. Give people a few moments to come up with their action step.
2. One at a time, the person calls into the circle, "I feel evolutionary." Everyone responds in unison, "Revolutionary!" The person then says, "I will . . . ," stating their commitment to action. Everyone responds, "Yes you will!"
3. To close, chant the words of Assata Shakur. Do this entire chant three times, progressing from soft voices to loud, emphatic shouts.

Call: It is our duty to fight for our freedom.
Response: It is our duty to fight for our freedom.
Call: It is our duty to win.
Response: It is our duty to win.
Call: We must love each other and support each other.
Response: We must love each other and support each other.
Call: We have nothing to lose but our chains.
Response: We have nothing to lose but our chains.

"It is our duty to fight for our freedom. It is our duty to win. We must love each other and support each other. We have nothing to lose but our chains." —Assata Shakur. Photo by Jonah Vitale-Wolff.

Soul Fire Farm Youth Food Justice Pledge

Inspired by the "The Black Child's Pledge" of the Black Panther Party for Self-Defense

I pledge allegiance to my Body, and will nourish myself with healthy food and outdoor play.

I pledge allegiance to the Earth, and will show my gratitude by taking care of her.

I pledge allegiance to my People, and will honor the memory and stories of my ancestors.

I pledge allegiance to the Community, and will stand up for the human rights and dignity of all people.

I pledge allegiance to my Mind, and will study diligently to gain knowledge and truth.

I pledge allegiance to Radical Love, and will do all I can to unite my people for justice.

CHAPTER FOURTEEN

Healing from Trauma

As Black farmers, we have to recognize the trauma that we are up against. Trauma is a grain of sand that gets into a clam. She can't cough it out, so she keeps covering up. The result looks beautiful and it's ours because it was passed down through the mothers, but what's inside is a lot of deep trauma. We must notice the pearl and how important it is for people to hold onto the pearl. That pearl becomes part of the clam and part of our story too. If we yank it out, we will kill her and we don't want to do that. Let's figure out how to live, how to care for ourselves connected to our whole sovereignty and our whole liberation.

—Chris Bolden-Newsome

My first day as a farmer at The Food Project in Boston, Massachusetts, was a homecoming for me. I carried a lot of pain and trauma in my 16-year-old body, and was burdened with both personal and ancestral violence and loss. I felt unsure whether I was worthy of the air that I inhaled and questioned whether there was a place for me on this green earth. This summer job was not an explicitly healing space, just a program to get urban and rural youth together to grow food and learn leadership skills. Still, the land worked her magic on me. My task that first day was to harvest cilantro for the farmers market. I had never interacted with this powerful plant before and the aromatic oils lingered in the creases of my fingers long after my train ride home, infiltrated my dreams, and called me to the present. The next eight weeks of farm labor awakened me to who I was meant to be.

While farming was initially healing for me, for many African heritage people, it is triggering and re-traumatizing. Almost without exception, when I ask Black visitors to Soul Fire Farm what they first associate with farming, they respond "slavery" or "plantation." As Chris Bolden-Newsome says, "The field was the scene of the crime." Hundreds of years of enslavement have devastated our sacred connection to land and overshadowed thousands of years of our noble, autonomous farming history. Many of us have confused the terror our ancestors experienced on land with the land herself, naming her the oppressor and running toward paved streets without looking back. We do not stoop, sweat, harvest, or even get dirty, because we imagine that would revert us to bondage. And yet we are keenly aware that something is missing, that a gap exists where once there was connection. This generation of Black people is becoming known as the "returning generation" of agrarian people. Our grandparents fled the red clays of Georgia, and we are now cautiously working to make sense of a reconciliation with land. We somehow know that without the land, we cannot return to freedom. In this chapter we bear witness to the racial atrocities committed against our

This generation of Black people is becoming known as the "returning generation" in terms of our relationship to land. Photo by Neshima Vitale-Penniman.

UPLIFT
Ruby Nell Sales

According to civil rights activist Ruby Sales, the Black church hymn "I love everybody. I love everybody in my heart," is an anthem of resistance against the supposed omnipotent power of the white enslaver. It says that even though white people may control our external lives, we are in charge of our internal lives, and we decide not to hate. No matter what is done to us, we will not hate. Martin Luther King Jr. echoed this sentiment: "I have decided to stick with love. Hate is too great a burden to bear."

In addition to marching from Selma to Montgomery and dedicating her career to human rights activism, Sales is now mentoring younger activists in the Black Lives Matter movement. At an Atlanta BLM convening, she apologized to the younger people in the room, saying, "I am sorry for the ways we abandoned you." We all have a hunger to be claimed by our elders, and as a result of the destruction of our intergenerational communities, we feel incomplete. She said, "One of the greatest trigger-fingers of the empire, is to destroy intimacy, to destroy how we know each other. And that the Black community has been under this assault ever since enslavement where Black people's families were sold away from each other. We've had to constantly fight to maintain that intimacy."

Black Lives Matter has been our outcry since the moment of our captivity. We have resisted dehumanization at every turn, through escape, armed rebellion, litigation, nonviolent direct action, art and music, religious ritual, boycott, and countless other ingenious strategies. Even as we have fought for recognition of our humanity in material form—jobs, housing, freedom—we have also known that Black Lives Matter is about reclaiming our own sense of value and identity in the context of intergenerational Black community and in the spirit of love.[1]

people, reflect upon the ways we have internalized this trauma, and explore strategies for personal healing and resistance so that we can reclaim our sacred belonging to land and self.

Historical Trauma: An Annotated Time Line

A first step in the healing process is to grieve. We need to look with wide-open eyes at the atrocities committed against our ancestors and our people, to feel the pain of these events, and to mourn our losses. Later in this chapter we explore ways to compost our pain into wholeness, but first we need to be with that pain. What follows is a selective time line of traumatic events experienced by African people due to European enslavement and colonization.

1455–Present: Doctrine of Discovery

Pope Nicholas granted Christian nations the authority to loot and enslave non-Christian nations, saying, "Invade, search out, capture, vanquish, and subdue all Saracens and pagans," take their possessions, and "reduce their persons to perpetual slavery." This decree justified European settlers in their genocide against the Indigenous people of what became the United States, killing 90 percent of a population of 20 to 100 million, displacing those who survived, and stealing their land. It also set the stage for the transatlantic slave trade.[2] The Doctrine of Discovery was upheld in US Supreme Court in 1823, which ruled that European discovery of land grants title annuls Native Americans' right to "occupy" land. In 2005, the US Supreme court referenced the doctrine of discovery in its ruling that denied the Oneida Nation of New York tribal sovereignty over their original lands.

1526–1857: Transatlantic Slave Trade

Twelve and a half million Africans were kidnapped by Europeans to work the agricultural fields of the Americas in the transatlantic slave trade.[3] The horrid conditions on the slave ship resulted in a mortality rate around 15 percent. "The space was so low that [the Africans] sat between each other's legs and [were] stowed so close together that there was no possibility of their lying down or at all changing their position by night or day . . . the heat of these horrid places was so great and the odor so offensive that it was quite impossible to enter them."[4]

1619–1865: Slavery in the United States

Six to 7 million enslaved African people labored in the tobacco, cotton, indigo, and sugar plantations of the American South, generating $6.5 to $10 trillion of wealth, in today's dollars, for their enslavers.[5] Enslaved people were subject to rape, torture, beatings, and murder, and prohibited from freedom of worship, learning to read, marrying, or moving about independently. Virginia laws allowed for the dismemberment of "unruly slaves," prohibited racial intermarriage, and mandated that white churches seize all possessions belonging to slaves. After Congress abolished the African slave trade in 1808, the internal slave trade flourished, devastating 30 percent of Black families. "I had a constant dread that Mrs. Moore, her mistress, would be in want of money and sell my dear wife," a freedman wrote, reflecting on his time in slavery. "We constantly dreaded a final separation. Our affection for each was very strong, and this made us always apprehensive of a cruel parting."[6] The 1857 Supreme Court decision in the *Dred Scott v. Sanford* case declared that Black people were "so far inferior, that they had no rights which the white man was bound to respect; and that the negro might justly and lawfully be reduced to slavery for his benefit."[7]

1704–Present: Racialized Police Brutality

North Carolina formed the nation's slave patrol force in 1704, a forerunner to the modern policing system.[8] The system was nationalized in 1793, when Congress

Members of Capital Area Against Mass Incarceration prepare for an immigrant rights demonstration by creating a large banner that reads NO ES MI PRESIDENTE. Photo by Sun Angel Media.

passed the fugitive slave laws allowing police, vigilantes, and dogs to hunt down both free and escaped Black people across state lines and drag them back into enslavement. At all points in US history, police have disproportionately detained, beaten, and even killed Black people as compared with white people. In 2002 Black people accounted for 13 percent of the US population, but 31 percent of the police killing victims.[9]

1862

The Homestead Act provided federal land grants to western settlers, a mechanism for transferring 270 million acres of Native American land to white people. While Black people were included in the legislation, de facto discrimination prevented most from participating. At the same time, the Morrill Act granted 30,000 acres of federal land to each state, the proceeds of which would fund agricultural universities that excluded Black students.

1865–77: Black Codes

The 13th Amendment abolished slavery except for when people were convicted of crimes. The South created laws called Black Codes to label African Americans as criminals and keep them working and living in neo-slavery conditions. The defining feature of the codes was a sweeping vagrancy law that allowed police to arrest Black people for unemployment, loitering, or failure to pay taxes, and force them into contract labor. "Law enforcement agencies and white farmers systematically colluded in arresting African American men via sweeping and groundless incarceration every harvest season in order to press them into unpaid field labor."[10] The codes also prevented landownership, congregating in churches, attending school, voting, bearing arms, or moving freely through public spaces. One law stated, "All freedmen, free negroes and mulattoes in this State, over the age of eighteen years, found on the second Monday in January, 1866, or thereafter, without lawful employment or business, or found unlawfully assembling themselves together, either in the day or night time . . . shall be deemed vagrants, and on conviction thereof shall be fined . . . and imprisoned." A forerunner to the modern social service system, the law also allowed the state to seize custody of children whose parents were not "industrious and honest" and send them to be "apprenticed" to their former owners.[11]

1865–1941: Convict Leasing

In the 1840s states began leasing out prisoners to private employers looking for cheap labor, but this practice exploded across the South after Emancipation. Black Codes were used to lock up formerly enslaved African Americans and force them to labor in farming, railroad construction, mining, and logging. In 1898, 73 percent of Alabama's entire state revenue came from leasing out its convicts, 90 to 95 percent of whom were Black. Working conditions were miserable, and death rates were high. At the Coalburg Prison Mine in Alabama, 90 men per 1,000 prisoners died during their sentence. The practice of convict leasing was not legally abolished until President Franklin D. Roosevelt issued a circular in 1941.[12]

1865–1940s: Sharecropping

Reverend Garrison Frazier and 20 other Black Baptist and Methodist ministers met with Union General William T. Sherman to request allotments of "40 acres" of land for each freed Black family in an area independent from whites. In a radical act of land reparations, Sherman issued Field Order #15 deeding "40 acres and a mule" over to Black families on the South Carolina and Florida coasts. However, President Andrew Johnson reversed the policy and land was seized from Black families and returned to their former enslavers.[13] Most Black farmers were unable to save enough money to purchase land on their own, so they remained in a high-poverty debt peonage system of sharecropping or tenant farming, where they paid white landowners for land, tools, seeds, and supplies. Many Black farmers ended up owing the white landowners more at the end of the season than at the beginning, compelling them to remain on the land or face legal consequences. Over 3 million Americans succumbed to pellagra, a disease resulting from niacin deficiency and associated with extreme poverty; Black women sharecroppers were hardest hit. Eddie Earvin was a spinach picker who fled Mississippi in 1963 after being made to work at gunpoint. "You didn't talk about it or tell nobody," Earvin said. "You had to sneak away."[14]

1877–1950 Terror Campaign

More than 4,000 African Americans were lynched in total between 1877 and 1950. In the two-year period from 1900 through 1901, known as the "Terror Campaign," 214 lynchings of Black people were reported in the South. Black landowners were specifically targeted for not "staying in their place"—that is, not settling for life as sharecroppers. Black people were also attacked for not removing their hats, for refusing to hand over a whiskey flask, for disobeying church procedures, for "using insolent language," for disputing labor contracts, and for refusing to be "tied like a slave." Other attacks were intended to simply "thin out the niggers a little." In 1921 a white mob destroyed Tulsa's "Black Wall Street," devastating independently owned Black businesses. In 1923 another white mob leveled the Black town of Rosewood, Florida, with impunity.[15]

1881–1965: Jim Crow Laws

These laws revived the principles of the Black Codes by enforcing racial segregation throughout the South in neighborhoods, schools, restrooms, restaurants, workplaces, juries, drinking fountains, and all public accommodations. By 1896 the *Plessy v. Ferguson* case put the federal stamp of approval on Jim Crow. The Jim Crow laws helped spur racist hysteria, lynching, rioting, and the rise of the Ku Klux Klan. "Jim Crow formalized second-class citizenship status for Black Americans."[16]

1908–Present: Theft of Black-Owned Land

In one detailed investigation by the Associated Press alone, white people violently stole at least 24,000 acres of land from 406 Black people, depriving them of tens of millions of dollars and often at the cost of their lives. For example:

> *After midnight on Oct. 4, 1908, 50 hooded white men surrounded the home of a black farmer in Hickman, Ky., and ordered him to come out for a whipping. When David Walker refused and shot at them instead, the mob poured coal oil on his house and set it afire, according to contemporary newspaper accounts. Pleading for mercy, Walker ran out the front door, followed by four screaming children and his wife, carrying a baby in her arms. The mob shot them all, wounding three children and killing the others. Walker's oldest son never escaped the burning house. No one was ever charged with the killings, and the surviving children were deprived of the farm their father died defending. Land records show that Walker's 2½-acre farm was simply folded into the property of a white neighbor. The neighbor soon sold it to another man, whose daughter owns the undeveloped land today.*[17]

Recently, a participant in our Black Latinx Farmers Immersion program shared that her family's land in Virginia has been gradually seized by white neighbors who showed up uninvited to her grandmother's house demanding that she sign papers "or else." Families seeking redress in court today are often told that the "statute of limitations has expired."

1910–Present: Heir Property Exploitation

Black landowners often do not have access to legal services in order to create wills, so their property is inherited in common by their descendants, who become legal co-owners. Heir property is usually not eligible for mortgages, home equity loans, USDA programs, or government housing aid, tying the hands of property owners to invest in their land. Heir property is also vulnerable to corrupt lawyers and predatory developers because they only need to convince one heir to sell in order to force the sale of the entire property, known as a partition sale. Developers hunt down distant relatives, often in other states, and offer them cash for their share of the land, then force sale of the entire property at auction. It is estimated that over 50 percent of Black land loss since 1969 was due to partition sales. "If we don't have our land, we don't have our family," says Queen Quet, chieftess of the Gullah/Geechee Nation. "This is the battle we're in now."[18]

1933–Present: Federal Discrimination Against Black Farmers

Throughout the South, USDA agents withheld crucial loans, crop allotments, and technical support services from Black farmers as well as excluded them from USDA county committees. For example, when Mississippi farmer Lloyd Shaffer went to the USDA office to apply for the programs to which he was entitled, "On three separate occasions, the white FHA loan officer took Lloyd Shaffer's loan application out of his hand and threw it directly into the wastebasket. Once Lloyd was kept waiting eight hours, from the time the office opened until after it closed at night, while white farmers came and went all day long, conducting business."[19] By the 1950s USDA programs had been "sharpened into weapons to punish civil rights activity." During the 1962 Greenwood Food Blockade, the White Citizens Council also weaponized the Federal Surplus Food Commodity Program to punish sharecroppers who organized for civil rights, cutting off food and

Founding members of the National Black Food & Justice Alliance at the Penn Center in St. Helena, South Carolina. NBFJA is committed to reclaiming land sovereignty for Black farmers. Photo courtesy of NBFJA.

driving 20,000 Black farmers to the edge of starvation.[20] The founder of the Citizens' Council drew up a plan to remove 200,000 African Americans from Mississippi by 1966 through "the tractor, the mechanical cotton picker . . . and the decline of the small independent farmers." Black farmers who held on to their land used their independence to support civil rights workers, which often made them targets for lynch mobs and local elites.[21] In 1965 the US Commission on Civil Rights, an independent agency created by the Civil Rights Act of 1957 to investigate and report on a broad spectrum of discriminatory practices, released a highly critical study revealing how the ASCS, the FHA, and the Federal Extension Service bitterly resisted demands to share power and resources with African American farmers, leading to a precipitous decline in Black landownership.[22] In 1983 President Reagan pushed through budget cuts that eliminated the USDA Office of Civil Rights—and officials admitted they "simply threw discrimination complaints in the trash without ever responding to or investigating them" until 1996, when the office reopened. In 1920 there were 925,000 Black farmers owning 16 million acres of land, 14 percent of the US farmland. By 1970, 90 percent of the farmers and the farmland were lost to the Black community.[23] "It was almost as if the earth was opening up and swallowing black farmers," wrote scholar Pete Daniel in his book *Dispossession*.[24]

1934–77: Redlining

The National Housing Act of 1934 institutionalized pre-existing racism and segregation within the housing industry. The Federal Housing Administration (FHA) created "residential security maps" that ranked neighborhoods from A to D, listing them from the most desirable to least desirable for lending. The

D neighborhoods were predominantly Black communities and were outlined in red, labeled too risky for mortgage support. These maps were used by public and private lenders to deny mortgages to Black people. Further, the FHA's Manual of 1936 advocated deed restrictions to "prevent the infiltration of inharmonious racial groups" and to "prohibit the occupancy of properties except by the race for which they are intended." Redlining led to lower property values, abandonment, vacancy, and decline in Black neighborhoods. When the GI Bill was enacted during World War II, veterans who wanted to buy homes in their own redlined neighborhoods were denied the zero-interest mortgages to which they were entitled. Consequently, fewer than 100 of the 67,000 mortgages insured by the GI Bill supported homes purchased by people of color. Black veterans were forced to turn to predatory lenders for "on contract" homeownership, where the seller kept the deed until the contract was paid in full. If the buyer missed a single payment, they would forfeit the down payment, the monthly payments, and the property itself. In the 1960s, 85 percent of all Black home buyers in Chicago bought on contract.[25] As recently as 2015 three US banks settled charges of systematically rejecting mortgage applications from people living in neighborhoods predominantly of color.[26]

1935–Present: Exclusion from New Deal

The National Industrial Recovery Act (1933), Agricultural Adjustment Administration (1933), Social Security Act (1935), National Labor Relations Act (1935), and Fair Labor Standards Act (1938) were enacted after the Great Depression to protect worker rights and bolster the economy, but categorically excluded agricultural and domestic workers, most of whom were Black. The statutory exclusion of agricultural and domestic employees was well understood as a race-neutral proxy for excluding Blacks as a compromise with southern Democrats intent on preserving white supremacy. As a result Black workers earned lower wages, no overtime pay, and no retirement or disability benefits; they did not have the right to unionize. Today Section 152(3) of the National Labor Relations Act still excludes agricultural and domestic workers, who are mostly Black and Latinx, from key protections. Consequently, farmworkers can be fired for seeking to unionize, and do not receive overtime benefits. If they work on a farm with fewer than seven employees, they are not even entitled to the federal minimum wage.[27] Domestic workers are also excluded from Title VII of the Civil Rights Act of 1964, the 1971 Occupational Safety and Health Act, the 1993 Family and Medical Leave Act, the 1990 Americans with Disabilities Act, and the Age Discrimination in Employment Act. Although migrant workers arrive in the US healthier, on average, that their peers at home, their life expectancy is only 49 years.[28]

1949–70s: Urban Renewal

The Housing Act of 1949 and the Federal Aid Highway Act of 1956 gave federal, state, and local governments the power and funding to displace residents through eminent domain. Cities condemned and cleared low-income neighborhoods, which they designated as "slums," to build highways and entice new development. Black and Latinx residents were uprooted, and vibrant neighborhoods were disrupted. At the same time, cities used federal funds to build high-rise public housing towers to concentrate the urban poor, increasing class and race segregation. Urban renewal destroyed 2,000 communities and displaced 300,000 families from their homes. Approximately half of urban renewal's victims were Black, prompting James Baldwin to observe, "Urban renewal means Negro removal."[29]

1971–Present: Mass Incarceration

"The Nixon campaign in 1968, and the Nixon White House after that, had two enemies: the antiwar left and black people . . . We knew we couldn't make it illegal to be either against the war or black, but by getting the public to associate the hippies with marijuana and blacks with heroin, and then criminalizing both heavily, we could disrupt those communities," explained former Nixon domestic policy chief John Ehrlichman in defense of Nixon's 1971 War on Drugs.[30] In the years following the War on Drugs, prison population skyrocketed to its current height of over 2.2 million with an additional 4.7 million Americans on probation or parole, giving the US the dubious distinction of incarcerating a higher percentage of our citizens than any other country in the world. A 2014 report by the ACLU on the criminal justice system documented that "racial disparities result from disparate treatment of Blacks at every stage of the criminal justice system, including stops and searches, arrests, prosecutions and plea negotiations, trials, and sentencing."[31] Incarceration rates are at least five times higher for Black males than for white males, and two times higher for Black women than white women. Harsh discipline policies in schools funnel Black students into the "school to prison pipeline," where suspensions and expulsions increase the likelihood of imprisonment. Studies show that Black students are disciplined at higher rates than white students when exhibiting the same behaviors. Black youth are sent to juvenile detention center for status offenses like truancy, running away, and incorrigibility at a rate 4.6 times higher than white children.[32]

Today's Wealth Gap

In 2014 the Pew Research Center found that white households had 13 times the median wealth of Black households in 2013, up from 8 times the wealth in 2010, and that disparity continues to increase.[33] Further, 80 percent of wealth is inherited, often traceable back to slavery. In the seven "cotton" states, one-third of all white income was derived from slavery. The book *Slavery's Capitalism*, edited by historians Sven Beckert and Seth Rockman, demonstrated that the capital accumulated by Lehman Brothers, Berkshire Hathaway, Aetna, Wachovia, and JPMorgan Chase can all be traced to slave labor.[34] Education, income, and employment disparities fail to explain current racial wealth differences, as "by far, the largest factors explaining these differences are gifts and inheritances from older generations: a down payment on a first home, a debt-free college education, or a bequest from a parent."[35] White adults who do not complete high school, have children before getting married, and do not work full-time still have much greater median wealth than Black and Latinx adults who are married, more educated, and work longer hours.[36] This wealth disparity extends to landownership. Black people own approximately 1 percent of rural land in the country, with a combined value of $14 billion. White people own more than 98 percent of US rural land, over 856 million acres valued at more than $1 trillion.[37]

Today's Income Gap

Black households earn only 59 cents for every dollar of white median household income. Black Americans are 2.5 times more likely to live in poverty than white Americans and twice as likely to be unemployed at all levels of education. Income disparities can largely be explained by discrimination in employment and education. In 2014, 48 percent of all Black children attended high-poverty schools, as compared with only 8 percent of white children. In 2003 scientists at the National Bureau of Economic Research submitted thousands of otherwise identical résumés that had been randomly assigned "white-sounding" names (like Brendan) and "Black-sounding" names (like Jamal). The former elicited 50 percent more callbacks. Later studies upheld these findings of "implicit bias" in the hiring process. Further, Black applicants with no criminal record were offered jobs at a rate as low as white applicants who had criminal records. Employers are more likely to promote white workers into skilled and high-paying jobs, and channel Black workers into back office positions.

Today's Food Access Gap

White neighborhoods have an average of four times as many supermarkets as predominantly Black communities. As a result of these food apartheid conditions, incidences of diabetes, obesity, and heart disease are highest for Black, Indigenous, and Latinx people. Black Americans are also three times more likely to go hungry than white Americans.[38]

Today's Power Gap

A deep disparity exists in the power structure of Ferguson, Missouri, where Michael Brown was shot and killed by Officer Darren Wilson. African Americans make up two-thirds of the city's population, but whites serve as mayor, five of six city councilors, six of seven school board members, and 50 of 52 police officers. The nation at large does not fare much better in terms of sharing power with Black people. Only four of the CEOs of Fortune 500 companies are Black; that's 2 percent and declining.[39] The media and Hollywood are overwhelmingly under the control of white men, with only 4.8 percent Black television writers. Congress is more diverse than ever, but Black people still only comprise 9 percent of elected representatives and 4 percent of the lobbyists. In the nonprofit world around 95 percent of the leadership is white, even though 60 percent of nonprofits purport to serve communities of color.[40] To this day, the people who wield power in the courtrooms and boardrooms are overwhelmingly white.

Today's Environmental Gap

It's not just in Flint where Black families experience disproportionate environmental burdens. A preponderance of research shows that environmental racism

Structural racism decreases Black people's access to nature, which increases their chances for depression, learning disabilities, obesity, diabetes, heart diseases, academic underperformance, stress, and social anxiety.

is alive and well in America. Environmental toxins such as lead, polychlorinated phenols, volatile organic compounds, and organophosphate pesticides are differentially concentrated in areas where Black and poor children live and go to school. Predominantly Black communities are located closer to hazardous waste facilities than white communities. These exposures lead to a 47 percent higher asthma rate for Black people as compared with whites, as well as increased cancer risks.[41] Environmental benefits are also experienced unequally. While Black neighborhoods were bulldozed to create parks across the nation during urban renewal, Black people enjoy less time in those parks and wild spaces. The nation's original parks explicitly excluded people of color, and today Black people are more likely to attend schools without green spaces, live in neighborhoods without parks, and lack the resources to travel to "pristine" wild spaces for recreation. Children's diminished contact with nature increases their chances for depression, learning disabilities, obesity, diabetes, heart diseases, academic underperformance, stress, and social anxiety.[42] Further, a history of sundown towns, lynching, and other racist violence in rural spaces has engendered a generational fear of wilderness.[43]

Internalized Racism

If all of our somatic cells have identical DNA, how is it possible that liver cells do not end up in our brain? The answer is epigenetics, the complex matrix of proteins that turn certain genes on and off, impacting how they express. Epigenetics turns on the genes that make nerve cells show up in the brain and epithelial cells show up in the skin. Our environment, including factors such as stress, sleep, diet, and exposure to toxins, alters our epigenetics and that of our offspring. When researchers investigated the health records of 3,000 Finish people who evacuated to Sweden during World War II, they found that the offspring of Finnish children who were displaced were four times more likely to suffer from depression and other mental illnesses than those whose parents were not displaced. Similarly, a study found that children of Holocaust survivors had higher levels of methyl groups associated with the gene that produces the stress hormone cortisol.[44] Studies in mice show that trauma can be passed down through multiple generations by altering epigenetics.[45] The uplifting news is that positive life experiences that contradict the trauma can correct negative epigenetic marks, improving outcomes for the individual and for their offspring.[46] Science is showing that both trauma and healing can be passed down genetically.

Trauma and healing can also be passed down culturally. Even if we do not experience the trauma of forced servitude and near-complete subordination to the whims of another, that trauma may still inform our personal and collective identity.[47] Internalized racism is the adoption, by people of color, of racist attitudes and stereotypes toward members of our own ethnic group, including ourselves. As Marlene Watson wrote in the introduction to her book *Facing the Black Shadow*, "What is the black shadow? It's the running inner dialogue we have with ourselves all day long about our fears of being inferior as black people. It is our internalization of the white man's lie that blacks are inferior to whites—the very lie that was the foundation of our ancestors' enslavement. The black shadow is more than simply internalized racism; it's also our complex feelings of fear and despair about being black, and consequently our longing to be less black."[48]

Internalized racism can manifest as subordinate behavior, negative self-image, denial of pain, victim identity, and violence. Some scholars believe that even if white people were to magically halt all white supremacist actions today, we would still need to heal for several generations before we stopped enacting white supremacy on one another. You are invited to consider the some examples of behaviors connected to internalized racism and reflect upon how they show up in you and your community:

Subordination. Having internalized our presumed inferiority to white people, we show deference to them, assume that their ideas are correct, and look to them to confirm our opinions and decide the proper course of action. Having internalized

UPLIFT

Black Is Beautiful Movement

"Black Is Beautiful" was born out of the Black Power movement of the 1960s, as a reclamation of the inherent worth and dignity of Black people and a rejection of white colonial beauty standards. It spread throughout the world, including to South Africa, where it manifested in that country's Black Consciousness Movement. Under the anthem "Black Is Beautiful," our people wore their hair in natural styles, adorned their bodies with colorful, elegant textiles, and shunned bleaching creams. Don Cornelius spread the "Black Is Beautiful" movement through his television show *Soul Train*, possibly the first non-minstrel black entertainment on the national stage. As Black music, dance, and fashion were celebrated on the screen, children were able to witness positive reflections of self. "I don't mean it lightly when I say that self-love was probably the most important lesson that was taught on the show," remembered Questlove.[49]

"Black Is Beautiful" is a reclamation of the inherent worth and dignity of Black people and a rejection of white colonial beauty standards. Photo by Capers Rumph.

that our safety and survival are dependent upon pleasing white people, we behave submissively, particularly with law enforcement and bosses. We learn how to anticipate and predict the wishes and desires of white people to win their approval and evade their punishment.

Victim identity. Having internalized our presumed powerlessness, we withhold our viewpoints and opinions, staying silent in the face of oppression. At meetings, we position ourselves on the sidelines or in the back of the room, not taking up too much space. We complain privately about injustice but assume that we have no authority to enact change. We assume that external forces are responsible for our life, such as "the system" or even "God." We fear for our lives even when there is no immediate danger. Having internalized suffering, we claim it as a badge of our identity and believe that we are not authentically Black if we are not struggling. Further, if other Black people succeed, we put them down, calling them sell-outs.

Negative self-image. Having internalized that we purportedly have no history outside of our interaction with Europeans, we do not learn about our African ancestors. We reject our traditional religions as barbaric, devil worship, or inherently corrupt. We emulate the dress, food, religion, educational practices, language, and mannerisms of white people. We straighten our hair with chemicals and hot irons, weave in other people's hair to make ours appear longer, rub bleaching creams into our skin, and cover our dark brown eyes with blue contact lenses. We believe that Black people with lighter skin and straighter hair are more beautiful, smart, and desirable. We believe that we are not intelligent enough to succeed in rigorous academics and not wise enough to take leadership in our organizations. We settle.

Violence. Having internalized the violence of the plantation and the Jim Crow South, where looking a white person in the eye could result in a death sentence, we inflict violence upon one another. We use corporal punishment on our children, admonishing them to "stay in their place." We "play the dozens," hurling insults at one another while remaining cool and unemotional. Having internalized genocide, we indulge addictions to alcohol, nicotine, drugs, sex, and sugar. We beat and murder our intimate partners and members of our community, numbing ourselves to the pain of ubiquitous early death.

Denial. Having internalized the experience of our ancestors discovering that no help would come to relieve them of their suffering, we give up hope that our struggle matters to other people. We go silent and deny our present and past pain. Elders in my own family have asked me to delete portions of the family history I was writing when it told of life under slavery. My elders in Ghana have responded to my questions about the slave trade by admonishing, "These are things we do not discuss." Instead we learn to put on masks of contentment and seek to carefully control our expressions and our environment. We caretake the feelings of others while carefully avoiding any probing into our own hearts, lest we explode.

Healing Ourselves

"I need to go dig a hole," I declared to my family as I hastily changed from city clothes back to my more natural farm-y attire. I was just home from a day of facilitating an "anti-racism training" for a mostly white organization and was reeling from an onslaught of well-meaning but still white supremacist comments and microaggressions. Rather than wipe my distress all over my family, I decided to immediately go to the forest and give my pain to the Earth. I grabbed a shovel and vigorously hacked a small hole in the fragrant ground, just big enough for my head. As expected, the Land was able to absorb my tears, screams, and curses and compost them into calm. Once my sorrow was released, I filled up the hole, thanked her, and returned to my family feeling centered.

We have all the tools we need within ourselves and our communities to heal from racial trauma and reclaim our wholeness. In this section, we identify specific strategies for releasing pain and attaining personal liberation.

Ancestor Work

Each one of us has innumerable ancestors who have endured suffering and emerged intact. Our ancestors are rooting for us, loving us, and attempting to share their wisdom with us. Our job is simply to listen. Opening ourselves to a connection with our ancestors can be as simple as holding a photograph or artifact of a beloved forebear and saying out loud, "I love you and I am listening to you. Please be with me." Others may desire to deepen their ancestor practice by establishing an ancestor altar. On a shelf

Making regular prayers and offerings strengthens your relationship with your ancestors.

UPLIFT
Haitian Stone Art

Komye, Leogane, is the stone carving and stone balancing capital of the beautiful island nation of Haiti. Elder stoneworkers mentor younger artists in this delicate form, gathering soft stone from the riverbed and liberating the form within. Reginald, a highly regarded stone balancer, explains, "Stone balancing is a training for the mind. You have to become very calm and patient until you can hear the stone. You must feel its exact center and then it will balance for you. For one hour or two hours you may have to listen to a single stone without faltering." In a similar way, carving the stone requires "seeing what the stone wants to become," according to Reginald. This dynamic meditation practice represents both a livelihood for residents of Komye as well as a healing practice. The artists believe that working with stone helps them to "let go of the stress and troubles" that they carry.[50]

Youth use Haitian stone balancing as a mindfulness tool. Photo by Neshima Vitale-Penniman.

or small table, place a bowl of water and artifacts that remind you of your ancestors, such as a list of names, photographs, family heirlooms, or stones from their burial site. Whenever you eat a meal, place a portion of food on the altar. Making regular offerings increases the power of the connection between you and your ancestors. You can pray to your ancestors at this altar, inviting them to fortify you with their love. You may experience this support in powerful dreams, magical synchronicities, or the strengthening of your intuition.[51]

Healing Partners and Circles

One of our foundational human needs is to be listened to with unconditional positive regard. We can make an agreement with a friend or group of people to take turns actively listening to each other and reminding each other that we are not alone. Reevaluation counseling offers tools and training for peer-to-peer counseling, including a body of work specifically on healing trauma as African heritage people. However, you do not need any formal training to get started with your healing partner, just an open heart and a timer. Remember that the release of emotional pain through tears, laughing, yawning, and trembling is part of the healing process and not a sign that something is "wrong." Here are some prompts to consider for your first few sessions together. In responding to these prompts, prioritize paying attention to the feelings rather than trying to answer each question comprehensively.

Members of this healing circle take turns listening to one another's stories and offering support. Photo by Neshima Vitale-Penniman.

- What is your life story as a Black person? What memories do you have related to Blackness? Start with the earliest and move through time.
- What do you love about being Black? What are you most proud of about Black people?
- What is challenging about being Black? In what ways do you believe racist messages about yourself and your ancestors? What hurts have you experienced because of your identity? In what ways are you holding yourself back because of your Black identity?
- What do you long for as a Black person? Do you have a yearning for more information, connection, authenticity related to your Blackness? Where is the emptiness for you?
- What is your relationship to land and nature in general? How does your Blackness inform this relationship?

Dynamic Meditation

Jun San, an elder Japanese Buddhist monk and our beloved neighbor, has dedicated her life to peace and justice activism, primarily in solidarity with Indigenous people. I once asked her, "Jun San, can you give me any tips for sitting meditation? I get antsy and can't focus?" She laughed and tossed back the cloth of her orange robe. "I no meditate! Too boring! I beat drum, chop wood, carry water," she responded. While sitting meditation is a unique and powerful tool, our indigenous African traditions often engage dynamic meditation, including drumming, long-distance running, chanting, singing, candle gazing, and stone carving or balancing. What differentiates meditation from just doing activities is the focused attention on a singular point in the present moment. While drumming, our entire focus may be on the sensation between the palm and the goat skin. While carving the stone, our entire focus may be on the cadence of our breath. While chanting, our entire focus may be on the echo of our song back to us from the mountains. When the mind wanders, we gently bring it back to the singular focus. Studies show that dynamic meditation can reduce stress, anxiety, depression, and substance addiction.

Caucus Spaces

While we embrace the texture and vibrancy of integrated spaces, at times it is important for us to caucus with others from similar backgrounds. Entering into spaces that are exclusively for Black people, people of color, or any other identity group central to our understanding of self can engender a feeling of safety and freedom of authentic expression. We need spaces where Black people can laugh and cry together, show our messy rage and uncertainty, and engage in our rituals unapologetically. You may convene your own caucus space, or reach out to one of these existing groups.

- Women of Color in Solidarity Conference, Bluestockings, New York, New York
- Kindred Southern Healing Justice Collective, Atlanta, Georgia
- Joy and Justice Workshops, Whole/Self Liberation, New York and online
- Harriet's Apothecary Healing Village, Brooklyn, New York and national
- Movement for Black Lives Healing Justice Spaces, national
- Generative Somatics workshops by Lisa Thomas-Adeyemo or Adaku Utah

Community Dance

The African conception of healing incorporates community, mind, and body into holistic rituals of transformation. Illness is viewed as a disruption in the natural order of humans' interactions with the spirit world, something that is external to the individual and does not define who they are. As such, healing must take place in the context of a witnessing community and in connection with the spirit world. Across the Diaspora, community dance rituals are

Across the Diaspora, community dance rituals are central therapeutic practices. Photo by Jonah Vitale-Wolff.

central therapeutic practices. For example, *Ndeup* is a Senegalese dance ritual used to heal a person by connecting to the spirits who established a contract with the community's original ancestors. Ndeup is a multiday ritual involving sabar drumming, nonchoreographed dance and free movement, offerings to the ancestors, and spirit possession. Similarly, *Zar* is a spirit possession deity system that originated in Ethiopia and is now found throughout parts of northeast Africa and the Middle East. Women lead this dance ceremony, which provides a container for participants to witness the patient's distress. Krump dance (Kingdom Radically Uplifted Mighty Praise) evolved from clown dance in the aftermath of the Los Angeles riots of 1992. This free, energetic, fast-paced, improvisational, aggressive dance form is used by African Americans as a way to release anger and frustration nonviolently.[52] Dance is a healing art that each of us can access. Whether in a West African dance class or house dance party, use the transformative movements to release the emotions that do not serve you.

Plant Medicine

Plant spirit medicine is available to us for trauma healing. A comprehensive discussion of herbs is offered in chapter 10. Two African plants are especially powerful in relieving anxiety and depression connected to trauma. *Solenostemon monostachyus* is used across West Africa as a tea to soothe the nervous system and end panic attacks. Megbezorli, *Dysphania ambrosioides*, is an epazote of West Africa used to treat anxiety, nightmares, and shortness of breath. Macerate the fresh leaves into a paste, mix with water, and take three times daily.[53] Vodou practitioners anoint their foreheads with an infusion of rosemary, vinegar, red wine, and honey before bed in order to ward off nightmares and welcome healing, prophetic dreams.[54] In all cases it is essential to make a small offering to the plant spirit and ask permission to use it for your healing work.

The times when it is most important that I engage one of the aforementioned healing practices are often the most difficult times to remember to do so. At the height of a triggering situation or in the midst of a panic attack, I often do not remember that I have tools for self-healing. For this reason, I prominently display a list of healing practices categorized by how long they take. My list includes 10 people I can call at any time, affirmations to say out loud, quick actions I can take to shift my energy, and more involved healing practices. I ask those in my family to help hold me accountable to using my list and claiming power in my healing process.

Of course, when we are living with a boot on our neck, there is only so much we can do internally to find equanimity. We also need to organize, resist, and embody our power as change agents. By first focusing on internal fortification, we arrive to movement work with the necessary sturdiness to face opposition without depleting ourselves. In the next chapter we explore organizing strategies to uproot racism in the food system and beyond.

CHAPTER FIFTEEN

Movement Building

With all due respect, King T'Challa, what can a nation of farmers have to offer the rest of the world?

—from the film *Black Panther*

We are part of an inexorable web of connection that binds our lives to those in our community, nation, and world. Even as we engage in the sacred work of tucking seeds into soil, we must also leverage our resources to change the structural conditions that keep our people in bondage. As my elders in Kroboland, Ghana, taught me, "*Late ete no no daa*," which means there need always be three stones in place for the cooking pot to stand firm. I believe that movement work is most successful when we engage three equally crucial strategies: (1) protest and direct action to resist oppression; (2) working within the system to evolve policies and practices toward justice; and (3) building alternative institutions and creating models of the world we want to see. When these strategies are in balance, our movement work is effective and transformative.

One of our models for integrated strategy is the work of southern Black farmers during the civil rights movement era. As noted in the introduction, our mentor, civil rights veteran Baba Curtis Hayes Muhammad, explained to us that "without Black farmers, there would have been no Freedom Summer—in fact, no civil rights movement." Muhammad leaned forward on his hand-carved wooden cane, eyes creased in a gentle but determined smile. He paused to make sure we were listening attentively, heeding the wisdom of the activists who preceded us.

Baba Muhammad explained the central role that Black farmers had played during the civil rights movement, coordinating campaigns for desegregation and voting rights as well as providing food, housing, bail money, and safe haven for activists and displaced tenant farmers. Independent Black farmers were their "own bosses," unlike sharecroppers or domestic workers, and as such could not be fired by retaliatory white bosses for having the audacity to attend a meeting or register to vote. This modicum of independence catapulted Black farmers into a leadership role in the civil rights movement. The percentage of landowners involved in the movement surpassed tenant participation by very large margins.[1] These farmers employed the "three cooking stone" approach; they maintained institutions (farms) to offer material support to their communities, engaged in system reform (voting and petitions), and led civil disobedience (marches and sit-ins).

At Soul Fire Farm we ask ourselves how we can emulate our elders and ancestors who selflessly gave their land, leadership, and resources to the broader movements for justice and dignity. So in addition to being "just" a farm, we also offer our space for

The Northeast Regional Farmers of Color network is an informal alliance of Black, Latinx, Indigenous, and Asian farmers making our lives on land in New England and upstate New York. Photo courtesy of NEFOC.

meetings, trainings, retreats, meals, and safe haven to dozens of activist groups each year, from the New York State Prisoner Justice Network to Miracle on Craig Street community center.

We also organize ourselves on a regional and international level. Our primary coalition is the Freedom Food Alliance, which Black farmer and prison abolitionist Jalal Sabur helped to start in 2009. We are a collective of farmers, political prisoners, and organizers in upstate New York who are committed to incorporating food justice to address racism in the criminal punishment system.

One of the Freedom Food Alliance's central efforts is the Victory Bus Project, a program that reunites incarcerated people with their loved ones while increasing access to farm-fresh food. Together with other local farmers, Soul Fire Farm contributes produce toward food packages, which families of prisoners can purchase using SNAP. Once they purchase the food, families get a free round trip to visit their loved ones at correctional facilities in upstate New York. Families may choose to give the food to prisoners as a care package, take it home, or both. While on the bus, Jalal facilitates conversations about the prison-industrial complex and food justice, using texts such as Michelle Alexander's *The New Jim Crow*.

Our other regional coalition is the Northeast Farmers of Color Network. We provide mutual aid and love for one another as rural Black and Brown farmers, and coordinate the "Reparations Map for Black-Indigenous Farmers," which matches people with resources to farmers seeking land and wealth. We also run Uprooting Racism trainings for hundreds

Farmers in the Dessalines Brigade work together on seed saving, composting, and reforestation projects. Photo by Neshima Vitale-Penniman.

of food justice groups interested in adopting a racial equity lens and engaging in reparations work.

Nationally we organize with the US Food Sovereignty Alliance, National Black Food & Justice Alliance, Black Urban Growers, Agricultural Justice Project, HEAL Food Alliance, and other formations to advance a collective policy agenda.

Committed to a food system that is globally just, not simply fair for people within our borders, we have worked in solidarity with other campesinos and peasant farmers in Haiti, Puerto Rico, Ghana, Mexico, and Brazil. As of this writing, we just completed a seven-year solidarity cycle with the farmers in Komye, Haiti, who are part of the Dessalines Brigade, an alliance of peasant farmers committed to food sovereignty. Together we planted thousands of mango trees, installed irrigation, repaired homes after the 2010 earthquake and 2015 hurricane, and established composting on every farm in the community. In 2013 the Global Food Sovereignty Prize was awarded to the Dessalines Brigade in Haiti for their cooperation to save Creole seeds and support peasant agriculture.[2] The group earned international attention when they set fire to hybrid seeds Monsanto donated after the 2010 earthquake. These seeds were dumped on the Haitian market at the time of the rice harvest, threatening to outcompete and undermine the local smallholder agricultural economy. Flavio Barbosa of Brazil, representing the Group of Four / Dessalines Brigade, explained, "Haiti is a country that everyone talks about helping because it has a lot of needs, but in the 21st century, Haiti has been recolonized." These modern-day colonizers include the Haitian government, NGOs, and biotech companies like Monsanto, whose "help" for the Haitian people has resulted in monocropping, deforestation, the destruction of Haitian markets, hunger, and poverty.

The challenges before us are monumental. We are not obligated to complete the task of repair, but we are required to act at the intersection of our capability and what the world needs. To maintain silence is to cast our vote for the status quo, to passively endorse a racist and exploitative food system, and to deny ourselves agency over the destiny of our community. This chapter offers strategies to challenge the Empire's undermining of Black land sovereignty. Building on the legacy of our ancestors and elders who successfully sued the federal government, founded universities, drove their tractors to DC in protest, escaped slavery to form maroon communities, and organized massive boycotts of corporations, we explore diverse resistance tactics that will result in the reclamation of our rightful place on land. Each section will highlight a distinct resistance strategy, including current efforts that you can join or emulate.

Litigation

While the criminal *in*justice system is rightfully notorious for conspiring harm against our people, it is sometimes possible to leverage legal processes to defend and uphold our rights. Most legal action occurs outside of the courtroom. The Black Belt Justice Center is one example of a legal nonprofit that supports African American farmers to retain and increase landownership, organize into cooperatives and entrepreneurial businesses, and navigate heir property law.[3] Law schools also have community law clinics that offer free and low-cost legal services to individuals and organizations. The law is not simple to navigate on your own, so having legal assistance can support you in exercising your rights. Here are some questions that lawyers can help you address:

UPLIFT
Pigford v. Glickman

For decades the USDA denied loans and relief to Black farmers while providing these entitlements to white farmers, driving the loss of over 12 million acres of Black-owned farmland. For example, a white loan officer took Black farmer Lloyd Shaffer's loan application out of his hand and threw it directly into the wastebasket on three occasions. At least 25,000 Black farmers had similar experiences of discrimination and banded together to sue the federal government in the *Pigford v. Glickman* lawsuit. During the deliberations, Black farmers organized demonstrations and civil disobedience, enduring arrest when they attempted to enter the agriculture building in Washington and speak with then Secretary of Agriculture Dan Glickman. In November 1999 the class-action suit was settled out of court for around $500 million, the largest civil rights settlement in history. In some ways this victory was largely symbolic, as each farmer took home an average of $50,000, not enough for a new tractor, never mind to buy back the farmland lost during the previous decades.[4] The USDA refused to admit that they had discriminated against Black farmers, despite the government's own civil rights reports in 1965, 1970, 1982, 1990, and three in 1997 documenting systemic racism.[5] Black farmers continue to fight for fair compensation through Pigford II payouts and cy pres distributions to farmer organizations.[6]

Civil rights law. Are my civil rights being violated? Is this agency or company enacting illegal discrimination against me? For example, was I denied an FSA loan for my farm or was I denied access to a farm incubator program, possibly because of my race, gender, religion, or disability?

Business law. How do I form the correct business or legal entity to help me attain my farm goals?

Contract law. How do I write a contract to ensure that I am getting a fair deal in my business with other companies or individuals? How do I write a lease agreement? What do I do if a contract is violated?

Inheritance law. How do I write a will that ensures that my children, or others whom I designate, inherit my land and resources? What can I do with the land I have inherited as an heir-in-common?

Property law. How do I prevent foreclosure on my farm or home? How do I navigate land-use and zoning regulations in my region?

Labor law. How do I comply with labor laws for my employees? How do I make sure I have fair working conditions and wages at my farm job?

Should you determine that your legal rights are being violated, as with the USDA's discrimination against Black farmers, you can sue for discrimination. In the case of discrimination by the USDA, begin by filing a complaint with the USDA Office of the Assistant Secretary for Civil Rights. If you suspect discrimination by an employer, the first step is to file an administrative charge with the federal Equal Employment Opportunity Commission and the corresponding state agency. In your filing you need to have either direct evidence against your employer or circumstantial evidence, including proof that you are a member of a protected class, that you are qualified for your position, that your employer took an adverse action against you, and that someone who is not in a protected class replaced you or was favored above you.

When an institution is systemically discriminating in disseminating its jobs, scholarships, land, training, or other resources, you may want to seek the help of a civil rights nonprofit to take your case to court. The American Civil Liberties Union, Southern Poverty Law Center, Lawyers' Committee for Civil Rights Under Law, and Legal Aid Society focus on "impact litigation" where they champion landmark cases that set precedent for upholding the rights of the general population.

Education

A university education in agriculture may be inaccessible for aspiring Black farmers. I've talked to many Caribbean growers who spent their whole lives farming their family land, immigrated to the United States, and sought to continue their agricultural careers—only to soon learn that agricultural universities were often far from their families and communities, prohibitively expensive, and culturally isolating. We need to expand programs like the USDA 1890 National Scholars Program that provide scholarships allowing farmers of color to complete agricultural degrees at no cost. To address geographic barriers, we need satellite "campuses" in our communities and on land cultivated by existing Black farmers. These degree programs must explicitly address racism in the food system and provide support for healing from land-based trauma.

Knowing that we cannot rely exclusively on formal education to meet our needs for technical advancement, we have created educational networks for peer-to-peer learning. The Southeastern African American Farmers' Organic Network (SAAFON) was founded in 2006 by Cynthia Hayes and Dr. Owusu Bandele to increase organic and sustainable practices among Black farmers. At the time there were no African American certified organic farmers in several southern states, yet the trainings led by SAAFON indicated there was an overwhelming interest in becoming certified. They now have 121 members, 50 of whom are USDA-certified organic.[7] They also work with Black farmers in the Caribbean on organic practices and certification.

Back in the early 2000s, I attended the summer conference of the Northeast Organic Farming Association with the goal of connecting to other farmers of color. There were only about 12 of us who presented

UPLIFT
Tuskegee University

Lewis Adams, a Black community leader, tinsmith, shoemaker, and harness maker, had a vision for an institution of higher learning for Macon County. Despite having no access to formal schooling under enslavement, Adams taught himself to read and write, and valued education dearly. He struck a deal with a white politician, W. F. Foster, to organize the Black vote in his favor if the politician would persuade the state of Alabama to fund a school for Black people. This came to pass and on July 4, 1881, Tuskegee University opened the doors of its one-room schoolhouse to the inaugural class of 30 students. Booker T. Washington was the first teacher, and helped grow the school into one of the most respected historically Black universities in the nation. Tuskegee students constructed the early campus buildings as part of their work-study. The institution promoted self-reliance and believed that physical labor was not only practical, but beautiful and dignified. Tuskegee trained generations of Black farmers and included in its faculty such outstanding teacher-scholars as George Washington Carver and Booker T. Whatley, who brought us models including "regenerative agriculture," community-supported agriculture (the CSA), and pick-your-own farming.[8]

The Black Farmers & Urban Gardeners Conference brings together hundreds of growers using the "each one teach one" model of resistance. Photo by Warren Cameron.

as people-of-color in the space, so I went up to each person and handed them a little slip of paper with a meeting time and place. Every single person came, including Karen Washington, who would become a mentor and dear friend. She proclaimed at that meeting, "One day, we will have a conference of our own." True to her word, Karen Washington started the Black Farmers & Urban Gardeners Conference in 2010. Hundreds of farmers, activists, and chefs from across the nation come together every year to share knowledge in a container that feels more like a family reunion than a conference.

When we come together for a conference, provide apprenticeships for one another, teach workshops, write manuals, or otherwise share knowledge within our community, we are engaging in powerful resistance work. "Each one teach one" is one of our people's proverbs. During slavery, Black people were denied education, so when someone learned how to read, it became their duty to teach someone else. This duty persists.

Direct Action

Direct action is the deliberate violation of an unjust law or policy, an exercise of collective power, and a demonstration of principled nonviolence. Direct action includes sit-ins, workplace occupations, camps, strikes, blockades, hacktivism, and sabotage. Direct action may be individual or collective, discreet or public. For example, an employer may discreetly decide to give wages under the table to a person without legal immigration documents, and deny their presence when ICE arrives. In contrast, when refugee Salim Rambo was being deported back to the Democratic Republic of Congo from the UK, an activist stood up on the flight and boldly refused to sit down until the asylum seeker was removed from the plane and allowed to stay in the country. This action was successful and Salim Rambo was not deported.[9] Any of us at any time can choose noncompliance in support of justice.

We can thank our international comrades for their models of collective direct action. In 2014 the Afro-Colombian Solidarity Network organized tens of thousands of strikers to shut down one of the country's most important international ports, demanding potable water, sewage, and basic services for the Black community.[10] Also in Colombia, the Black Women's Movement in Defense of Life and the Ancestral Territories marched peacefully from their home of Cauca to Bogotá to demand the restoration of their ancestral mining rights and the cessation of megaprojects that displace their people.[11] The Proceso de Comunidades Negras (Black Communities Process, PCN) works collaboratively with the aforementioned Colombian formations to protect the right of Black women in southwestern Colombia, through direct action and direct services. Their founding grandmothers, such as Doña Paulina Balanta, taught them that "the territory is life, and life does not have a price" and that "the territory is dignity and it does not have a price."

Professor Wangari Maathai founded the Green Belt Movement in 1977, organizing women farmers in Kenya to plant over 51 million trees to date. While she was awarded the Nobel Peace Prize in 2004, her movement was not always met with acclaim. When Maathai denounced President Daniel arap Moi's plan to build a skyscraper in the middle of Nairobi's largest park, security forces visited her home in the night. Undeterred, she went on to lead a rally for the release of political prisoners. She and other Green Belt women were beaten and threatened with genital mutilation to force them to behave "like women should." Members of Parliament suggested that the Green Belt Movement be banned as a subversive organization. The women carried on, occupying a church adjacent to the park for nearly one year. As many of the soldiers were Christian, they refused to break into the church to arrest the women. Of her pro-democracy, pro-environment direct action, Maathai said, "Fear is the biggest enemy you have. I think you can overcome your fear when you no longer see the consequences . . . you must have courage."[12]

In 1993 grassroots activists from Colombia and Kenya joined forces with the Haitian Peasant Movement and 200 million other small-scale landworkers on four continents to form La Via Campesina, the

UPLIFT

NYC Community Gardens Movement

Former New York City mayor Rudolph Giuliani waged war on the city's community gardens in 1998, placing all of the 700 urban oases up for disposition to private interests. These gardens were primarily the work of Black and Latinx growers who reclaimed the vacant lots left behind from urban renewal and the disinvestment brought on by redlining. The gardeners got organized quickly and began lawsuits, protests, and civil disobedience. In the Esperanza garden on the Lower East Side, gardeners chained themselves to cement blocks as the bulldozers arrived to raze their 22 years of devoted investment in that tiny piece of earth. They also engaged spiritual resistance, erecting a large sculpture of a *coqui*, which is said to repel attackers in Puerto Rican legend. The police arrived and cut the chains, arresting 31 protesters for trespassing and obstructing justice, and detaining them overnight. A work crew with a backhoe, bulldozer, and chain saws set to work destroying the garden. As they were carted away, the protesters vowed, "We're going to haunt Giuliani like the Furies from Greek mythology." The battle was lost, but the war was arguably won. The community gardeners attracted national attention and were able to save 500 community gardens, 200 of which were designated as city parks and others of which became land trust property.[13]

international peasant movement.[14] Arguably the world's largest social movement, Via Campesina members have staged direct action in the streets of Cancun, Seattle, Quebec City, and wherever else institutions such as the World Trade Organization, World Bank, and UN meet to discuss food and agriculture. Via Campesina stands against corporate control of the food system under the guise of "free trade" and supports community-controlled sustainable food systems and fair trade. Via Campesina activists have brought international attention to the global food crisis.

In the United States, the Black farmer organizations taking the lead on direct action are the National Black Food & Justice Alliance and the Black Land & Liberation Initiative. On Juneteenth,* 2017, they occupied 40 acres' worth of vacant lots across the nation demanding reparations of land for Black people. The occupants offered free bag lunches to children and held a celebratory barbecue, demonstrating what productive land use should look like. The National Black Food & Justice Alliance (NBFJA) brings together dozens of farming organizations to act together to protect Black-owned farmland, including the recently foreclosed land of Eddie and Dorothy Wise in North Carolina. NBFJA's strategies also include organizing, building visibility, reframing narratives, and building institutions.[15]

As Reverend Martin Luther King Jr. wrote in 1963, "Nonviolent direct action seeks to create such a crisis and foster such a tension that a community which has constantly refused to negotiate is forced to confront the issue. It seeks so to dramatize the issue that it can no longer be ignored . . . The purpose of our direct action program is to create a situation so crisis packed that it will inevitably open the door to negotiation." As

* Juneteenth, or "Freedom Day," commemorates the abolition of slavery in Texas on June 19, 1865—and more generally the emancipation of African Americans throughout the former Confederacy.

Occupation of land or buildings can be an effective resistance strategy. Here, activists occupy city hall to protest the incarceration of our children. Photo by Sun Angel Media.

we confront unjust policies, we can consider direct action as a means to open negotiation. Consider organizing an occupation, blockade, or other visible noncompliance to call attention to your struggle.

Land Defense

Land and freedom cannot be disentangled. As early as 1650, kidnapped African people escaped to the Seminole territories of Florida, the Great Dismal Swamps of North Carolina, and the shores of Lake Borgne in Louisiana to form maroon communities. Free African Americans also started their own towns, like Lyles Station, Indiana, a community of Black farmers whose population peaked at 800 and persists to this day. "The old men were smart men, and they taught us the land was important," says Stanley Madison, a Lyles Station resident.[16]

Today the struggle for Black land continues, led by the Federation of Southern Cooperatives Land Assistance Fund, founded in 1967. The federation's 20,000 members engage in direct action, litigation, and education in defense of land. They sponsored the 1992 Caravan of Black and Native American Farmers to Washington, DC, to demand reparations for USDA discrimination. Together with Land Loss Prevention Project and other allies, they fought and won legislative battles to fund the Outreach and Assistance for Socially Disadvantaged Farmers and Ranchers Program (the 2501 program) and the 1890 Land Grant College scholarships.[17] Most recently, the federation is on the frontlines of the heirs' property battle.

Due to lack of financial resources and legal assistance, 81 percent of Black landowners in previous generations did not make wills. Their descendants inherited the property without clear title and were consequently limited in what they could do with the land. So-called heirs property is not eligible for mortgages, home equity loans, FEMA aid, USDA programs, or any of the loans or conservation

This Indigenous elder in the Triqui region of Mexico has successfully resisted many corporate and governmental attempts to take her family's land.

programs that keep many rural farmers in business. Additionally, any of the co-owners has the legal right to sell their share or to bring the entire parcel to court-ordered auction. Predatory developers take advantage of this legal loophole and entice faraway relatives to sell their share for peanuts, then flip the property for huge profits. The dispossession crisis among the Gullah people has been driven largely by heirs property exploitation.

The Federation of Southern Cooperatives, Land Loss Prevention Project, Black Family Land Trust, and other advocacy groups are working to pass the Uniform Partition of Heirs Property Act, which would add due process protections for families, such as notice, appraisal, and the right of first refusal. These organizations also provide free legal support to farmers so that they can create wills to protect their property and save their land from foreclosure and other forms of theft.[18] As Savi Horne, executive director of Land Loss Prevention Project, reminds us, "Never forget that food justice requires land justice." Currently, leaders in South Africa are considering comprehensive and radical land reform that would transfer certain farmland from white ownership to Black ownership without compensation. This model encourages us to think creatively and expansively about what is possible in terms of reclaiming land sovereignty.

Many of the graduates of Soul Fire Farm's Black Latinx Farmers Immersion are going "home" to the South to revive land that is still in their family. They are part of a returning generation of Black farmers, of which farmer Ben Burkett commented, "Some of them have great passions and dreams, some have acquired achings and fears, and all of them have acknowledged, to varying degrees, the ways in which a people can feel bound to their land. They're coming home, and they're bringing with them the skills and strategies to capitalize on the organizing the farmers did during the civil rights movement and their recent lawsuit. They're helping build networks and coalitions that grow, using structures and strategies

UPLIFT

Sojourner Truth

Sojourner Truth was born in bondage in 1797 in Ulster County, New York, and escaped to freedom with her infant daughter in 1828. In addition to her acclaimed contributions as an abolitionist and women's rights advocate, Truth was arguably one of the first Black American activists to champion landownership as a means to self-determination for her people. In 1870 she met with President Ulysses S. Grant at the White House, and asked him to give formerly enslaved people land grants in the West so that they could attain economic independence and be free from southern white racism. She persisted in her land grant quest for seven years, offering several public speeches on the matter, but the government refused.[19]

Truth was echoing the call of Garrison Frazier and other Black ministers in the South who designed the idea of "40 acres and a mule," and met with Union general William T. Sherman to negotiate implementation. Sherman agreed to this radical land redistribution plan, but President Jackson later reversed the policy. Despite the broken promises of "40 acres and a mule," Black people purchased 120,738 farms by 1890. By 1910 black farmers had accumulated 218,972 farms and nearly 15 million acres, 14 percent of the nation's farmland.

that establish them locally as people to be contended with. It's true that you can't go home again, because home is a time as well as a place. But you can return to a particular piece of earth that's in your blood and your heart."[20]

Policy Change

"We will not let their pens write us out of existence!" proclaimed Lindsey Lunsford of Tuskegee University at the 2017 Black Farmers Conference in Atlanta, Georgia. We were co-facilitating a policy workshop with Dara Cooper of the National Black Food & Justice Alliance, sharing strategies for grassroots political organizing. Lunsford explained passionately how crucial it is that we understand and advocate for legal protections, which can endure for generations and benefit people beyond our immediate community.

For example, the federal Farm Bill is the most important piece of legislation regulating the food system, governing nutrition assistance, farm credit, conservation, research, and trade. Until 1990 there were no provisions in the Farm Bill to address the needs of farmers of color; in fact, the programs of the Farm Bill were used as discriminatory tools against Black farmers. As a result of the advocacy in our communities, the "2501 program" for farmers of color and special funding set-asides in conservation programs and beginning farmer training grants were enacted.[21] Even today, 50 percent of white farmers and only 31 percent of black farmers receive funding through a USDA program.[22] We must persist in demanding our fair share of this public trust.

Changing policies requires organizing and thorough comprehension of the policy you want to enact or change. Corporations banded together to create ALEC, the American Legislative Exchange Council, which writes "model bills" that become laws favoring corporate interests. ALEC makes it easy for politicians to just sign on the dotted line without having to do any of their own research or thinking. We can and must be more organized than the corporate interests.

UPLIFT
Vision for Black Lives Policy Platform

The Movement for Black Lives (MBL) is as old as white racism, but coalesced in its current incarnation in 2013, in response to the acquittal of Trayvon Martin's murderer, George Zimmerman, and under the leadership of Black women—Patrisse Cullors, Alicia Garza, and Opal Tometi. MBL is now a global network of 40 organizations that intervene in violence inflicted on Black communities by the state and vigilantes. MBL convened in 2015 to write a unifying policy platform called "A Vision for Black Lives: Policy Demands for Black Power, Freedom and Justice." The platform has six "demands": (1) End the war on black people; (2) reparations; (3) invest–divest; (4) economic justice; (5) community control; and (6) political power. It contains over 30 specific policies at the local, state, and federal levels that should be implemented to meet the demands. Of these beautifully crafted policies, several pertain to the plight of Black farmers, demanding the restoration of land, free education including in agriculture, and the end to the war on immigrants.[23]

Farmers gather under a mango tree in Komye, Haiti, to discuss their policy platform. Photo by Neshima Vitale-Penniman.

Northeast Farmers of Color Network Policy Demands

Policy Action Steps to End Racism in the Food System

Please contact your elected officials and encourage them to support the following policies:

1. Real Food for Our People.
    a. Fully fund SNAP and WIC, eliminating barriers to access. Make EBT/SNAP easier for farmers to use by allowing online payment and automatic deduction. Expand healthy, sustainable, culturally appropriate options within these programs.
    b. Fund real food access in community institutions like schools, hospitals, day cares, prisons, and senior centers. Strengthen the Child Nutrition Reauthorization.
    c. Provide capital, credit, tax breaks, and training to worker and community-owned cooperative food enterprises that generate wealth for our people. Strengthen the Healthy Food Financing Initiative.
    d. Include agriculture and food systems science in the public school curriculum.
    e. End marketing of unhealthy food and food brands to children, including in schools. End subsidies for junk food marketing by closing the tax loophole that allows corporate write offs for marketing.
    f. Treat junk food and beverage companies like tobacco companies: hold companies liable for health impacts, and include visible warning labels, restricted advertising, barriers to purchase, and raise taxes that are re-invested in community.

2. Dignity for Farm Workers.
    a. Equalize all labor and wage laws so that farm and food workers have a living wage, a day of rest, health insurance, overtime, workers compensation, and collective bargaining rights. Update the Fair Labor Standards Act and National Labor Relations Act to afford farmworkers equal protection under the law.
    b. End penal farms, where incarcerated people are enslaved for food production.
    c. Create supportive pathways for (migrant, seasonal) farmworkers to become land-owning farmers running their own businesses, owner-operators. Create pathways to legalization for all undocumented people, included pathways to citizenship for all those that want it, and end deportations until a comprehensive policy is in place.
    d. Support smaller and independent producers so that they can pay a living wage to farmworkers.
    e. Replace the indentured servitude of the H2A visa program with the North American Agricultural Work Visa, and uphold all provisions of the Universal Declaration of Human Rights.

3. Community-Based Farmer Training.
    a. Include urban farmers in the USDA farming census as a unique category and provide technical support to these farmers. Pass the Urban Agriculture Act.
    b. Provide funding for farmer training programs led by people of color, that address trauma and history, and offer strategies for navigating in the racist food system. These programs should take place *in the community* but be

credit-bearing through partnerships with land grant universities.

c. Create and host an online portal for new farmers of color to find farms run by farmers of color and get training there.

d. Secure and protect land access and nonpredatory credit and capital for independent producers, particularly producers of color.

4. Economic Viability for Farmers.

a. Use public funds to pay farmers for preserving and enhancing ecosystem services and guarding the public trust (water purification, carbon sequestration, pollination, genetic diversity). This should be paid for by a tax on industrial agriculture for their externalities: dead zones in aquatic ecosystems, aquifer depletion, killing pollinators. Conventional agricultural should require a "certification" for its practices, rather than have that burden placed on organic farmers. Fully fund the Environmental Quality Incentives Program and Conservation Reserve Program, with set-asides for farmers of color.

b. Offer price supports and price parity for farm products to ensure that income from crop sales covers the expenses of producing those crops. Include non-commodity, heritage, and cultural crops in these programs.

c. End the practice of unfair contracts siphoning earning from farmers to enrich corporations. Pass the Producer Protection Act.

d. Equalize and expand access to crop insurance, technical assistance, non-GMO seed, equipment sharing, low interest credit, and technical assistance for independent producers, particularly

Some of the strategies that grassroots activists use to influence policy include:

Food policy councils. Join these local or state coalitions that bring together advocates, businesses, farmers, food workers, and citizens to analyze the food system as a whole and make policy recommendations.

USDA county committees. Run for election or vote for someone in your community to serve on these committees, which determine the type of programs the county will offer to deliver FSA resources at a local level.

People's assemblies. Bring people to testify at town hall meetings, or hold your own people's assembly and invite politicians to come hear your testimony.

Allies with influence. Ask influential lobbying organizations to adopt and champion your policy agenda. We have had early successes with the National Young Farmers Coalition, National Sustainable Agriculture Coalition, and Northeast Sustainable Agriculture Working Group.

Elections. Run for office or work on the campaigns of other changemakers of color, as the organizers in Jackson, Mississippi, did for Chokwe Antar Lumumba.

Regardless of your strategy, the first step is to clarify your demands and the entity in charge of the decision to meet or reject those demands. The Northeast Farmers of Color Network met to write our policy demands, which were inspired by the work of the HEAL Food Alliance and *Strengthening Our Rural Roots* by Oleta Garrett Fitzgerald and Sarah Bobrow-Williams.

producers of color. Include free legal and accounting clinics. Fully fund the Office of Advocacy and Outreach at the USDA.

e. Increase access to markets for farmers of color through food hubs, processing centers, farmers markets, and farm-to-institution programs.

f. Reduce paperwork burden for federal and state farming grants. Increase government staff support of application process. Make grants accessible to small farmers who are not incorporated. Move the application period to winter season. Eliminate matching funds requirements. Fully fund Outreach and Assistance for Socially Disadvantaged Farmers (2501).

5. Reparations for Stolen Land and Wealth.
   a. Reparations are necessary in the form of land and wealth redistribution to those who had land and wealth stolen from them: African American, Latinx, Indigenous people. Establish a commission to study reparations and propose a comprehensive redistribution of wealth and land.
   b. Enforce a moratorium on foreclosures of Black land; create a national trust or community-based organization that absorbs Black farmland and transfers it within the Black community. Implement the Uniform Partition of Heirs Property Act in all states.
   c. Create and implement farmer debt forgiveness programs in cases of discrimination. USDA should refinance loans for Black farmers.
   d. End "tied aid" policies that flood international markets with surplus commodities, and undermine smallholder farmers. Ban corporate land grabbing domestically and abroad.

Consumer Organizing

Farmworker rights organizations continue to be at the frontlines of consumer organizing. The Coalition of Immokalee Workers (CIW) of Florida began resisting declining wages for Mexican, Guatemalan, and Haitian tomato pickers in 1993, using hunger strikes, marches, and work stoppages. They then launched the first-ever farmworker boycott of a major fast-food company, calling for Taco Bell to take responsibility for the human rights abuses in the fields where its produce was harvested. After years of pressure Taco Bell agreed to meet all of CIW's demands to improve working conditions and wages for Florida tomato pickers. CIW turned to other major purchasers of tomatoes, winning campaigns with Burger King, Whole Foods, Bon Appétit Management Co., Compass Group, Aramark, Sodexo, Trader Joe's, and Chipotle Mexican Grill. By 2010 CIW had transformed over 90 percent of the Florida tomato industry by convincing the Florida Tomato Growers Exchange to sign on to its Fair Food Program, which combines a strict code of conduct with a complaint resolution system, health and safety program, worker-to-worker education, and independent auditing to ensure compliance. CIW also uncovered human trafficking operations on the farms and helped liberate over 1,200 enslaved workers. CIW's tactics focus on pressuring corporations to change behavior through boycotts and other direct action, rather than depending on laws to change.

Inspired by the tactics of CIW, dairy workers in Vermont put the pressure on their major buyer, Ben & Jerry's ice cream. There are 1,200 to 1,500 workers

UPLIFT

Delano Grape Strike

In 1965 Filipino farmworkers demanded a raise to $1.40 per hour for picking grapes. Their union, the Agricultural Workers Organizing Committee (AFL-CIO), was denied. When the farmworkers met in a packed hall their president, Larry Itliong, yelled out, "I want those in favor of a strike to stand up with your hand raised." Every last person stood and every last person walked off the job on September 8.

Itliong invited the Mexican farmworkers in the region to join their strike. By September 16 the National Farm Workers Association, under the leadership of Cesar Chavez and Dolores Huerta, joined the effort. The two unions merged to form the United Farm Workers, and over 2,000 pickers joined the strike. They organized an international consumer boycott of grapes, which brought the industry to its knees. By July 1970 the UFW had succeeded in reaching a collective bargaining agreement with the table-grape growers, affecting in excess of 10,000 farmworkers.

Dr. Martin Luther King Jr. saw the farmworkers movement and the civil rights movements as one. He wrote to Chavez in 1966, "As brothers in the fight for equality, I extend the hand of fellowship and good will and wish continuing success to you and your members. The fight for equality must be fought on many fronts—in the urban slums, in the sweatshops of the factories and fields. Our separate struggles are really one—a struggle for freedom, for dignity and for humanity." As he worked to build the Poor People's Movement, Dr. King met with Chicano leaders including Chavez, Bert Corona, Corky Gonzales, and Reies Tijerina.[24]

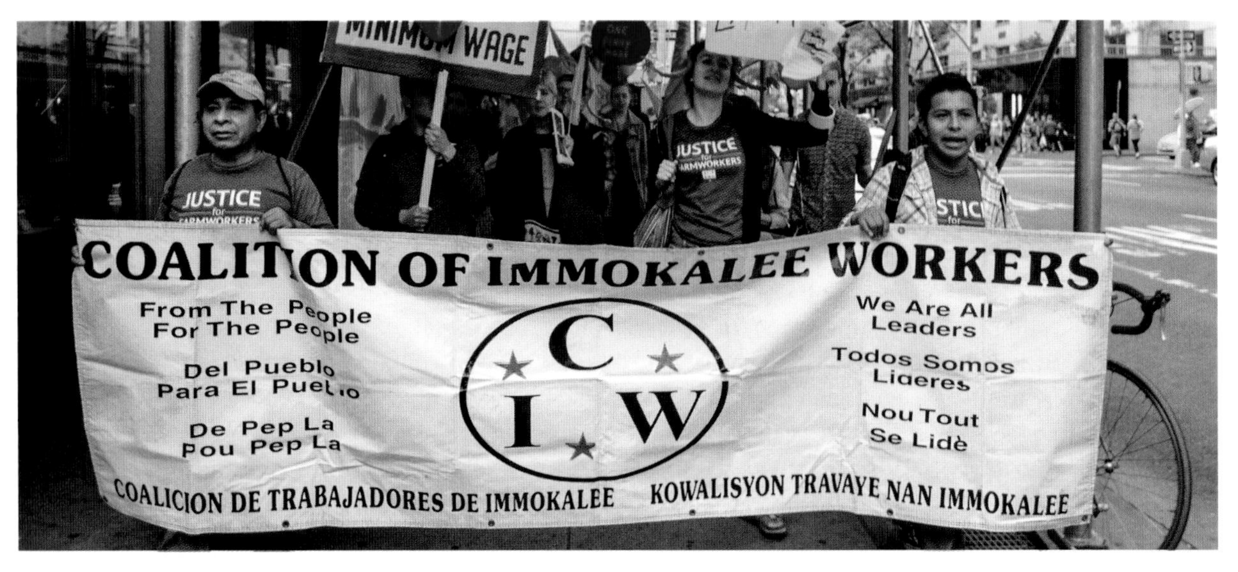

The Coalition of Immokalee Workers is a farmworker rights group on the front lines of consumer organizing. Photo courtesy of CIW.

in Vermont's dairy industry, many of whom worked seven days per week, had schedules that prevented sleeping more than a few hours, earned low wages, and had substandard housing. Their Milk With Dignity campaign demanded that workers earn the state minimum wage and have one day off in seven, eight hours of rest between shifts, and housing that included a bed, electricity, and clean running water. They demanded that Ben & Jerry's pay a premium on their milk to cover the costs of these basic dignities. In 2017 the campaign reached an agreement with the corporation, transforming the lives of thousands of farmworkers across the state.

The Agricultural Justice Project (AJP) and Domestic Fair Trade Association (DFTA) are working to take CIW's strategy to the national level. AJP convinces farmers and suppliers to voluntarily adhere to labor standards and fair trade practices in their supply chain, which affords them the right to use the "food justice certified" label. Soul Fire Farm is currently working through the AJP certification process, which guarantees workers freedom of association, living wages, safe and adequate housing, health and safety protections, medical care, sick leave, and family leave. You can contact AJP for support launching or joining a market-based campaign for human rights or environmental justice.

Mutual Aid and Survival Programs

"We are about a collective way of doing things," explained Xavier Brown, a DC-based urban farmer with Soilful City. "We work how ants work. One ant finds food and thousands of ants pull it back to the nest. We work to replicate how root systems work, passing information and food around. My elders have taught me that our theory of change is to mimic nature in how we work together. I could do it myself but that would be taxing and less effective."

Xavier Brown was echoing the strategic philosophy of generations of Black farmers, and the central organizing principle in our communities today. The Federation of Southern Cooperatives Land Assistance Fund (FSC-LAF) are our elders and guides in terms of this mutual aid strategy. Founded in 1967 FSC-LAF brings together 70 active cooperative groups with a combined membership of over 20,000

UPLIFT

Colored Farmers' National Alliance and Cooperative Union

Beginning in 1886 Black farmers organized for mutual aid and self-defense in Texas. They were targeted by land sharks, merchants, horse thieves, cattle ranchers, and repressive Black codes. These farmers formed the Colored Farmers' National Alliance and Cooperative Union (CFNACU), which peaked at 1.2 million members in 1891. CFNACU provided vocational training, discount purchasing depots, and marketing support to its members. CFNACU called for a general strike of Black cotton pickers to demand a wage increase from 50 cents to $1 per 100 pounds of cotton. The strike was vehemently opposed by the white Farmers Alliance and was crushed by local vigilantes in the Arkansas delta, who lynched several of the striking workers. After the failure of the general strike, CFNACU began its decline and disbanded in 1896. However, the legacy of mutual aid lived on in the 154 businesses that were part of the 1918 Negro Cooperative Guild, and later in the Federation of Southern Cooperatives, Freedom Farm, New Communities, and other Black farmer unions.[25]

The Victory Bus Project gives people living in New York City a way to visit their loved ones incarcerated upstate, while providing fresh food packages from member farms. Photo courtesy of Michael Rudin.

families, primarily in Mississippi, Alabama, Georgia, and South Carolina. They formed the Rural Training & Research Center in Sumter County, Alabama, which teaches agroforestry, credit union management, business planning, cooperative strategy, property law, marketing, value added products, and agricultural skills.[26] They also put their money and bodies on the line for one another in defense of land sovereignty. For this powerful work, they earned the 2015 Global Food Sovereignty Prize.

In the Northeast the Freedom Food Alliance engages in another form of mutual aid. The alliance formed out of the political prisoner Herman Bell's Victory Gardens Project in Maine, where volunteers cultivated land and distributed no-cost food in urban areas.* In 2009 Black farmers and food justice activists convened at the Growing Power Conference in Milwaukee and launched the Victory Bus Project. The project gives people living in New York City a way to visit their loved ones incarcerated upstate, while providing fresh food packages from member farms.

The Freedom Food Alliance exemplifies a crucial component of movement work: the survival program. Too often our generation centers "armchair activism," which may consist of nothing more than online pontification or defeatist headshaking about the inadequacies of others' efforts. Meanwhile our people are deprived of land, food, and the means of survival. The Black Panthers understood the need to animate their theoretical framework with direct action and direct service. To the extent that our resources and physical ability permit, we, too, are obligated to get our hands dirty in the work of direct mutual aid.

* At the time of writing, Herman Bell was released on parole after 45 years of incarceration.

CHAPTER SIXTEEN

White People Uprooting Racism

But all our phrasing—race relations, racial chasm, racial justice, racial profiling, white privilege, even white supremacy—serves to obscure that racism is a visceral experience, that it dislodges brains, blocks airways, rips muscle, extracts organs, cracks bones, breaks teeth. You must never look away from this. You must always remember that the sociology, the history, the economics, the graphs, the charts, the regressions, all land, with great violence, upon the body.

—TA-NEHISI COATES

At the Woodbourne Correctional Facility in upstate New York, guest teachers are not allowed to use red markers, and all images must be approved by the administrators in advance. Jonah and I worked within these and dozens of other regulations as we prepared our "farming and food justice" workshop for the incarcerated individuals who were earning their college degree through the Bard Prison Initiative.

On our way to the prison, we drove through one economically depressed rural Hudson Valley town after another, each one less resourced than the previous one. A few houses dotted the road, laundry hanging over sagging porches. We felt the painful irony that the Woodbourne Correctional Facility sat, quite literally, on land that once produced food. The land was now perverted as a tool for tearing apart our communities and draining its resources.

We passed through razor wire and triple locks, fingerprinting and white guards, peeling paint and 1970s encyclopedias, and caged humans. Thirty men, almost all Black or Latinx, attended our class. They told stories of gardening with their grandmothers, climbing mango trees at home in Puerto Rico, and running their family farms in the South. Bolstered by these memories, we got to work on business plans for the farm and food enterprises they planned to create once they were out.

Upstate New York was once home to thriving trade routes and prosperous dairy farms. The US Department of Agriculture's destruction of parity pricing and its "get big or get out" policies since the 1950s undermined the small-farm economy and incentivized consolidation. As agriculture declined in the area, the prison industry rose to take its place. New York State Department of Corrections is the state's largest agency, with a 2017 budget of $3.182 billion.[1] Politicians representing the towns where prisons are located vehemently protect this bloated industry and the jobs it provides in their mostly white communities.

These remote upstate New York prisons are filled with people of color from New York City, despite evidence that keeping close ties to one's community

Jalal Sabur prepares to bring vegetables to people incarcerated at the Green Haven Correctional Facility. Photo courtesy of Michael Rudin.

and family decreases recidivism.[2] While the majority of drug users and sellers in the state are white, 90 percent of people locked up for drug offenses are Black and Latinx. The result is a dysfunctional relationship between racist criminal justice policy and economic stability. The prison system harms rural communities because it ignores economic devastation and sidesteps development policy. It harms urban communities by kidnapping their people and reinforcing institutional racism. It harms us both by pitting our basic needs—for economic survival and freedom—against each other.

Working-class Black and white people have long understood that we are natural allies, and the owning class has done everything in their power to invent and promote racist ideas to divide us. Bacon's Rebellion in 1676 occurred when Black and white indentured servants took up arms against the landed gentry of the Virginia House of Burgesses. This interracial militia captured Jamestown and burned it to the ground. Word of the rebellion spread far and wide, and several more uprisings ensued. The planter elite were alarmed and deeply fearful of alliance between their workers, so they enacted laws that permanently enslaved Virginians of African descent and gave poor white indentured servants new rights and status. The white rebels were pardoned and the Black rebels were punished, further cementing the racial divide.[3]

Despite violent attacks by elite forces, poor whites and Blacks continued to organize together. They formed the Southern Tenant Farmers Union, an interracial alliance demanding their fair share of subsidies and profits, and improved working conditions. Harrison George of the Communist Party remarked in 1932, "The impoverished farmers are on the march. We cannot order them to retreat, even if we desired."[4]

Our generation must carry on this march. White supremacy erodes our humanity and is our common enemy. The white elite created white supremacy, a "historically based, institutionally perpetuated system of exploitation and oppression of continents, nations, and peoples of color by white peoples and nations of the European continent for the purpose of establishing, maintaining and defending a system of wealth, power and privilege."[5] White supremacy infuses all aspects of society including our history, culture, politics, economics, and entire social fabric, producing cumulative and chronic adverse outcomes for people of color. What can be created, can be destroyed. White people need to be active in the dismantling of white supremacy.

Black food and land sovereignty is the collective responsibility of the entire community, not just the purview of African American people. This chapter offers strategies for European heritage people to examine their privilege and take concrete action to uproot personal and institutional racism. We explore strategies for enacting reparations, forming alliances, transforming organizations, "calling in" oppression, and connecting with our personal histories. Recognizing that, as historian Howard Zinn once said, "you can't be neutral on a moving train," this chapter offers steps to actively work against the grain in your spheres of influence and start to undo the harms of racism.

Reparations

Enslaved Africans never received reparations for their unpaid labor or broken families. The promise "40 acres and a mule" during Reconstruction was retracted. In fact, some white plantation owners received reparations for their "lost property." If African American people were paid $20 per week for our agricultural labor rather than enslaved, we would have $6.4 trillion in today's dollars in the bank right now. This figure does not include reparations for denied credit and homeownership opportunities, exclusion from the social safety net and education, or property theft and destruction.[6] There is a reason why the typical white household has 16 times the wealth of a typical black household: 80 percent of wealth is inherited, often traceable back to slavery times.[7] Existing policies reinforce and augment the wealth gap.

The most important action that white people can take to uproot racism is to enact reparations, to quite literally give back what was stolen. We are not talking about Georgetown University–style reparations, where the institution gave a formal apology and preferential admission to the descendants of people sold as slaves in 1838, but did not offer scholarships or include the descendants in the reparations committee. We are talking about true reparations, which have the following three characteristics:

1. "Nothing about us, without us": Black people get to define what reparations look like.
2. "No strings attached": Transfers of land and resources without oversight or conditionality.
3. "The whole pie": Give the land, money, and jobs away, even and especially when it entails personal sacrifice.

Take stock of your resources, including your job, assets, property, and power. Ask yourself what you can give away in a loving act of reparations. Can you job-share with a community member and then hand over the reins once that person learns

Farmer Dallas (*left*) won the first reparations map victory, earning funds for Harriet Tubman Freedom Farm in North Carolina. Photo by Neshima Vitale-Penniman.

the ropes? Can you give away your land and downsize your living arrangements? Can you eliminate superfluous expenses and give that money away to front-line organizations? Can you convince other people with money and power to align their actions with the priorities of front-line communities? Black and Indigenous members of the Northeast Farmers of Color Network and Soul Fire Farm alumni put together a reparations map (www.soulfirefarm.org/support/reparations) to channel these resources directly to people of color working on farming and food justice projects. Users can explore the map and contact one another directly to transfer resources.

Reparations also entails working for policy changes that undermine the systemic nature of the wealth gap. The following are among the policies for which we should advocate.

To be in allyship as a white person can mean doing "unglamorous" tasks in support of people-of-color-led projects. Photo by Capers Rumph.

- Pass HR 40: Commission to Study and Develop Reparation Proposals for African-Americans Act.
- Guarantee a "Universal Basic Income"—a minimum livable income to cover the costs of food, shelter, education, transportation, and health care.
- Provide free and universal education, from pre-K through university.
- Increase federal and state investments in historically Black colleges and universities.
- Eliminate the down payment for home buyers of color and provide foreclosure interventions.
- Enforce housing antidiscrimination laws.
- Cap the mortgage interest tax deduction.
- Tax all income for Social Security, not just the first $128,400, and raise the payout to low-income earners.
- Close loopholes like "step up" that allow people to avoid paying tax on property they inherit.
- Enact affirmative action in colleges and universities.
- Provide equitable funding for public schools across districts.
- Raise the minimum wage and Social Security payouts.
- Provide free and universal health care.
- Grant the right to unionize for all employees.
- Enact a progressive restructuring of tax codes at the local, state, and federal levels.
- Divest from the criminal *in*justice system, and invest in restorative justice, mental health services, and job programs.
- End money bail and capital punishment.

Finally, reparations demands that we release the frontier mentality that plagues progressive spaces. The frontier mentality is the erroneous idea that the way to solve existing problems is to create or grow an initiative led by white people, rather than support existing projects led by front-line communities.[8] This myth is rooted in the frontier mind-set of European colonizers who romanticized the "Wild West" as an

Black-Indigenous Led Organizations Working on Food, Land, and Freedom

- African People's Education and Defense Fund
- Black Belt Justice Center
- Black Family Land Trust
- Black Farmers and Agriculturalists Association
- Black Fraternal Organization of Honduras (Organizacion Fraternal Negra Hondureña Ofraneh)
- Black Immigration Network
- Black Urban Growers
- CATA—The Farmworker Support Committee (El Comite de Apoyo a Los Trabajadores Agricolas)
- Center for Heirs' Property Preservation
- Centro Campesino
- Color of Food
- Community to Community Development
- Cooperation Jackson
- Cosecha
- Detroit Black Community Food Security Network
- Dignity & Power Now
- Ella Baker Center
- Familias Unidas por la Justicia
- Farmworker Association of Florida
- Federation of Southern Cooperatives/ Land Assistance Fund
- Food Chain Workers Alliance
- Food Sovereignty Ghana
- Freedom Food Alliance
- HEAL Food Alliance
- International Indigenous Youth Council
- Land Loss Prevention Project
- Líderes Campesinas
- Malcolm X Grassroots Movement
- Movement for Justice in El Barrio
- National Black Farmers Association
- National Black Food & Justice Alliance
- National Black Growers Council
- National Coordination of Peasant Organizations of Mali (Coordination Nationale des Organisations Paysannes)
- National Domestic Workers Alliance
- National Women in Agriculture Association
- Operation Spring Plant
- PODER: People Organized in Defense of Earth and Her Resources
- Restaurant Opportunities Centers United
- Sierra Seeds
- Southeastern African American Organic Farmers' Network (SAAFON)
- Standing Rock Sioux
- Via Campesina, La; International Peasant Movement

"unsettled" area with unlimited free land and opportunity. This myth drove the brutal exploitation of Indigenous people, Mexican Americans, and Blacks and the exploitation of natural resources. To release the frontier mentality, we must follow the lead of people of color. Logically, those harmed directly by racism are the same people who know best how to remedy that harm.

Sometimes "following the lead" of people of color means doing unglamorous tasks behind the scenes that fortify the work but do not garner recognition. My partner, Jonah, is currently the only white-identified person working at Soul Fire Farm, and I watch him fix toilets, trap ants, repair leaks, organize paperwork, and order supplies, while others facilitate programs and give public talks to much acclaim.

He has quite literally dedicated his life to building the infrastructure to support a people-of-color-led liberation project.* Similarly, I witness the White Noise Collective cook food and prepare spaces for Harriet's Apothecary, a Black-led space. They disappear during the healing event and return when it finishes to clean up and give our guests rides to the train station.

A first step is to find out what Black-led organizations are working toward liberation in your area and see what you can do to support. An incomplete list of formations working on food, land, and liberation for people of color is provided here in a sidebar.

Forming Interracial Alliances

Adopting a listener's framework is the first step for white people who want to form interracial alliances. Rather than trying to "outreach" to people of color and convince them to join your initiative, find out about existing community work that is led by people directly impacted by racism and see how you can engage. Front-line communities have the long-term commitment, strategy, and expertise necessary to transform the system. Nine guidelines for interracial alliance building are provided here.

1. **Center relationships.** There is no shortcut to building relationships. The first step is to *show up*. Spend as much time as possible attending community meetings, getting involved in neighborhood associations, and lending your skills and time to projects that are important to members of the community. When you tangibly demonstrate your commitment to the community, people are more likely to invite you in as a partner in initiatives. When you only reach out to someone because you want their presence at your onetime event, that is tokenizing, not relationship building.
2. **Pass the mike.** While you might be savvy with social media, writing, or public speaking, that does not give you permission to control the narrative of other people. Rather than telling or interpreting the stories of people of color, work to amplify the voices of people directly. Offer your technical support so that people can tell their own stories.
3. **Welcome feedback.** Develop a no-retaliation system for community members to give you feedback on your role in the alliance. Demonstrate your receptivity to feedback by thanking those who have the courage to offer you suggestions and implementing those suggestions in a timely fashion. Circle back with your partners asking if the changes you have made are acceptable.
4. **Cultural deference.** Because white culture is so ingrained in our society, we can mistakenly assume that work habits like tight deadlines, written communication, and Robert's Rules of Order are universal norms. Take time to get to know the communication styles, conflict management strategies, pace, and other characteristics of your partner organization and work to fit into that culture rather than impose your own.
5. **Facilitate skills transfer.** White privilege often affords European heritage people with the opportunity to learn technical skills, such as bookkeeping, grant writing, web design, legal advocacy, and business management. As you volunteer your time to apply these skills to anti-racism work, invite community members to work side by side with you to gain these competencies.
6. **Pay well.** Leverage your resources so that people of color are paid for their time as they organize for a racially just society. If you are an employer, hire people who might otherwise be overlooked in the capitalist economy: people of color, formerly incarcerated people, those without documents, elders, and those with disabilities. Offer jobs with training built in, a living wage, health care, and job security. Fund-raise so that people of color can be paid for their time as consultants, speakers, and organizers. Never

* Jonah is a Sephardic Jew who presents as Latino and is proactive about naming his white privilege. Many scholars mark the 1940s as the era when Jews "became white" in the United States.

UPLIFT
White Anti-Racists

Elijah P. Lovejoy was the abolitionist editor of newspapers in Missouri and Illinois. His printing press was destroyed four times by pro-slavery mobs. On November 7, 1837, Lovejoy was gunned down defending his newspaper and buried in an unmarked grave with no funeral. Over 100 years later Andrew Goodman worked with James Chaney and Michael Shwerner on the Congress of Racial Equality (CORE) voter registration effort in Mississippi. On June 21, 1964, members of the Ku Klux Klan murdered the three activists with the collusion of the deputy sheriff. Lovejoy and Goodman are two of the hundreds of white people who have fought—and sometimes died—for Black freedom. We lift up their names along with Sarah and Angelina Grimké, James Zwerg, Joan Trumpauer Mulholland, Tim Wise, Viola Liuzzo, Margo Adair, Jessie Daniel Ames, Kirsten Anderson, Anne Braden, Marilyn Buggey, Marilyn Buck, Robin DiAngelo, Bernardine Dohrn, Virginia Foster Durr, Ruth Frankenberg, Heather Hackman, Heather Heyer, Naomi Jaffe, Selma James, Frances E. Kendall, Chelsea Manning, Peggy McIntosh, Juliette Hampton Morgan, Kathy Obear, Minnie Bruce Pratt, Adrienne Rich, Eleanor Roosevelt, Ann Russo, Mab Segrest, Ricky Sherover-Marcuse, Lillian Eugenia Smith, Fay Stender, Julie Rawson, Jack Kittredge, and Peggy Terry.[9]

assume that Black people should volunteer their time as part of alliance work.

7. **Make it easy.** Limit or eliminate the number of hoops that community members need to jump through to access the resources that you control. In the case of grantmaking, provide video/audio application processes instead of exclusively written applications. Offer multiyear grants for general operating expenses, rather than requiring innovating special projects each cycle. Coordinate with other funders to have a universal application and reporting process, and minimize the demands you make on people's time. Organizers should not have to feel obligated to show up at your events because you donated money or other resources to their project. Include community members in the committees and boards that guide grantmaking and resources allocation.
8. **Pass the oars.** At all times people of color should be setting the agenda and determining the priorities in your alliance work. Check and recheck that you are not dominating the space or steering the ship. Because we have all internalized white supremacy to some degree, it is easy to collude in the "white expert" narrative and defer to white people's ideas. White folks need to be proactive to interrupt this pattern.
9. **Ask why you want to work with people of color.** When organizations have the "diversity conversation," it invariably turns to the seemingly intractable problem of "attracting people of color to our organization." Perhaps a better initial question is, "How can we work with other white people to raise consciousness and shift policies that are causing the harm?" In response to an influx of white progressives traveling to work with the Zapatistas in Mexico, the latter issued a statement that read, "If you can remove the boot from our neck by stopping your society from funding our government who is doing [the harm] directly

Participants in Soul Fire Farm's Uprooting Racism training explore the history of racism and resistance in the food system. Photo by Neshima Vitale-Penniman.

to us, then boy, wouldn't that be a big relief? . . . Then, please go home and organize . . . not just against imperialism and massive military expenditures going down to support the war in Mexico, but against the shit that you need to reorganize as your own problems. Stop letting us distract you from the fact that your cities have third worlds in them, that racism and sexism, things that we are really beginning to get a grip on here, are rampant in your home. Go home and take care of that."[10]

Organizational Transformation

Given that fewer than 20 percent of nonprofit leaders are people of color, it is very likely that you will find yourself working in a white-led organization on social justice issues. There are several inherent problems with this arrangement, from the moral fallacy of the "white savior complex" to the exploitation inherent in "poverty pimping" to keep white do-gooders employed. Existential questions should also arise for predominantly white organizations as to whether the resources going into the organization could be allocated more effectively and justly by investing in Black-led projects. Even with these sticky and challenging questions, many white-led organizations take the courageous step toward internal transformation. For example, both the National Young Farmers Coalition and the Groundswell Center for Local Food & Farming developed equity statements and corresponding equity practices as first steps toward uprooting racism in their organizations.

In order to transform your organization from one that is complicit in white supremacy to one that actively resists white supremacy, there are a number of actions you can take.

Decision Making and Power

Organizations working to end racism should have people of color in leadership at all levels, including on the staff, board, trustees, and volunteer leadership. Transforming an organization toward power sharing means first ensuring that everyone in the organization understands how power is distributed, how decisions are made, and how they can increase their decision-making power. Make training and mentorship available, along with clear steps for advancement open to everyone in the organization, including program participants. Too often, those with the most power in an organization or business are white, while those with the least decision-making influence are people of color. White people are responsible for making space so that people of color can lead.

Budget and Finances

Implement a cap on the wage and benefits gap between the highest- and lowest-paid person in the organization, so that there is equitable distribution of resources for everyone involved. Everyone in the organization should be able to see the budget and balance sheet at any time and have input on budget priorities. White-led organizations can work to actively transfer their resources to front-line organizations by contributing money, meeting space, customers, and staff time as well as making introductions to press and funders.

Accessibility

Work to increase accessibility of your programs and events to people of color. Provide transportation, childcare, food, wheelchair accessibility, all-gender bathrooms, and interpretation services. In some cases it is also appropriate to provide stipends for attendance. Review whether your decorations, music, and menu selections reflect a white-dominant culture. Consider hiring a guest lecturer, DJ, or caterer from the community you "serve" to bring their expertise to your events.

UPLIFT

Haymarket Fund

The Haymarket Fund was founded in 1974 to honor the Haymarket affair, a working people's uprising in 1886 that paved the way for the eight-hour workday. The Haymarket Fund has donated money to almost every major social justice movement in New England. However, for decades they were a white-led organization distributing money to mostly white-led organizations. In 1998 they embarked on a deliberate journey to uproot the racism ingrained in every aspect of the organization. With the guidance of the People's Institute for Survival and Beyond, they updated their mission, changed organizational policies, hired new leadership, and shifted organizational culture. While they will always be engaged in a learning process, Haymarket Fund is now led by people of color at all levels of the organization. People work together across racial lines to build relationships rooted in trust and accountability. Haymarket Fund published a manual detailing their transformation process, called *Courage to Change*.

Culture and Commitment

Implement an equity statement and safer space practices that explicitly address racism, sexism, transphobia, ableism, classism, and other oppressions. Invest organizational resources into transforming the culture from white-dominant to culturally inclusive. This may mean devoting more time for self-care, reflection, and collaboration, shifting work pace, updating definitions of success, and investing in more

Equity Statements by Predominantly White Farming Organizations

National Young Farmers Coalition, "Ending Violence Against People of Color in Food and Farming": www.youngfarmers.org/ending-violence-against-people-of-color-in-food-and-farming

Groundswell Center for Local Food & Farming, "Equity Statement": groundswellcenter.org/equity-statement

Northeast Sustainable Agriculture Working Group, "Statement of Intent on Race and Equity": nesawg.org/node/230882

National Sustainable Agriculture Coalition, "Statement on Race and Equity": sustainableagriculture.net/about-us/mission-goals/nsac-statement-on-racial-equity

time for training and support. Make space for people to bring elements of our cultures—music, stories, food, dance—to the organization. Be willing to name racism and directly address oppressive behavior.

Programs

Your organization's programs should be designed to build and share power with people of color, not to "serve" or "save" us. Address the root cause of problems, recognizing that this will eventually mean your project becomes obsolete. The people directly affected by oppressions should be the people involved in planning and designing those initiatives, and should have a pathway to take on leadership roles in the organization. Provide anti-racism training to all of your staff and infuse anti-racism topics into your community programs. If you are unsure what race has to do with your mission or goals, hire an anti-racism consultant to help you develop that equity lens.

Narrative

Update the narratives you tell about the work to uplift the contributions of people of color in the field. For example, many farming organizations omit the contributions of Fannie Lou Hamer, Booker T. Whatley, George Washington Carver, and the other visionaries highlighted in previous chapters. Actively participate in regional and national networks that are led by people of color. Use your social media, newsletter, and press platforms to promote the campaigns and stated priorities of people of color.

Behavior

Check your own white supremacist behavior and that of other white people in the organization. Racist structures are perpetuated through the accrual of seemingly minor exclusive and biased acts. Here are some common white supremacist patterns to challenge:

- Speaking first, more often, or interrupting
- Unilaterally setting the agenda
- Assuming white people are more capable
- Trivializing the experience of others
- Dismissing the content of what people of color say, because of disliking the "tone" or "attitude"
- Speaking on behalf of others
- Assuming one person of color speaks for the group
- Unilaterally controlling the organization's resources
- Reducing power struggles to personality conflicts
- Assuming that misunderstandings are the root of problems in organization
- Demanding proof or justification for perspectives of people of color
- Appropriating cultural elements of people of color
- Expecting gratitude and praise
- Defending mistakes because of "good intentions"

Participants in Soul Fire Farm's Uprooting Racism training create action plans for transforming their organizations. Photo by Neshima Vitale-Penniman.

- Assuming that everyone has the same options and access to resources
- Expecting people of color to educate white people about oppressions
- Expecting to be trusted
- Assuming that high-performing people of color are exceptional
- Expecting emotional comfort

Calling In

One of the highest forms of love is accountability. Rather than dismissing, shunning, or shaming people who make mistakes, known as "calling out," we can invite people into awareness. Rather than compounding a cycle of trauma by acting as persecutors of those who harm, we can invite one another to change our behavior for the better.[11] The process of loving accountability is known as "calling in." When someone in your organization or community engages in oppressive behavior, you have the opportunity to interrupt the harm. The steps for calling in are discussed below, adapted from the work of Mel Mariposa's "A Practical Guide to Calling In."

Preparing for the Conversation

Clarify goal. Ask yourself, "What is my goal? What am I hoping will change as a result of this conversation?" Clarify exactly what behavior you are hoping to shift as a result of this conversation.

Personal assessment. Ask yourself, "Am I the right person? What is our personal relationship? Is the subject matter something that might trigger me?

UPLIFT
Ruby Sales

Black elder activist Ruby Sales tells her oppressors, with a firm and unyielding love in her voice, "You cannot make me hate you."[12] Mama Sales was born in 1948 in Alabama and attended segregated schools. She was arrested at age 17 for picketing a whites-only store, jailed for six days, and threatened by a shotgun-wielding construction worker upon release. Jonathan Daniels, a white Episcopal seminarian and fellow activist, pushed her out of the way and took the bullet meant for her, dying instantly. Sales dedicated the rest of her life to human rights work and currently directs the Spirit-House project in Washington, DC. She is the living embodiment of Booker T. Washington's commitment to love in the face of unspeakable horror. As he said, "I will permit no man to narrow and degrade my soul by making me hate him." Sales and Washington challenge us to love even when we are justified in our hatred, acknowledging that hatred diminishes our humanity.

If so, who might be able to support me, or to have the conversation on my behalf? Am I someone they will listen to?" If this conversation is about an incident involving a third party, make sure you have consent to discuss it with others. Do not assume that another person wants you to champion their cause.

Role-play. Before having the conversation, it can be helpful to role-play with a trusted friend. Generally speaking, organizations can implement calling-in role-plays using fabricated scenarios to normalize the process and increase the likelihood of accountable conversations.

Having the Conversation

Centering. Before you enter into a courageous conversation, ensure that you are centered and grounded. Take a few minutes to breathe deeply and call upon the support of your ancestors, the Earth, and other higher powers that you feel connected to. You may also choose to meditate, take a walk, or drink a cup of tea to get grounded.

Privacy. Arrange for a private space to have the conversation to ensure confidentiality. Ask, "Can we talk for a few minutes in private?" Have the conversation in person. Never call in via email and definitely not on social media. Tragically, public call-outs have compounded mental health challenges and led to suicides in many communities. You do not want to cause more harm through humiliation.

Vulnerability. Start by sharing any fears or apprehensions you have around the conversation, and identifying anything that could be a potential obstacle. This has a disarming effect on the listener and reminds you both about your humanity.

Hopes. Share your hopes for the outcome of the conversation, and why you care enough to have this conversation with them. Let them know that you see accountability as a form of love.

Specific behaviors. Share the feedback about the specific behavior. Get clear around what specific behavior this person engaged in, rather than any assumptions, projections, or judgments around their motivations or character. The more specific you can be about the behavior and how the behavior affected you and/or others, the more you're going to be able to give this person feedback that can help them learn and grow.

Participants in Soul Fire Farm's Uprooting Racism training use role-play to practice courageous conversations and "calling in." Photo by Neshima Vitale-Penniman.

Empathy. If applicable, share about a time when you made a mistake that harmed someone and remind the person that this behavior does not mean they are a bad person.

Ask questions. Asking someone whose behavior is having unintended consequences what effect they'd like to have, and working to understand why they have behaved the way they have, can create more space for compassion, and can further support them in growing and changing.

Next steps. Ask the person to propose appropriate next steps to remedy the harm. It may be as simple as stopping the behavior. In some cases an apology may be necessary. Encourage the person to offer an unconditional apology, even if they do not "agree" that they caused harm.

Resources. Offer resources for further learning, such as readings, training opportunities, and people with relevant experiences.

Following Up

Setting limits. As much as we desire every accountability conversation to end with decisive behavior change, this is not always the case. If you are part of an organization with this person, you can work to enact organizational policies that have consequences for oppressive behavior. If someone is not willing to change a behavior, it may be time to take corrective action or ask the person to leave the organization.

Ongoing support. For individuals engaging in a change process, regular check-ins can be helpful. Ask the person how they see their behavior shifting, what questions they have, and what support they need. For example, a person who caused harm may decide that they want to apologize and make amends, and need support figuring out how to approach the conversation. If you have the emotional capacity to be there for them, do.

Don't unfriend, engage. While it can be tempting to distance yourself from white people with problematic behavior, so you can be labeled the "good white ally," it is actually more helpful for you to stay engaged with these struggling individuals. When anti-racist white people dismiss conservative whites, it not only places the burden of accountability on people of color, but can lead to disastrous outcomes like the election of "45." Stay engaged in your circles of white people, continually offering the courageous voice of reason and justice.

Personal Development

What is your history? Who are your people? How did you become white? All of us have a family history that predated the invention of race. Part of healing from racial trauma is to know your family story and to sit with its implications. Your ancestors and your family members likely caused unspeakable harm. Your ancestors also likely created beauty, loved deeply, and acted with profound generosity of heart. We need to sit with those contradictions, find love for ourselves, and act to repair the harms inflicted by our people.

As seed keeper and activist Owen Taylor explains on the Table Underground podcast:

> *This whole project of whiteness in this country is pretty new and devastating, really, to everybody. Obviously, the biggest victims are not white people, but we've lost so much as well in [terms of] losing pieces of our culture. I know that for my Italian ancestors . . . becoming "American" is what they would call it, was so important because they would belong, they would find more material success. There were so many incentives to lose your language, lose your religion, lose your food culture, in exchange for all the privileges and benefits of being considered white in America. It's something I talk about with my partner, who is Black . . . and I know for him it's really frustrating to be in mixed groups or with white folks who will say something like "Oh I have no culture," which is something I grew up thinking and believing too . . . which is, first of all unfair to those who came before us, [and] second of all totally untrue. But it's also what allows white supremacy to flourish, this concept that there is such a thing as whiteness, and that we're buying into it . . . There is so much power in realizing that . . . there's so much to find there, that is there [in our lineage], that you don't have to recreate or take on someone else's [culture]."*[14]

When white people are plagued by guilt about their unearned privilege, they often attempt to tuck away that privilege by ignoring their whiteness and appropriating the cultural elements of people of color. "I'm not the oppressor, I'm 'down' with POC," they say wordlessly through yoga mats and dreadlocks. According to *Who Owns Culture?* by lawyer and author Susan Scafidi, taking intellectual property, traditional knowledge, cultural expressions, or artifacts from someone else's culture without permission is cultural appropriation. This can include

UPLIFT
Traces of the Trade

Katrina Browne learned from her grandmother, at age 28, that her ancestors were the most successful slave-trading family in American history, responsible for the kidnap and transport of over 10,000 African people. The DeWolfs were part of a "web of broad-based Northern complicity in slavery." The North prides itself on its abolitionist history and sidesteps its role in the exploitation of black people. Katrina spent the years from 1998 to 2008 making a documentary about the implication of her family's leadership in the slave trade, called *Traces of the Trade: A Story from the Deep North*. She remarks, "None of us want to feel implicated. Because then we would have to feel responsible for problems we don't want to feel responsible for—deep, old, intractable divides . . . [this] has been instrumental in helping us see that we benefit disproportionately from systems that were set up to serve us, even when we aren't intending to do harm."[13]

unauthorized use of another culture's dance, dress, music, language, folklore, cuisine, traditional medicine, and religious symbols. It's most likely to be harmful when the source community is a minority group that has been oppressed or exploited in other ways or when the object of appropriation is particularly sensitive, such as sacred objects. "Currently, the commodification of differences promotes paradigms of consumption wherein whatever difference the Other inhabits is eradicated, *via* exchange, by a consumer cannibalism that not only displaces the Other but denies the significance of the Other's history through a process of removing context," remarks bell hooks.[15]

In the song "White Privilege II," Macklemore challenges, "We take all we want from black culture, but will we show up for black lives?" To begin untangling from the pervasive practice of cultural appropriation, you can convene a discussion group and ask yourselves the following questions, compiled by Larisa Jacobson of Soul Fire Farm.

Undoing Cultural Appropriation Dialogue Questions

1. What is something that gives you life in your own cultural identity?
2. When have you observed something that is culturally significant for you be appropriated by another person (for their gain, without consent)? (Note that an inability to answer this question is also an answer.)
3. What is something you benefit from—socially, spiritually, economically, or otherwise—that comes from a culture not represented in your own identity? Would you call that cultural appropriation? Why or why not?
4. What are ethical ways of borrowing?
5. How do we know if we have the "blessing" to borrow from someone else's culture?
6. What does compensation or reciprocity look like?
7. What are ways that you check yourself (or not) when engaging with something from another culture?
8. What does it mean to live by the precepts of the Honorable Harvest as explained by Robin Wall Kimmerer: "To take only what is given, to use it well, to be grateful for the gift, and to reciprocate the gift"?[16]

Once we have detached from stealing the culture and identity of others, we are liberated to wrestle with our own privileged and targeted identities.

Embracing one's own cultural heritage is part of uprooting racism, exemplified here in this Jewish mikveh ceremony. Photo by Neshima Vitale-Penniman.

Table 16.1. Matrix of Intersectionality

Social Identity Categories	Privileged Social Groups	Border Social Groups	Targeted Social Groups	Ism
Race	White people	Mixed-race people (with recent white ancestry)	Asian, Black, Latinx, Indigenos people	Racism
Color	White-presenting people	People with light brown, tan skin	People with dark skin	Colorism
Sex	Men	—	Women, intersex	Sexism
Gender	Gender-conforming cis men and women	Gender-ambiguous cis men and women	Transgender, nonbinary people	Transgender oppression
Sexual orientation	Heterosexual people	Bisexual people	Queer, lesbian, gay people	Heterosexism
Class	Rich, owning, upper-middle-class people	Middle-class people	Working-class, poor people	Classism
Ability/ disability	Temporarily able-bodied people	People with temporary disabilities	People with disabilities	Ableism
Religion	Protestants, secular	Roman Catholics	Jews, Hindus, Muslims, Indigenous religions	Religious oppression
Age	Adults	Young adults	Elders, children	Ageism, adultism
Language	Fluent English-only speakers	Bilingual speakers of English	Speakers of languages other than English	Language oppression
Education	College-educated people	High-school-educated people	People without formal high school education	Elitism
Documentation	US citizen	US resident	Undocumented	Xenophobia

Adapted from *Teaching for Diversity and Social Inclusion*, 2nd ed., Routledge, 2007.

Almost all of us embody a complex matrix of identities, including some that provide social advantage and others that limit access. Explore the "Matrix of Intersectionality" (table 16.1) and record your identities in the appropriate categories. Then share stories related to the identities that are most difficult for you to claim. For example, as an able-bodied person, it is easy for me to take that identity for granted as "normal" and ignore the ways that this privilege opens doors for me that are closed for others. I do not bristle when sidewalks are crusted over with snow or bathrooms are not accessible, because I can effortlessly navigate. I ask myself, "How can I have the same urgency around disability rights as I do about race and gender equality?" Individuals make up the system, so this work of challenging our limited perspectives on identity can transform institutions and power structures. While it is just a small step, my reflective work with this matrix resulted in Soul Fire Farm raising money for and installing a wheelchair-accessible bathroom. What action can you catalyze by bringing your privileged identities into awareness?

In her poem "On the Pulse of Morning," Maya Angelou said, "History, despite its wrenching pain / Cannot be unlived, but if faced / With courage, need not be lived again." We are never finished with the work of facing our history with courage. Continue to educate yourself, engage in conversations, and catalyze action toward a racially just world.

Recommended Reading on Racial Justice

- *White Privilege: Unpacking the Invisible Knapsack* by Peggy McIntosh (online)
- *Dismantling Racism: A Resource Book* by Western States Center (online)
- *Examining Whiteness: An Anti-Racism Curriculum* by Reverend Doctor William J. Gardiner (online)
- *The Case for Reparations* by Ta-Nehisi Coates (Spiegel & Grau, 2015)
- *The New Jim Crow: Mass Incarceration in the Age of Colorblindness* by Michelle Alexander (The New Press, 2012)
- *Rewriting the Racial Rules: Building an Inclusive American Economy* by Andrea Flynn et al. (online)
- *Opportunities for White People in the Fight for Racial Justice* by Jonathan Osler (online)
- *Curriculum for White Americans to Educate Themselves on Race and Racism—from Ferguson to Charleston* by Jon Greenberg (online)

Recommended Training Programs for Dismantling Racism

- People's Institute for Survival and Beyond
- AORTA Collective
- Catalyst Project
- Training for Change
- Beyond Diversity 101
- White Noise Collective
- Center for Study of White American Culture
- Interaction Institute for Social Change
- Soul Fire Farm's Uprooting Racism in the Food System

To free ourselves, we must feed ourselves. Are you ready to get free together?

AFTERWORD

A few months back, one hundred Black farmers gathered in the gymnasium of a church in Atlanta, Georgia, to discuss what to do next now that the Pigford lawsuit against the USDA for discrimination was settled and the disbursements were complete. I was honored to be in the space alongside movement elders like Shirley Sherrod, John Boyd, Barbara Norman, Dorathy Barker, John Zippert, and Ralph Paige (in blessed memory). These were the leaders who had the audacious courage to ride their tractors onto the mall in Washington, DC, demanding reparations for the land that had been snatched from them as a result of institutionalized bigotry. These were the leaders who made the largest civil rights settlement in US history possible.

As my heart was filled with humility and inspiration, it also held alarm. At age 38, I was the "youth" in the room. The agricultural census said that the average age of Black farmers was 62 years, but the median age in that room was even higher. I felt quite sure that if we did not figure out how to pass on the legacy of our agricultural traditions, this art of living in a sacred manner on land would go extinct for our people. Then, the KKK, the White Citizens Council, and Monsanto would be rubbing their hands together in glee, saying, "We convinced them to hate the earth and now it is all ours. The water, the minerals, and the soil—all ours." And then the soil herself would don sackcloth and grieve, saying, "The ones who I birthed have forgotten and foresaken me, worshipping instead the works of human hands."

We will not let the colonizers rob us of our right to belong to the Earth and to have agency in the food system. We are Black Gold—our melanin-rich skin the mirror of the sacred soil in all her hues. We belong here, bare feet planted firmly on the land, hands calloused with the work of sustaining and nourishing our community.

BLACK GOLD

by Naima Penniman

I am evidence of love under fingernails
knee caps stained from kneeling to pray
sacred remains of yesterday
fertile with future

I am the musk after rain
soft firm unconditional embrace
bosom of returning

I am the earthen floor
slapped with souls of feet
origin story and sorcery

I am vast and cosmic compost
sand, silt, stone
composition of decomposed bones
and primordial ferns

I am sun
turned to rot
particles of stardust
heaped in terra cotta pots

This thing the living clings to
sorghum seed and black eyed pea
nascent seedling
ancient tree

This melanin rich thickness
dripping mineral and mystery

I am galactic blackness
humus of fecundity

I am mud clung to cassava root
in the lower holds of slave ships
I am the soggy clay that raised up sugarcane
and cotton profits of enslavement

I am arid Trails
damp with Tears of dislocation
earth snatched from Japanese farmers
thrust into concentration

I am raped
for modern day colonization
for global market chocolate
whose growers face starvation

I am the ground under Harriet's Railroad
the smell of promise past plantation
the cracks in pavement post Migration
that spawned the shoots that pushed through subjugation

I am loam lush with nitrogen in Carver's legume fields
the calcium in Black Panthers free breakfast meals
I am cooperative acres Fannie Lou Hamer made real
the toxic lots Hattie Carthan's garden plots healed

I am ashes of burned Monsanto seed
stamped under the feet of Haitian revolutionaries

I am flush with tomatoes from Taco Bell boycotts
fertilized by the fruit Immokalee farmers left to rot

I am proof of life after death
I am dawning from decay
my belly of mass graves
my open palms of gardens

I am transmuter of toxins
cauldron of embryos
cradle of coffins
atmospheric alchemist
sequestering carbon

I am dust
bleached, stripped, strained
by purveyors of pipelines
and mountain top explosions

Erosion of homelands
washed to ocean
metropolis encroaching
I am choking

I am swept up and stepped upon
guzzled by the gluttonous

I am paradise paved over
poison drenched
drained of sustenance

I buttress monocultures of monopolies
to feed cultures of injustices

Silent utterance
I am thunderous

I am umber, ochre
saffron, sienna
crimson, cinnamon
cocoa brown and ebony

I am gold, gold, Gold

You are soiled
filthy
black
dirt
rich

You are soul, soul, Soul

Take me in your palms
breathe in my memory

Remember me

Fall soft where you belong, my seed
I need you

The future depends on me

RESOURCES

African People's Education and Defense Fund (apedf.org)
Afro-Colombian Solidarity Network (afrocolombian.org)
AgPlan (agplan.umn.edu)
Agricultural Justice Project (www.agriculturaljusticeproject.org)
Agriculture and Land-Based Training Association (ALBA) (www.albafarmers.org)
Alabama A&M University College of Agricultural, Life and Natural Sciences (www.aamu.edu/Academics/alns/Pages/default.aspx)
Alliance for Food Sovereignty in Africa (afsafrica.org)
American Farmland Trust Directory (farmlandinfo.org/directory)
Ancestral Apothecary School (ancestralapothecaryschool.com)
AORTA Collective (aorta.coop)
Appetite for Change "Grow Food" video (https://youtu.be/PqgU3co4vcI)
ATTRA (attra.ncat.org); ATTRA directory of Sustainable Farming Internships and Apprenticeships (attra.ncat.org/attra-pub/internships)
Baker Creek Heirloom Seeds (www.rareseeds.com)
Beyond Diversity 101 (www.beyonddiversity101.org)
Black Belt Justice Center (blackbeltjustice.org)
Black Dirt Farm Collective (www.facebook.com/blackdirtfarmcollective)
Black Family Land Trust (bflt.org)
Black Farmers and Agriculturalists Association (www.bfaa4us.org)
Black Fraternal Organization of Honduras (Organizacion Fraternal Negra Hondureña Ofraneh) (www.ofraneh.org/ofraneh/index.html)
Black Immigration Network (blackimmigration.net)
Black Land & Liberation Initiative (blacklandandliberation.org)
Black Oaks Center for Sustainable and Renewable Living (www.blackoakscenter.org)
Black Organizing for Leadership and Dignity (boldorganizing.org)
Black Urban Growers (www.blackurbangrowers.org)
Blue Otter School of Herbal Medicine (www.blueotterschool.com)
California FarmLink (californiafarmlink.org)
Capital Area Against Mass Incarceration (www.caami.org)
Carts Vermont (www.cartsvermont.com)
CATA—The Farmworker Support Committee (El Comite de Apoyo a Los Trabajadores Agricolas) (cata-farmworkers.org)
Catalyst Project (collectiveliberation.org)
Center for Heirs' Property Preservation (www.heirsproperty.org)
Center for the Study of White American Culture (www.euroamerican.org)
Centro Ashé (www.centroashe.org)
Centro Campesino (centrocampesino.org)
Climbing Poetree (www.climbingpoetree.com)
Color of Food (thecolorofood.com)
Community Foundations of the Hudson Valley (communityfoundationshv.org)
Community Seed Network (www.communityseednetwork.org)
Community to Community Development (www.foodjustice.org)
Cooperation Jackson (www.cooperationjackson.org)
Cornell Small Farms Program (smallfarms.cornell.edu/online-courses)
Cornerstone Farm Ventures (www.cornerstone-farm.com)
Cosecha (movimientocosecha.com)
D-Town Farm (www.d-townfarm.com)
Detroit Black Community Food Security Network (www.dbcfsn.org)

Dignity & Power Now (dignityandpowernow.org)
Domestic Fair Trade Association (DFTA) (www.thedfta.org)
Eagletree Herbs (www.eagletreeherbs.com)
Earthseed Land Cooperative (earthseedlandcoop.org)
East New York Farms! (ucceny.org/enyf)
Ella Baker Center (ellabakercenter.org)
Environmental Justice League of Rhode Island (ejlri.org)
Equal Employment Opportunity Commission (www.eeoc.gov)
Equity Trust (equitytrust.org)
Familias Unidas por la Justicia (familiasunidasjusticia.org/en/home)
Family. Agriculture. Resource. Management. Services. (F.A.R.M.S.) (www.30000acres.org)
Farmland Access (farmlandaccess.org)
Farm School NYC (www.farmschoolnyc.org)
Farm School NYC & Sustainable Flatbush (sustainableflatbush.org)
Farms to Grow, Inc. (www.farmstogrow.com/about)
Farmworker Association of Florida (floridafarmworkers.org)
Fedco Seeds and Fedco Trees (www.fedcoseeds.com)
Federation of Southern Cooperatives (www.federationsoutherncoop.com); Federation of Southern Cooperatives Rural Training & Research Center (www.federationsoutherncoop.com/fscprograms/rtccenter.htm)
Field & Forest Products (www.fieldforest.net)
Flats Mentor Farm (www.worldfarmers.org/programs/#flats)
Food Chain Workers Alliance (foodchainworkers.org)
Food Hub Collaboration (www.wallacecenter.org/foodhubcollaboration)
Food Project, The (thefoodproject.org)
Food Sovereignty Ghana (foodsovereigntyghana.org)
Foot Print Farms (footprintfarmsms.com)
Freedom Food Alliance (freedomfoodalliance.wordpress.com)
Fruition Seeds (www.fruitionseeds.com)
Global Village Farms (www.globalvillagefarms.org)
Gold Water Alchemy (www.goldwateralchemy.com)
Good Food Jobs (www.goodfoodjobs.com)
Green Belt Movement (www.greenbeltmovement.org)
Griffin Greenhouse Supplies (www.griffins.com)
Groundswell Center for Local Food and Farming—Incubator Program (groundswellcenter.org)
Grow Where You Are (www.growwhereyouare.farm)
Harmony Homestead & Wholeness Center (unveilwholeness.com)
Harriet's Apothecary (www.harrietsapothecary.com)
Hattie Carthan Farm (www.hattiecarthancommunitymarket.com)
HEAL Food Alliance (healfoodalliance.org)
Herbal Tea House (greengreenscrub.com)
High Mowing Organic Seeds (www.highmowingseeds.com)
Homowo African Heritage Seed Collection at Old Salem Museums and Gardens (www.oldsalem.org/garden/gardens-landscape/african-american-seed-collection)
Hudson Valley Farm Hub ProFarmer Training Program (hvfarmhub.org/programs/farmer-training)
Hudson Valley Seed Company (hudsonvalleyseed.com)
Indigenous Remedies (irherbs.com)
Indigenous Seed Initiative (www.indigenousseedinitiative.org)
Interaction Institute for Social Change (www.interactioninstitute.org)
International Coalition to Commemorate African Ancestors of the Middle Passage (ICCAAMP) (www.remembertheancestors.com)
International Indigenous Youth Council (indigenousyouth.org)
Intervale Center (www.intervale.org)
Johnny's Selected Seeds (www.johnnyseeds.com)
Kiva (www.kiva.org)
Lakeview Organic Grain (lakevieworganicgrain.com)
Land Loss Prevention Project (www.landloss.org)
Land Stewardship Project (landstewardshipproject.org)
Land Trust Alliance Directory (findalandtrust.org)
Lawyers' Committee for Civil Rights Under Law (lawyerscommittee.org)
Líderes Campesinas (www.liderescampesinas.org/english)

Malcolm X Grassroots Movement (https://mxgm.org)
Mayflor Farms (mayflorfarms.com)
MESA (mesaprogram.org)
mindbodygreen (www.mindbodygreen.com)
Minnesota Food Association (www.mnfoodassociation.org)
Minnesota Food Association Big River Farms Farmer Education Program (www.mnfoodassociation.org/training-program)
Movement for Black Lives (policy.m4bl.org)
Movement for Justice in El Barrio (www.facebook.com/Movement-for-Justice-in-El-Barrio-54775959685)
Movement Ground Farm (movementgroundfarm.wordpress.com)
Mudbone Farm (www.facebook.com/pg/MudBoneFamilyFarm)
Natasha Bowens's Color of Food Map (thecolorofood.com/cof-map)
National Black Farmers Association (www.nationalblackfarmersassociation.org); Let's Get Growing Program
National Black Food & Justice Alliance (www.blackfoodjustice.org)
National Black Growers Council (nationalblackgrowerscouncil.com)
National Community Land Trust Network (cltnetwork.org)
National Coordination of Peasant Organizations of Mali (Coordination Nationale des Organisations Paysannes) (www.cnop-mali.org)
National Credit Union Administration (www.mycreditunion.gov/about-credit-unions/pages/how-to-start-a-credit-union.aspx)
National Domestic Workers Alliance (www.domesticworkers.org)
National Hmong American Farmers (www.nhaf.org)
National Incubator Farm Training Initiative (nesfp.org/national-and-state-networks/national-incubator-farm-training-initiative)
National Sustainable Agriculture Coalition (sustainableagriculture.net)
National Women in Agriculture Association (www.nwiaa.org)
National Young Farmers Coalition (www.youngfarmers.org); Land Link directory (www.youngfarmers.org/landlinks)
Natural Choices Botanica School of Herbalism & Holistic Health (www.naturalchoicesbotanica.com/school.html)
New Entry Sustainable Farming Project (nesfp.org)
Nolt's Produce Supplies (www.noltsproducesupplies.net)
North East Community Center (NECC) (www.neccmillerton.org)
Oldways Preservation Trust (oldwayspt.org)
Olympia Food Co-op (olympiafood.coop)
Operation Spring Plant (operationspringplant.blogspot.com)
Organic Seed Alliance (seedalliance.org)
Osain Yoruba Herbal Medicine, Nile River Medicine (www.nilevalleymedicine.net/products/yoruba-course-in-herbal-medicine)
People's Institute for Survival and Beyond (www.pisab.org)
PODER: People Organized in Defense of Earth and Her Resources (www.poder-texas.org)
Proceso de Comunidades Negras (Black Communities Process, PCN) (renacientes.net)
Queen Afua Wellness Center (queenafua.com)
Queering Herbalism, Herbal Freedom School (queerherbalism.blogspot.com)
Rain-Flo Irrigation (www.rainfloirrigation.com)
Real Seed Catalogue (www.realseeds.co.uk/seedcleaner.html)
Restaurant Opportunities Centers United (rocunited.org)
RID-ALL Green Partnership (www.greennghetto.org)
Rock Steady Farm & Flowers (www.rocksteadyfarm.com)
Rocky Acres Community Farm (www.facebook.com/rockyacrescommunityfarm)
Rooftop Gardens Canada (www.alternatives.ca or www.rooftopgardens.ca)
Roots of Resistance (www.instagram.com/rootsofresistance)
Rootwork Herbals (www.rootworkherbals.com/herbalism-school)

Sacred Roots Wellness (www.sacredrootswellness.earth)
Sacred Vibes Apothecary (www.sacredvibeshealing.com)
SCORE Business Plan Resources (www.score.org/content/business-plan-resources)
Seed Farm, Pennsylvania (www.theseedfarm.org/farm-incubator)
Seed Keepers (tierranegrafarms.org/seed-keepers)
Seed Savers Exchange (www.seedsavers.org)
Seedlibraries.org (communityseednetwork.org)
Seeds for Change UK (www.seedsforchange.org.uk)
Sierra Seeds (sierraseeds.org)
Sitting Bull College (sittingbull.edu)
Soilful City, Washington DC (soilfulcity.com)
Soul Fire Farm (www.soulfirefarm.org); Soul Fire Farm reparations map (www.soulfirefarm.org/support/reparations)
Soul Flower Farm (www.soulflowerfarm.com)
Southeast Wise Women (www.sewisewomen.com)
Southeastern African American Farmers' Organic Network (SAAFON) (www.saafon.org)
Southern Exposure Seed Exchange (www.southernexposure.com)
Southern Poverty Law Center (www.splcenter.org)
Southern University Agricultural Research and Extension Center (SUAREC) (http://sualgc.susenterprisecms.com/page/small-farmers)
Southside Community Land Trust (www.southsideclt.org)
Sow True Seed (sowtrueseed.com)
Standing Rock Sioux (www.standingrock.org)
Stinging Nettle (www.thestingingnettlesonoma.com)
Third Root Community Health Center (thirdroot.org)
Three Part Harmony Farm (threepartharmonyfarm.org)
Training for Change (www.trainingforchange.org)
Truelove Seeds (trueloveseeds.com)
Truly Living Well (trulylivingwell.com)
Tuskegee University (www.tuskegee.edu)
Tuzini Farms (www.facebook.com/tuzinifarms)
Urban Agricultural Legal Resource Library (a project of the Sustainable Economies Law Center) (www.urbanaglaw.org)
Urban Farm Institute (urbanfarminginstitute.org)
Urban Growers Collective (urbangrowerscollective.org)
US Food Sovereignty Alliance (usfoodsovereigntyalliance.org)
USDA 1890 National Scholars Program (www.outreach.usda.gov/education/1890)
USDA Census of Agriculture (www.agcensus.usda.gov)
USDA Office of the Assistant Secretary for Civil Rights (https://www.ascr.usda.gov/complaint-resolution)
Via Campesina, La; International Peasant Movement (viacampesina.org)
White Noise Collective (www.conspireforchange.org)
Wildseed Community Farm and Healing Village (www.wildseedcommunity.org)
Women of Color in Solidarity (www.wocsolidarity.org)
Working World, The (www.theworkingworld.org)
Yisrael Family Farm (yisraelfamilyfarm.net)

NOTES

Introduction: Black Land Matters

1. Jeffrey Jordan et al., *Land and Power: Sustainable Agriculture and African Americans* (College Park, MD: Sustainable Agriculture Research and Education, 2007), www.sare.org/content/download/50650/665630/file/landandpower.pdf.
2. Booker T. Whatley, *Whatley's Handbook on How to Make $100,000 Farming 25 Acres: With Special Plans for Prospering on 10 to 200 Acres* (Kutztown, PA: Rodale Institute, 1987).
3. *Arc of Justice: The Rise, Fall, and Rebirth of a Beloved Community*, Open Studio Productions, 2016, http://openstudioproductions.com/store/arc-of-justice-the-rise-fall-and-rebirth-of-a-beloved-community.
4. Michelle Ver Ploeg et al., *Access to Affordable and Nutritious Food: Measuring and Understanding Food Deserts and Their Consequences*, report to Congress for the USDA Economic Research Service (Collingdale, PA: Diane Publishing, 2009).
5. Maulana Karenga, "Principles and Practices of Kwanzaa: Repairing and Renewing the World," *Los Angeles Sentinel*, December 24, 2009, http://www.officialkwanzaawebsite.org/documents/PrinciplesandPracticesofKwanzaa_000.pdf.
6. Jess Gilbert, Gwen Sharp, and M. Sindy Felin, "The Loss and Persistence of Black-Owned Farms and Farmland: A Review of the Research Literature and Its Implications," *Southern Rural Sociology* 18, no. 2 (2002): 1–30, http://journalofruralsocialsciences.org/pages/Articles/SRS%202002%2018/2/SRS%202002%2018%202%201-30.pdf.
7. Mark D. Hersey, *My Work Is That of Conservation: An Environmental Biography* (Athens: University of Georgia Press, 2011).
8. Y. Y. Liu, *Good Food and Good Jobs for All: Challenges and Opportunities to Advance Racial and Economic Equity in the Food System* (New York: Race Forward, 2012), https://www.raceforward.org/research/reports/food-justice.
9. Toni Morrison, *Song of Solomon* (New York: Alfred A. Knopf, 1977).

Chapter 1: Finding Land and Resources

1. Meissha Thomas et al., "What Is African American Land Ownership?," Federation of Southern Cooperatives Land Assistance Fund, 2004, http://www.federationsoutherncoop.com/aalandown04.htm; *2012 Census of Agriculture: Black Farmers* (Washington, DC: USDA National Agricultural Statistics Service, 2014), https://www.agcensus.usda.gov/Publications/2012/Online_Resources/Highlights/Black_Farmers/Highlights_Black_Farmers.pdf.
2. Dara Cooper, *Reframing Food Hub Discourse: Food Hubs, Racial Equity, and Self-Determination in the South* (New York: Center for Social Inclusion, 2006), https://www.centerforsocialinclusion.org/publication/reframing-food-hubs.
3. John Emmeus Davis, "The Backstory: Historical Background for Events Featured in the *Arc of Justice*," in *Arc of Justice: Rise, Fall and Rebirth of a Beloved Community*, Open Studio Productions, 2016, https://static1.squarespace.com/static/574610bb20c647ad3c3c2daf/t/583c8c659de4bb595172a9b2/1480363113503/AoJ_backstory_v4a.pdf.
4. "Conservation," Black Family Land Trust, http://www.bflt.org/conservation.html.
5. *Finding Farmland: A Farmer's Guide to Working with Land Trusts* (Hudson, NY: National Young Farmers Coalition, 2015), http://www.youngfarmers.org/wp-content/uploads/2015/01/NYFC-Finding-Affordable-Farmland.pdf.
6. Jessica Gordon Nembhard, *Collective Courage: A History of African American Cooperative Economic Thought and Practice* (University Park: Pennsylvania State University Press, 2014).
7. "Preserving Farms for Farmers," Equity Trust, http://equitytrust.org/farms-for-farmers.
8. *2012 Census of Agriculture: Black Farmers.*
9. "About Us," California FarmLink, http://www.californiafarmlink.org/about-us.
10. "National Incubator Farm Training," New Entry Sustainable Farming Project, https://nesfp.org/food-systems/national-incubator-farm-training-initiative.

11. Tasha M. Hargrove et al., "A Case Study Analysis of a Regional Food System: The Sustainable Agriculture Consortium for Historically Disadvantaged Farmers Program," *Professional Agricultural Workers Journal* 1, no. 2 (2014): 1–11.
12. Appiah Sarpong and Michael Addusei, "Multinomial Logistic Analysis of 'Susu' Contribution in Ghana," *Journal of Economic and Social Development* 1, no. 1 (2014): 96–105.
13. Joelle Cruz, "Memories of Trauma and Organizing: Market Women's Susu Groups in Postconflict Liberia," *Organization* 21, no. 4 (2014): 447–62.
14. *Growing Opportunity: A Guide to USDA Sustainable Farming Programs* (Washington, DC: National Sustainable Agriculture Coalition, 2017), http://sustainableagriculture.net/wp-content/uploads/2017/02/FSA-Guide-Final.pdf.
15. "Minority and Women Farmers and Ranchers," USDA Farm Service Agency, https://www.fsa.usda.gov/programs-and-services/farm-loan-programs/minority-and-women-farmers-and-ranchers/index.
16. *Guide to Farming in New York State: What Every Ag Entrepreneur Needs to Know* (Ithaca, NY: Cornell Small Farms Program, 2015), https://ecommons.cornell.edu/handle/1813/40229.
17. Tracy Dunn and Jeff Neumann, "40 Acres and a Mule Would Be at Least $6.4 Trillion Today—What the US Really Owes Black America," *Yes! Magazine*, May 14, 2015, http://www.yesmagazine.org/issues/make-it-right/infographic-40-acres-and-a-mule-would-be-at-least-64-trillion-today.
18. "Supporting Black-Led, Black Liberation," Resource Generation, http://resourcegeneration.org/what-we-do/supporting-black-led-black-liberation.
19. Nembhard, *Collective Courage*.
20. Tim O'Brien, "Albany's Historic Rapp Road Neighborhood Has Roots in Southern Migration," *Times Union*, August 25, 2015.

Chapter 2: Planning Your Farm Business

1. Dennis Austin and Robin Luckham, *Politicians and Soldiers in Ghana 1966–1972* (London: Frank Cass, 1975).
2. Charles Reagan Wilson, "Mississippi Rebels: Elvis Presley, Fannie Lou Hamer, and the South's Culture of Religious Music," *Southern Quarterly* 50, no. 2 (2013): 9–30.
3. "Fannie Lou Hamer Founds Freedom Farm Collective," Student Nonviolent Coordinating Committee, 2017, https://snccdigital.org/events/fannie-lou-hamer-founds-freedom-farm-cooperative.
4. Jessica Gordon Nembhard, *Collective Courage: A History of African American Cooperative Economic Thought and Practice* (University Park: Pennsylvania State University Press, 2014).
5. Oluwo Ifakolade Obafemi, *Ile Ifa International: Orunmila's Healing Spaces* (Xlibris US, 2011).
6. Margaret Lund, *Solidarity as a Business Model: A Multi-Stakeholder Cooperatives Manual* (Kent, OH: Cooperative Development Center at Kent State University, 2011), http://www.uwcc.wisc.edu/pdf/multistakeholder%20coop%20manual.pdf.
7. Booker T. Whatley, "The Small Farm Plan by Booker T. Whatley," interview in *Mother Earth News*, May–June 1982, http://www.motherearthnews.com/homesteading-and-livestock/small-farm-plan-zmaz82mjzkin?PageId=7.
8. Barbara Seeber, "The Producer," *Utne Reader*, October–November 1984, 100.
9. Alyssa Battistoni, "America Spends Less on Food than Any Other Country," *Mother Jones*, February 1, 2012, http://www.motherjones.com/food/2012/02/america-food-spending-less.
10. Alana Rhone et al., *Low-Income and Low-Supermarket-Access Census Tracts, 2010–2015* (Washington, DC: USDA Economic Research Service, 2017), https://www.ers.usda.gov/webdocs/publications/82101/eib-165.pdf?v=42752.
11. Dara Cooper, *Reframing Food Hub Discourse: Food Hubs, Racial Equity, and Self-Determination in the South* (New York: Center for Social Inclusion, 2006), https://www.centerforsocialinclusion.org/publication/reframing-food-hubs.
12. Dennis Derryk, personal communication to the author, 2016.
13. *Regional Food Hub Resource Guide* (Washington, DC: USDA Agricultural Marketing Service, 2012), https://www.ams.usda.gov/sites/default/files/media/Regional%20Food%20Hub%20Resource%20Guide.pdf.
14. Marylynn Steckley, "Eating Up the Social Ladder: The Problem of Dietary Aspirations for Food Sovereignty," *Agriculture and Human Values* 33, no. 3 (2016): 549–62.
15. Christian N. Vannier, "Rational Cooperation: Situating Konbit Labor Practice in Context," *Journal of Haitian Studies* 15, no. 1–2 (2009): 333–49.
16. Nikki Giovanni, *Racism 101* (New York, NY: Quill, 1995), 154–55.

17. Dana Martin and Melissa Fery, *Growing Farms: Successful Whole Farm Management Planning Book: Think It! Write It!* (Corvallis: Oregon State University Extension Service, 2011).

Chapter 3: Honoring the Spirits of the Land

1. Alik Shahadah, "Religions in Africa Today," African Holocaust Society, March 23, 2017, http://www.africanbelief.com.
2. Wande Abimbola, "Religion, World Order, and Peace: An Indigenous African Perspective," *Cross Currents* 60 (2010): 307–09.
3. "Ifa Divination System," United Nations Education, Scientific and Cultural Organization, https://ich.unesco.org/en/RL/ifa-divination-system-00146.
4. Ayò Salami, *Ifá: A Complete Divination* (Lagos: NIDD Publishing and Printing, 2002).
5. John Henley, "Haiti: A Long Descent into Hell," *Guardian*, January 14, 2010, https://www.theguardian.com/world/2010/jan/14/haiti-history-earthquake-disaster.
6. Asia Austin Colter, "Azaka, the Loa," in *Encyclopedia of African Religion*, vol. 1, ed., Molefi Kete Asante and Ama Mazama (Thousand Oaks, CA: SAGE Publications, 2009), 82–83.
7. Charlotte Hammond, "'Children' of the Gods: Filming the Private Rituals of Haitian Vodou," *Journal of Haitian Studies* 18 (2012): 64–82.
8. Jacob Kẹhinde Olupona and Terry Rey, *Òrìṣà Devotion as World Religion: The Globalization of Yorùbá Religious Culture* (Madison: University of Wisconsin Press, 2008).
9. Marcel Carty. *Vodou: The Next Stage* (Xlibris US, 2010).
10. Awo Olumide Achaba, *Orisha Oko—Orisha of Fertility, Progress and Evolution: A Guide to Accessing the Divinity's Blessings* (Amazon Digital Services, LLC, 2014).
11. Milo Rigaud, *Secrets of Voodoo* (San Francisco: City Light Books, 1969).
12. Terry Rey and Alex Stepick, *Crossing the Water and Keeping the Faith: Haitian Religion in Miami* (New York: NYU Press, 2013).
13. Boakye Agyarko, "The Manya Krobo Annual Ngmayem Festival," October 30, 2015, https://www.modernghana.com/news/652710/the-manya-krobo-annual-ngmayem-festival.html.
14. Rose Mary Allen, "The Harvest Ceremony Seú as a Case Study of the Dynamics of Power in Post-Emancipation Curaçao (1863–1915)," *Caribbean Quarterly* 56 (2010): 13–29.
15. Jeffrey E. Anderson, *The Voodoo Encyclopedia: Magic, Ritual, and Religion* (Santa Barbara, CA: ABC-CLIO, 2015).
16. Fred Opie, *Zora Neale Hurston on Florida Food: Recipes, Remedies, and Simple Pleasures* (Mount Pleasant, SC: Arcadia Publishing, 2015).
17. "Carol Beckwith and Angela Fisher African Ceremonies Collection," JSTOR World Heritage Sites Africa, 1980–1999, https://www.aluka.org/heritage/collection/BFAC.
18. Benjamin Hebbkethwaite and Joanne Bartley, *Vodou Songs in Haitian Creole and English* (Philadelphia: Temple University Press, 2011).
19. Hebbkethwaite and Bartley, *Vodou Songs*.
20. Hebbkethwaite and Bartley, *Vodou Songs*.
21. "Slave Work Songs," Colonial Williamsburg, http://www.history.org/history/teaching/enewsletter/february03/worksongs.cfm; *Hoe, Emma, Hoe*, video, 1:17, https://www.youtube.com/watch?v=SIoWRVE-H58.
22. Zora Neale Hurston, *Halimufack*, audio, 2:07, 1939, https://www.loc.gov/item/flwpa000014.
23. Ann Hoog, *Zora Neale Hurston: Recordings, Manuscripts, Photographs, and Ephemera*, in the Archive of Folk Culture and Other Divisions of the Library of Congress, 2014, https://www.loc.gov/folklife/guides/Hurston.html.
24. Toshi Reagon, *A Sower Went Out to Sow Her Seed: Parable of the Sower—An Opera*, video, 2016, https://vimeo.com/158102311.
25. Jephté Guillaume, *Ibo Lele*, audio, 7:18, http://music.spirituallifemusic.net/track/ibo-lele.

Chapter 4: Restoring Degraded Land

1. Steven E. Hollinger, *Midwestern Climate Center Soil Atlas and Database* (Champaign: Illinois State Water Survey, 1995), http://www.isws.illinois.edu/pubdoc/c/iswsc-179.pdf.
2. "370.8 Ranking of Mineral Soils," in *Official Compilation of Codes, Rules and Regulations of the State of New York* (Toronto: Thomson Reuters Westlaw, 2017).
3. "Lead at Superfund Sites: Human Health," US Environmental Protection Agency, 2016, https://www.epa.gov/superfund/lead-superfund-sites-human-health.
4. M. R. O'Connor, "The World's Favorite Disaster Story," *Vice News*, October 13, 2016, https://news.vice.com/story/one-of-the-most-repeated-facts-about-deforestation-in-haiti-is-a-lie.
5. Brian Wheeler, "Gullah Geechee: Descendants of Slaves Fight for Their Land," *BBC News*, December 5, 2016, http://www.bbc.com/news/magazine-37994938.

6. "Hazard Standards for Lead in Paint, Dust and Soil (TSCA Section 403)a," US Environmental Protection Agency, May 8, 2017.
7. Michael V. Mickelbart, Kelly M. Stanton, Steve Hawkins, and James Camberato, "Commercial Greenhouse and Nursery Production," Purdue University, 2017.
8. Andreas D. Peuke and Heinz Rennenberg, "Phytoremediation," *Science and Society* 6, no. 6 (2005): 497–501, doi: 10.1038/sj.embor.7400445.
9. David E. Stilwell and John F. Ranciato, *Use of Phosphates to Immobilize Lead in Community Garden Soils* (New Haven: Connecticut Agricultural Experiment Station, 2008), http://www.ct.gov/caes/lib/caes/documents/publications/bulletins/b1018.pdf.
10. Carl J. Rosen and Peter M. Bierman, "Using Manure and Compost as Nutrient Sources for Fruit and Vegetable Crops," University of Minnesota Extension, https://www.extension.umn.edu/garden/fruit-vegetable/using-manure-and-compost.
11. Emmanuel Krieke, *Environmental Infrastructure in African History: Examining the Myth of Natural Resource Management in Namibia* (New York: Cambridge University Press, 2013).
12. Edda Fields-Black, *Deep Roots: Rice Farmers in West Africa and the African Diaspora* (Bloomington: Indiana University Press, 2008).
13. *Lead-Safe Yard Manual: A Do-It-Yourself Guide to Low-Cost Soil Remediation and Safe Greening* (Worcester, MA: Worcester Roots Project, 2017), http://www.worcesterroots.org/wp-content/uploads/2011/08/DIY_leadsafe_landscapingFinalDraft7-29-11sm.pdf.
14. Mark D. Hersey, *My Work Is That of Conservation: An Environmental Biography* (Athens: University of Georgia Press, 2011), 126.
15. Julianna White, "Terracing Practice Increases Food Security and Mitigates Climate Change in East Africa," Research Program on Climate Change, Agriculture and Food Security, 2016, https://ccafs.cgiar.org/blog/terracing-practice-increases-food-security-and-mitigates-climate-change-east-africa#.WJTUEhsrLIU.
16. Mark Hertsgaard, *Hot: Living Through the Next 50 Years on Earth* (New York: Mariner Books, 2011).
17. Diane Toomey, "Exploring How and Why Trees Talk to Each Other," *Yale Environment 360*, 2016, https://e360.yale.edu/features/exploring_how_and_why_trees_talk_to_each_other.
18. Jack Kittredge, *Soil Carbon Remediation: Can Biology Do the Job?* (Barre, MA: Northeast Organic Farming Association/Massachusetts Chapter, 2015), http://www.nofamass.org/sites/default/files/2015_White_Paper_web.pdf.
19. Mark Schonbeck, "What Is 'Organic No-Till,' and Is It Practical?," *eOrganic*, July 20, 2015.
20. Mark Schonbeck and Ron Morse, "Choosing the Best Cover Crops for Your Organic No-Till Vegetable System: A Detailed Guide to Using 29 Species," Rodale Institute, January 29, 2004, http://www.newfarm.org/features/0104/no-till/chart.shtml.

Chapter 5: Feeding the Soil

1. C. V. Cole et al., "Analysis of Historical Changes in Soil Fertility and Organic Matter Levels of the North American Great Plains," in *1989 Soils and Crops Workshop Proceedings* (Saskatoon, Canada: University of Saskatchewan, 1989).
2. K. W. Flach et al., "Impact of Agriculture on Atmospheric CO_2," in *Soil Organic Matter in Temperate Agroecosystems: Long-Term Experiments in North America*, ed. E. A. Paul et al. (Boca Raton, FL: CRC Press, 1997).
3. Ann Lewandowski, "Organic Matter Management," University of Minnesota Extension, 2002, https://www.extension.umn.edu/agriculture/soils/soil-properties/soil-management-series/organic-matter-management.
4. "Guide to Texture by Feel," USDA Natural Resources Conservation Service, Soils, https://www.nrcs.usda.gov/wps/portal/nrcs/detail/soils/edu/?cid=nrcs142p2_054311.
5. A. Jones et al., *Soil Atlas of Africa* (Luxemburg: Publications Office of the European Union, 2013), http://eusoils.jrc.ec.europa.eu/Library/Maps/Africa_Atlas/Download/49.pdf; Christien Ettema, *Indigenous Soil Classifications: What Is Their Structure and Function, and How Do They Compare to Scientific Soil Classifications?* (Athens: University of Georgia, 1994), http://www.css.cornell.edu/faculty/dgr2/Docs/Misc/IntroToEthnopedology.pdf.
6. James Fairhead et al., "Indigenous African Soil Enrichment as a Climate-Smart Sustainable Agriculture Alternative," *Frontiers in Ecology and the Environment* 14, no. 2 (2016): 71–76.
7. Paul Yeboah, email to research assistant, August 4, 2017.
8. Mae-Wan Ho, "Beware the Biochar Initiative," Permaculture Research Institute, November 18, 2010, https://permaculturenews.org/2010/11/18/beware-the-biochar-initiative.

9. Tom DeGomez et al., "Basic Soil Components," *eXtension*, January 21, 2015, http://articles.extension.org/pages/54401/basic-soil-components.
10. Jerry Minnich, *The Earthworm Book: How to Raise and Use Earthworms for Your Farm and Garden* (Emmaus, PA: Rodale Press, 1977).
11. Mohammadtaghi Vakili et al., "A Review on Composting of Oil Palm Biomass," *Environment, Development and Sustainability* 17, no. 4 (2015): 691–709.
12. Justice Cudjoe, email to research assistant, August 4, 2017
13. Demalda Newsome, personal communication to the author, February 5, 2017.
14. Lorinda Balfanz et al., "Vermicompost Tea," University of Minnesota Extension, http://www.extension.umn.edu/garden/yard-garden/soils/vermicompost-tea.
15. Thomas Björkman, "NY Cover Crop Guide," Cornell University College of Agricultural and Life Sciences, http://covercrops.cals.cornell.edu.
16. Mark D. Hersey, *My Work Is That of Conservation: An Environmental Biography* (Athens: University of Georgia Press, 2011).
17. Paul Yeboah, email to research assistant, August 4, 2017.

Chapter 6: Crop Planning

1. Robin Wall Kimmerer, *Braiding Sweetgrass: Indigenous Wisdom, Scientific Knowledge, and the Teachings of Plants* (Minneapolis: Milkweed Editions, 2013), 130.
2. James McCann, *Maize and Grace: Africa's Encounter with a New World Crop, 1500–2000* (Cambridge, MA: Harvard University Press, 2005), xiii, 289.
3. "Leading Causes of Death in Males, 2014," Centers for Disease Control and Prevention, 2014, https://www.cdc.gov/healthequity/lcod/men/2014/black/index.htm.
4. Judith Carney, *Black Rice: The African Origins of Rice Cultivation in the Americas* (Cambridge, MA: Harvard University Press, 2001).
5. Logan Kistlera et al., "Transoceanic Drift and the Domestication of African Bottle Gourds in the Americas," *Proceedings of the National Academy of Sciences of the United States of America* 111, no. 8 (2014).
6. *Homowo Harvest Collection Brochure* (Winston-Salem, NC: Old Salem Museums & Gardens, 2017).
7. G. L. Nesom, "Toward Consistency of Taxonomic Rank in Wild/Domesticated Cucurbitaceae," *Phytoneuron* 13 (2011): 1–33.
8. Catherine Zuckerman, "5 African Foods You Thought Were American," *National Geographic*, September 21, 2016, http://www.nationalgeographic.com/people-and-culture/food/the-plate/2016/09/5-foods-from-africa.
9. Edda Fields-Black, *Deep Roots: Rice Farmers in West Africa and the African Diaspora* (Bloomington: Indiana University Press, 2008).
10. Andrew Smith, *The Tomato in America: Early History, Culture, and Cookery* (Champaign: University of Illinois Press, 2001).
11. Camila Domonoske, "A Legume with Many Names: The Story of 'Goober,'" *NPR Code Switch*, April 20, 2014, https://www.npr.org/sections/codeswitch/2014/04/20/304585019/a-legume-with-many-names-the-story-of-goober.
12. Larisa Jacobson, personal communication to the author, 2018.
13. "Corn Growing Guide," Southern Exposure Seed Exchange, 2017, http://www.southernexposure.com/corn-growing-guide-ezp-52.html.
14. Gretchen Kell, "Millet Project Shows Grain Isn't Just for the Birds," *Berkeley News*, August 28, 2015, http://news.berkeley.edu/2015/08/28/the-millet-project.
15. Olga Linares, "African Rice (*Oryza glaberrima*): History and Future Potential," *Proceeding of the National Academy of Sciences* 99, no. 25 (2002).
16. Michael Twitty, "A Letter to the Newgrorati: Of Collards and Amnesia," *AfroCulinaria*, 2016, https://afroculinaria.com/2016/01/16/a-letter-to-the-newgrorati-of-collards-and-amnesia.
17. "Heirloom Herb & Vegetable Garden," Chesapeake Bay Maritime Museum, http://cbmm.org/pdf/Heirloom_Garden_Handout2012.pdf.
18. Variety information comes from 2017 interviews with Southern Exposure Seed Exchange, D. Landreth Seed Company (https://cottagegardenliving.wordpress.com/2015/02/07/african-american-collection-landreth-heirloom-seeds), Owen Taylor, and Chris Bolden-Newsome.
19. Akinola A. Agboola, "Crop Mixtures in Traditional Systems," United Nations University, 2017, http://archive.unu.edu/unupress/unupbooks/80364e/80364E08.htm.
20. Amanda Stone et al., *Africa's Indigenous Crops* (Washington, DC: Worldwatch Institute, 2011), http://www.worldwatch.org/system/files/NtP-Africa%27s-Indigenous-Crops.pdf.
21. Justice Cudjoe, email to research assistant, August 4, 2017.
22. Paul Yeboah, email to research assistant, August 4, 2017.

23. R. T. Martinez, *An Evaluation of the Productivity of the Native American 'Three Sisters' Agriculture System in Northern Wisconsin* (Stevens Point: University of Wisconsin, 2007).
24. George Kuepper and Mardi Dodson, "Companion Planting and Botanical Pesticides: Concepts & Resources," ATTRA, 2001, updated 2016, https://attra.ncat.org/attra-pub/viewhtml.php?id=72; B. Amoako-Atta, "Observations on the Pest Status of the Striped Bean Weevil Alcidodes leucogrammus Erichs. on Cowpea Under Intercropping Systems in Kenya," *International Journal of Tropical Insect Science* 11, no. 4 (1983): 351–56.
25. B. T. Kang, "Introduction to Alley Farming," Food and Agriculture Organization, 1992, http://www.fao.org/Wairdocs/ILRI/x5545E/x5545e04.htm.

Chapter 7: Tools and Technology

1. Josef Kienzle and Úna Murray, "Tools Used by Women Farmers in Africa," presented at the Food and Agriculture Organization of the United Nations Workshop on Food Security, September 1998.
2. Susan Ferris and Ricardo Sandoval, *The Fight in the Fields: Cesar Chavez and the Farmworkers Movement* (New York: Mariner Books, 1997).
3. "Effectiveness of ROPS for Preventing Injuries Associated with Agricultural Tractors," Centers for Disease Control and Prevention, 1993, https://www.cdc.gov/mmwr/preview/mmwrhtml/00019495.htm.
4. Owusu Bandele, "The Deep Roots of Our Land-Based Heritage: Cultural, Social, Political, and Environmental Implications," in *Land and Power: Sustainable Agriculture and African Americans*, ed. Jeffrey Jordan et al. (Waldolf, MD: SARE, 2009).
5. Remini Boualem et al., "The Foggara: A Traditional System of Irrigation in Arid Regions," *GeoScience Engineering* 60, no. 2 (2014): 32–39.
6. Randy Creswell and Dr. Franklin Martin, *Dryland Farming: Crops & Techniques for Arid Regions* (Fort Myers, FL: Echo, 1998).
7. Angela Lakwete, *Inventing the Cotton Gin: Machine and Myth in Antebellum America* (Baltimore: John Hopkins University Press, 2003).
8. Mary Schons, "African American Inventors II," *National Geographic*, June 21, 2011, https://www.nationalgeographic.org/news/african-american-inventors-19th-century.
9. Patricia Sluby, *The Inventive Spirit of African Americans: Patented Ingenuity* (Santa Barbara, CA: Praeger, 2004).

Chapter 8: Seed Keeping

1. Judith Carney, *Black Rice: The African Origins of Rice Cultivation in the Americas* (Cambridge, MA: Harvard University Press, 2009).
2. Beverly Bell, "Black Farmers Still Valiantly Fighting to Save Their Farms: Black Farmers' Lives Matter: Defending African-American Land and Agriculture in the Deep South," *Black Left Unity*, October 9, 2015, http://www.blunblog.org/2015/10/black-farmers-still-valiantly-fighting.html.
3. "Who Owns Nature? Corporate Power and the Final Frontier in the Commodification of Life," ETC Group, November 12, 2008, http://www.etcgroup.org/content/who-owns-nature.
4. Gloria Kostadinove, "Peasant Farmers Unite to Secure Food Sovereignty," *Borgen Magazine*, April 2, 2014, http://www.borgenmagazine.com/peasant-farmers-unite-secure-food-sovereignty.
5. Jared Metzker, "Haitian Farmers Lauded for Food Sovereignty Work," Food Sovereignty Prize, August 14, 2013, http://foodsovereigntyprize.org/haitian-farmers-lauded-for-food-sovereignty-work.
6. "What Is AFSA?," Alliance for Food Sovereignty in Africa, 2017, http://afsafrica.org/home/what-is-afsa.
7. "Crop Diversity: Use it or Lose It," Food and Agriculture Organization of the United Nations, 2010, http://www.fao.org/news/story/en/item/46803/icode.
8. Chris Bolden-Newsome, interview with the author, 2017.
9. Lee Buttala and Shanyn Siegel, *The Seed Garden: The Art & Practice of Seed Saving* (Decorah, IA: Seed Savers Exchange, 2015).
10. Jill Neimark, "A Lost Rice Variety—And the Story of the Free 'Merikins' Who Kept It Alive," *NPR The Salt*, May 10, 2017, http://www.npr.org/sections/thesalt/2017/05/10/527449714/a-lost-rice-variety-and-the-story-of-the-freed-merikins-who-kept-it-alive.
11. *Seed Saving for Home Use* (Mineral, VA: Southern Exposure Seed Exchange, 2017), http://www.southernexposure.com/growing-guides/saving-seeds-home-use.pdf.
12. Ira Wallace, interview with Juliet Tarantino, 2017.

Chapter 9: Raising Animals

1. Edda Fields-Black, *Deep Roots: Rice Farmers in West Africa and the African Diaspora* (Bloomington: Indiana University Press, 2008).
2. Oumarou Badini et al., "A Simulation-Based Analysis of Productivity and Soil Carbon in Response to

Time-Controlled Rotational Grazing in the West African Sahel Region," *Agricultural Digest* 94, no. 1 (2007): 87–96; Ann Adams, "South Africa and Intensive Short Duration Grazing," HMI, 2010, https://holisticmanagement.org/blog/south-africa-and-intensive-short-duration-grazing.

3. Anne D. Edwards and Donna K. Carver, "Keeping Garden Chickens in North Carolina," North Carolina Cooperative Extension Service, 2008, https://content.ces.ncsu.edu/keeping-garden-chickens-in-north-carolina.
4. Douglas Chambers and Kenneth Watson, *The Past Is Not Dead: Essays from the Southern Quarterly* (Jackson: University Press of Mississippi, 2012).
5. Bjørnstad Mwacharo et al., "The History of African Village Chickens: An Archaeological and Molecular Perspective," *African Archaeological Review* 30, no. 1 (2013): 97–114.
6. Andrew Lawler, "How the Chicken Built America," *New York Times*, November 25, 2014, https://www.nytimes.com/2014/11/26/opinion/how-the-chicken-built-america.html.
7. Leni Sorensen, "In Our Own Time," The Jefferson Monticello, 2005, https://www.monticello.org/site/plantation-and-slavery/our-own-time.
8. Lynn Bliven and Erica Frenay, "New York State: On-Farm Poultry Slaughter Guidelines," Cornell Cooperative Extension, 2012, http://smallfarms.cornell.edu/resources/guides/on-farm-poultry-slaughter-guidelines.
9. Enroue Halfkenny (Babalawo Onigbonna Sangofemi), personal communication to the author, 2016.
10. Franz C. M. Alexander, "Experience with African Swine Fever in Haiti," *Annals of the New York Sciences* 653, no. 1 (1992): 251–56.
11. Jean-Bertrand Aristride, "Globalization and Creole Pigs," *Earth Island Journal* 6, no. 2 (2001): 47.
12. Edward Cody, "Pigs Are Making a Comeback," *Washington Post*, June 24, 1998.
13. Philip Gaertner, "Whether Pigs Have Wings: African Swine Fever Eradication and Pig Repopulation in Haiti," *Stretch*, Fall 1990.
14. Dr. Stuart D. Wills, "The Kingdom of This World," https://msu.edu/~williss2/carpentier/part2/boiscaiman.html.
15. Mark D. Hersey, *My Work Is That of Conservation: An Environmental Biography* (Athens: University of Georgia Press, 2011).
16. Steve Huntzicker, Mahlon Peterson, and Dave Wachter, *Guide to Raising Healthy Pigs* (Madison: University of Wisconsin Extension, 2009), https://learningstore.uwex.edu/Assets/pdfs/A3858-03.pdf; Carol Ekarius, *Small-Scale Livestock Farming: A Grass-Based Approach for Health, Sustainability, and Profit* (North Adams, MA: Storey Publishing, 1999).
17. Sean Powers, "How Working the Land Is Helping US War Veterans to Heal," *Living on Earth*, Public Radio International, October 3, 2016, https://www.pri.org/stories/2016-09-24/can-working-farm-help-war-veterans-heal.
18. Amy Goodman, "North Carolina Hog Farms Spray Manure Around Black Communities; Residents Fight Back," *Democracy Now*, May 3, 2017, https://www.democracynow.org/2017/5/3/nc_lawmakers_side_with_factory_farms.
19. Bryan Walsh, "The Triple Whopper Environmental Impact of Global Meat Production," *Time*, December 16, 2013, http://science.time.com/2013/12/16/the-triple-whopper-environmental-impact-of-global-meat-production.

Chapter 10: Plant Medicine

1. M. Fawzi Mahomoodally, "Traditional Medicines in Africa: An Appraisal of Ten Potent African Medicinal Plants," *Evidence-Based Complementary and Alternative Medicine* (2013), https://www.hindawi.com/journals/ecam/2013/617459.
2. Pierre Lutgen, "Artemisia Against Malaria: Efficient but Banished," *MalariaWorld Journal*, 2015, https://malariaworld.org/blog/artemisia-against-malaria-efficient-banished.
3. Stephanie Mitchem, *African American Folk Healing* (New York: NYU Press, 2007).
4. Peter Burchard, *George Washington Carver: For His Time and Ours* (Washington, DC: National Park Service, US Department of the Interior, 2005).
5. Herbert Covey, *African Slave Medicinal: Herbal and Non-Herbal Treatments* (Lanham, MD: Lexington Books, 2007).
6. Katrina Hazzard-Donald, *Mojo Workin': The Old African American Hoodoo System* (Urbana: University of Illinois Press, 2013).
7. Crystel Aneira, "Herbal Riot," http://herbalriot.tumblr.com.
8. Jerome S. Handler and JoAnn Jacoby, "Slave Medicine and Plant Use in Barbados," *Journal of the Barbados Museum and Historical Society* 41 (1993).
9. Wanda Fontenot, *Secret Doctors: Ethnomedicine of African Americans* (Santa Barbara, CA: Praeger, 1994).

10. Andrew Chevallier, *The Encyclopedia of Medicinal Plants: A Practical Reference Guide to More than 550 Key Medicinal Plants and Their Uses* (London: DK Publishing, 1996).
11. Steven Foster and James A. Duke, *Peterson Field Guide to Medicinal Plants and Herbs of Eastern and Central North America*, 3rd ed. (Boston: Houghton Mifflin Harcourt, 2003).
12. Jessica Houdret, *Practical Herb Garden: A Comprehensive A–Z Directory and Gardener's Guide to Growing Herbs Successfully* (Leicester, U.K.: Anness Publishing, 2003).
13. Steve "Wildman" Brill and Evelyn Dean, *Identifying and Harvesting Edible and Medicinal Plants in Wild (and Not So Wild) Places* (New York: William Morrow, 1994).
14. Colin Fitzgerald, "African American Slave Medicine of the 19th Century," Bridgewater State University, 2016, http://vc.bridgew.edu/cgi/viewcontent.cgi?article=1376&context=undergrad_rev.
15. H. M. Burkill, "Solenostemon monostachyus (P Beauv.) Briq. [family LABIATAE]," Global Plants, JSTOR, http://plants.jstor.org/stable/10.5555/al.ap.upwta.3_63.
16. Liesl Van der Walt, "Mentha Longifolia," South African National Biodiversity Institute, 2004, http://pza.sanbi.org/mentha-longifolia.
17. Maisah B. Robinson and Frank H. Robinson Sr., "Slave Medicine: Herbal Lessons from American History," *Mother Earth Living*, 1998.
18. Robin Wall Kimmerer, *Braiding Sweetgrass: Indigenous Wisdom, Scientific Knowledge, and the Teachings of Plants* (Minneapolis: Milkweed Editions, 2013), 183.
19. Julie Doyle Durway, "Beyond the Railroad," *Appleseeds* 6, no. 7 (2004): 30; Roland Richardson, "From Sea to Shining Sea," *Parks and Recreation* 52, no. 4 (2017): 30–31.
20. Dann J. Broyld, personal communication to the author, July 11, 2017.
21. F. P. Porcher, *Resources of the Southern Fields and Forests, Medical, Economical and Agricultural* (Charleston, SC: Evans and Cogswell, 1863), 2, 411.
22. Steven Foster and James A. Duke, *A Field Guide to Medicinal Plants: Eastern and Central North America* (Peterson Field Guide Series) (Boston: Houghton Mifflin Company, 1990), 200.
23. Sarah Mitchell, "Bodies of Knowledge: The Influence of Slaves on the Antebellum Medical Community" (thesis, Virginia Tech, 1997), http://theses.lib.vt.edu/theses/available/etd-65172149731401/.
24. Hazzard-Donald, *Mojo Workin'*.
25. Robinson and Robinson, "Slave Medicine."
26. Mitchell, "Bodies of Knowledge."
27. Fitzgerald, "African American Slave Medicine."
28. George Washington Carver, "Letter of Carver's to the Montgomery Advertiser of February 14, 1940," in Burchard, *George Washington Carver*.
29. George Washington Carver, "Letter from George Washington Carver to Gertrude Thompson Miller, Huntington, WV, 23 October 1941," in Burchard, *George Washington Carver*, 74.
30. George Washington Carver, "Letter from George Washington Carver to Abbie Bugg, Torpedo, PA, 22 September 1942," in Burchard, *George Washington Carver*, 74.
31. Christine Andreae, "Slave Medicine," Monticello Library exhibit, https://www.monticello.org/library/exhibits/lucymarks/medical/slavemedicine.html.
32. George Brandon, "The Uses of Plants in Healing in an Afro-Cuban Religion, Santeria," *Journal of Black Studies* 22, no. 1 (1991): 55–76; "Hierbas de Oshas y Orishas," Proyecto Orunmila, http://www.proyecto-orunmila.org/hierbas-de-osha-y-orisha.

Chapter 11: Urban Farming

1. *Regaining Ground: Cultivating Community Assets and Preserving Black Land* (New York: Center for Social Inclusion, 2011), http://www.centerforsocialinclusion.org/wp-content/uploads/2014/07/Regaining-Ground-Cultivating-Community-Assets-and-Preserving-Black-Land.pdf.
2. *Homecoming*, video, directed by Charlene Gilbert (PBS, 1999).
3. "Historical Shift from Explicit to Implicit Policies Affecting Housing Segregation in Eastern Massachusetts," Fair Housing Center of Greater Boston, http://www.bostonfairhousing.org/timeline/1934-FHA.html; Federal Housing Administration, *Underwriting Manual: Underwriting and Valuation Procedure Under Title II of the National Housing Act with Revisions to April 1, 1936* (Washington, DC), Part II, Section 2, Rating of Location.
4. Ariana Arancibia, "Mapping the Effects of Redlining and Gentrification on Community Gardens in NYC," 2017, https://drive.google.com/file/d/1N4S8VwomZ6-SaZvZnfo2k1ncweZcuFVP/view.
5. Themis Chronopoulos, "African Americans, Gentrification, and Neoliberal Urbanization: The Case of Fort Greene, Brooklyn," *Journal of African American Studies* 20, no. 3–4 (2016): 294–322; Kristin Reynolds and Nevin Cohen, *Beyond the Kale: Urban Agriculture*

and Social Justice Activism in New York City (Athens: University of Georgia Press, 2016).

6. Hannah Wallace, "Malik Yakini of Detroit's Black Community Food Security Network," *Civil Eats*, December 19, 2011.
7. "Land Access," Urban Agricultural Legal Resource Library, Sustainable Economies Law Center, 2017, http://www.urbanaglaw.org/land-access.
8. Natalie Angier, "Researchers Find a Concentrated Anticancer Substance in Broccoli Sprouts," *New York Times*, September 16, 1997, http://www.nytimes.com/1997/09/16/us/researchers-find-a-concentrated-anticancer-substance-in-broccoli-sprouts.html.
9. C. Clouse, *Farming Cuba: Urban Agriculture from the Ground Up* (New York: Princeton Architectural Press, 2014); André Viljoen and Katrin Bohn, "Scarcity and Abundance: Urban Agriculture in Cuba and the US," *Architectural Design* 82, no. 4 (2012): 16–21.
10. Amélie Germain et al., "Guide to Setting Up Your Own Edible Rooftop Garden. Alternatives and the Rooftop Gardens Project," 2008, www.alternatives.ca or www.rooftopgardens.ca.
11. Sophie Greenbaum, "A Case Study Analysis of a Regional Food System: The Sustainable Agriculture Consortium for Historically Disadvantaged Farmers Program," University of Michigan, 2014, https://detroitenvironment.lsa.umich.edu/coleman-young-his-influence-on-urban-farming.

Chapter 12: Cooking and Preserving

1. "Farm Subsidy Primer," EWG, https://farm.ewg.org/subsidyprimer.php; Brad Plumer, "The $956 Billion Farm Bill in One Graph," *Washington Post*, January 28, 2014, https://www.washingtonpost.com/news/wonk/wp/2014/01/28/the-950-billion-farm-bill-in-one-chart.
2. Roberto A. Ferdman, "The Disturbing Way That Fast Food Chains Disproportionately Target Black Kids," *Washington Post*, November 12, 2014, https://www.washingtonpost.com/news/wonk/wp/2014/11/12/the-disturbing-ways-that-fast-food-chains-disproportionately-target-black-kids.
3. C. Galbete, "Food Consumption, Nutrient Intake, and Dietary Patterns in Ghanaian Migrants in Europe and Their Compatriots in Ghana," *Food and Nutrition Research* 61, no. 1 (2017): 1341809, https://www.ncbi.nlm.nih.gov/pmc/articles/PMC5510194.
4. S. J. O'Keefe, "Fat, Fibre and Cancer Risk in African Americans and Rural Africans," *Nature Communications* 6 (2015): 6342, https://www.ncbi.nlm.nih.gov/pubmed/25919227.
5. Robert Hall, "Africa and the American South: Culinary Connections," *Southern Quarterly* 44, no. 2 (2007): 19–52.
6. John T. Edge, *The Potlikker Papers: A Food History of the Modern South* (New York: Penguin, 2017).
7. "Burundi: Farmer Finds New Technique for Preserving Tomatoes," *Barzawire*, November 28, 2016, http://wire.farmradio.fm/en/farmer-stories/2016/11/burundi-farmer-finds-new-technique-for-preserving-tomatoes-15454.
8. R. D. Pace and W. A. Plahar, "Status of Traditional Food Preservation Methods for Selected Ghanaian Foods," *Food Reviews International* 5, no. 1 (1989): 1–12.
9. Norman F. Haard, "Fermented Cereals: A Global Perspective," Food and Agriculture Organization, 1999, http://www.fao.org/docrep/x2184e/x2184e00.htm#conm.
10. Leanne Brown, *Good and Cheap: Eat Well on $4/Day* (self-published, 2014), https://cookbooks.leannebrown.com/good-and-cheap.pdf.

Chapter 13: Youth on Land

1. Susan Strife and Liam Downey, "Childhood Development and Access to Nature: A New Direction for Environmental Inequality Research," *Organization & Environment* 22, no. 1 (2009): 99–122.
2. Cecily Maller et al., "Healthy Nature Healthy People: 'Contact with Nature' as an Upstream Health Promotion Intervention for Populations," *Health Promotion International* 21, no. 1 (2006): 45–54; Andrea Faber Taylor et al., "Coping with ADD: The Surprising Connection to Green Play Settings," *SAGE Journals* 33, no. 1 (2001): 54–77.
3. Richard Louv, *Last Child in the Woods: Saving Our Children from Nature Deficit Disorder* (New York: Workman Publishing Company, 2005).
4. Lauren Stanforth, "Youth Leader Yusuf Burgess, 64: He Imparted His Love of Nature to Children in Albany," *Times Union*, 2014.
5. "Youth Food Bill of Rights," https://www.youthfoodbillofrights.com.
6. Shirley Williams, "Black Child's Pledge," *The Black Panther*, October 26, 1968.
7. Shani Ealey, "Black Panthers' Oakland Community School: A Model for Liberation," Black Organizing Project, 2016, http://blackorganizingproject.org/black-panthers-oakland-community-school-a-model-for-liberation.

8. Cassady Rosenblum, "At Historic Black Panthers School, Black Teachers Were Key to Student Success," *Oakland North*, December 15, 2016, https://oaklandnorth.net/2016/12/15/at-historic-black-panthers-school-black-teachers-were-key-to-student-success.
9. Taryn Finley, "Colin Kaepernick Just Started a Black Panther–Inspired Youth Camp," *Huffington Post Black Voices*, November 1, 2016, https://www.huffingtonpost.com/entry/colin-kaepernick-black-panther-youth-camp_us_5818b31ee4b0390e69d2935f.
10. "Fast Food FACTS," Robert Wood Johnson Foundation, https://www.rwjf.org/en/library/research/2010/11/fast-food-facts.html.

Chapter 14: Healing from Trauma

1. Ruby Sales, "Where Does It Hurt?," *On Being*, 2016, https://onbeing.org/programs/ruby-sales-where-does-it-hurt.
2. The Bull *Romanus Pontifex* (Nicholas V), January 8, 1455.
3. Jamelle Bouie, "The Atlantic Slave Trade in Two Minutes." *Slate*, June 15, 2015, http://www.slate.com/articles/life/the_history_of_american_slavery/2015/06/animated_interactive_of_the_history_of_the_atlantic_slave_trade.html.
4. "Aboard a Slave Ship, 1829," EyeWitness to History, 2000, www.eyewitnesstohistory.com.
5. Andrea Flynn et al., *Rewrite the Racial Rules: Building an Inclusive American Economy* (New York: Roosevelt Institute, 2016).
6. Ta-Nehisi Coates, "The Case for Reparations," *Atlantic Magazine*, June 2014, https://www.theatlantic.com/magazine/archive/2014/06/the-case-for-reparations/361631.
7. Quote from Roger B. Taney, March 6, 1857.
8. Victor E. Kappeler, "A Brief History of Slavery and the Origins of American Policing," Eastern Kentucky University, http://plsonline.eku.edu/insidelook/brief-history-slavery-and-origins-american-policing.
9. German Lopez, "Police Brutality and Shootings in the US," *Vox*, May 6, 2017, https://www.vox.com/cards/police-brutality-shootings-us/us-police-shootings-statistics.
10. Shorlette Ammons et al., "A Deeper Challenge of Change: The Role of Land Grant Universities in Assessing and Ending Structural Racism in the US Food System," Inter-Institutional Network for Food, Agriculture, and Sustainability, February 22, 2018.
11. Douglas Blackmon, *Slavery by Another Name: The Re-Enslavement of Black Americans from the Civil War to World War II* (New York: Doubleday, 2008).
12. Christopher R. Adamson, "Punishment After Slavery: Southern State Penal Systems, 1865–1890," *Social Problems* 30, no. 5 (1983): 555–69.
13. Henry Louis Gates Jr., "The Truth Behind '40 Acres and a Mule,'" PBS, 2014, http://www.pbs.org/wnet/african-americans-many-rivers-to-cross/history/the-truth-behind-40-acres-and-a-mule.
14. Isabel Wilkerson, *The Warmth of Other Suns: The Epic Story of America's Great Migration* (New York: Penguin Random House, 2011).
15. "Lynching in America: Confronting the Legacy of Racial Terror," Equal Justice Initiative, 2017, https://lynchinginamerica.eji.org.
16. N. L. M. Brown and B. M. Stentiford, eds., *Jim Crow: A Historical Encyclopedia of the American Mosaic* (Santa Barbara, CA: Greenwood, 2014).
17. Todd Lewan and Dolores Barclay, "AP Documents Land Taken from Blacks Through Trickery, Violence, and Murder," Associated Press, 2011, http://nuweb9.neu.edu/civilrights/wp-content/uploads/AP-Investigation-Article.pdf.
18. Thomas Mitchell, "Restoring Hope for Heirs Property Owners: The Uniform Partition of Heirs Property Act," *American Bar Association* 40, no. 1 (2016), https://www.americanbar.org/publications/state_local_law_news/2016-17/fall/restoring_hope_heirs_property_owners_uniform_partition_heirs_property_act.html; Leah Douglas, "African Americans Have Lost Untold Acres of Land Over the Last Century: An Obscure Legal Loophole Is Often to Blame," *Nation*, June 26, 2017.
19. Carol Estes, "Second Chance for Black Farmers: After Decades of Discrimination, Black Farmers Are Struggling for Justice," *Yes! Magazine*, June 30, 2001, http://www.yesmagazine.org/issues/reclaiming-the-commons/second-chance-for-black-farmers.
20. https://www.southernfoodways.org/the-greenwood-food-blockade/
21. Nathan A. Rosenberg and Bryce Wilson Stucki, "The Butz Stops Here: Why the Food Movement Needs to Rethink Agricultural History," *Journal of Food Law and Policy* 13, no. 12 (2017).
22. Pete Daniel, "African American Farmers and Civil Rights," *Journal of Southern History* 73, no. 1 (2007).
23. Allison Alkon, "Paradise or Pavement: The Social Constructions of the Environment in Two Urban Farmers' Markets and Their Implications for Environmental Justice and Sustainability," *Local Environment* 13, no. 3 (2008): 271–89.

24. Pete Daniel, *Dispossession* (Chapel Hill: University of North Carolina Press, 2013), 21.
25. "Historical Shift from Explicit to Implicit Policies Affecting Housing Segregation in Eastern Massachusetts," Fair Housing Center of Greater Boston, http://www.bostonfairhousing.org/timeline/1934-FHA.html; Federal Housing Administration, *Underwriting Manual: Underwriting and Valuation Procedure Under Title II of the National Housing Act with Revisions to April 1, 1936* (Washington, DC), Part II, Section 2, Rating of Location.
26. Brentin Mock, "Redlining Is Alive and Well," CityLab, 2015, https://www.citylab.com/equity/2015/09/redlining-is-alive-and-welland-evolving/407497.
27. Juan F. Pereas, "The Echoes of Slavery: Recognizing the Racist Origins of the Agricultural and Domestic Worker Exclusion from the National Labor Relations Act," *Ohio State Law Journal* 72, no. 1 (2011).
28. Alicia Bugarin and Elias Lopez, *Farm Workers in California* (Sacramento: California Research Bureau, 1998).
29. "Urban Renewal Under Fire," *CQ Researcher*, August 21, 1963, http://library.cqpress.com/cqresearcher/document.php?id=cqresrre1963082100.
30. Dan Baum, "Legalize It All," *Harper's Magazine*, April 2016, https://harpers.org/archive/2016/04/legalize-it-all.
31. Flynn et al., *Rewrite the Racial Rules*.
32. Mariame Kaba, "Juvenile Justice in Illinois: A Data Snapshot," 2014, https://chiyouthjustice.files.wordpress.com/2014/04/juvenile_justice_in_illinois.pdf.
33. Rakesh Kochhar and Richard Fry, "Wealth Inequality Has Widened Along Racial, Ethnic Lines Since End of Great Recession," Pew Research Center, December 12, 2014, http://www.pewresearch.org/fact-tank/2014/12/12/racial-wealth-gaps-great-recession.
34. Sven Beckert and Seth Rockman, eds., *Slavery's Capitalism* (Philadelphia: University of Pennsylvania Press, 2016).
35. Flynn et al., *Rewrite the Racial Rules*.
36. Amy Traub et al., *The Asset Value of Whiteness* (New York: Demos, 2014), http://www.demos.org/publication/asset-value-whiteness-understanding-racial-wealth-gap.
37. Antonio Moore, "Who Owns Almost All America's Land?," *Inequality.org*, 2016, https://inequality.org/research/owns-land.
38. Alison Leff, "Race, Poverty and Hunger," *Poverty & Race*, July–August 2002, http://www.prrac.org/full_text.php?text_id=757&item_id=7796&newsletter_id=63&header=Race+%2F+Racism.
39. Richard L. Zweigenhaft, "Diversity Among CEOs and Corporate Directors: Has the Heyday Come and Gone?," Guilford College, 2013, http://www2.ucsc.edu/whorulesamerica/power/diversity_among_ceos.html.
40. Derwin Dubose, "The Nonprofit Sector Has a Ferguson Problem," *Nonprofit Quarterly*, December 5, 2014, https://nonprofitquarterly.org/2014/12/05/the-nonprofit-sector-has-a-ferguson-problem.
41. Jason Byrne and Jennifer Wolch, "Nature, Race, and Parks: Past Research and Future Directions for Geographic Research," *Progress in Human Geography* 33, no. 6 (2009): 743–65.
42. Susan Strife and Liam Downey, "Childhood Development and Access to Nature: A New Direction for Environmental Inequality Research," *Organization & Environment* 22, no. 1 (2009): 99–122.
43. Jaimee Swift, "It's Not Just Flint: Environmental Racism Is Slowly Killing Blacks Across America," *The Grio*, January 24, 2016, https://thegrio.com/2016/01/24/flint-water-environmental-racism-blacks.
44. Henry Bodkin, "Childhood Trauma Can Be Inherited by Future Generations—New Study," *Telegraph*, November 29, 2017, http://www.telegraph.co.uk/science/2017/11/29/childhood-trauma-can-inherited-future-generations-new-study.
45. Eric J. Nestler, "The Mind's Hidden Switches," November 22, 2011, in *Science Talk*, produced by *Scientific American*, podcast, audio, https://www.scientificamerican.com/podcast/episode/the-minds-hidden-switches-11-11-22.
46. Erika Beras, "Traces of Genetic Trauma Can Be Tweaked," April 15, 2017, in *60-Second Science*, produced by *Scientific American*, podcast, audio, https://www.scientificamerican.com/podcast/episode/traces-of-genetic-trauma-can-be-tweaked.
47. Ron Eyerman, *Cultural Trauma: Slavery and the Formation of African American Identity* (Cambridge, U.K.: Cambridge University Press, 2001).
48. Marlene F. Watson, *Facing the Black Shadow* (self-published, 2013).
49. Henry Louis Gates Jr., "The African Americans: Many Rivers to Cross," video, PBS, http://www.pbs.org/wnet/african-americans-many-rivers-to-cross/video/black-is-beautiful.
50. Reginald, personal communication to the author, 2016.
51. Kamari Maxine Clarke, *Mapping Yorùbá Networks: Power and Agency in the Making of Transnational Communities* (Durham, NC: Duke University Press, 2004).

52. Nicole Monteiro and Diana J. Wall, "African Dance as Healing Modality Throughout the Diaspora: The Use of Ritual and Movement to Work Through Trauma," *Journal of Pan African Studies* 4, no. 6 (2011): 234.
53. *Herb Guide and Vegetable Catalogue* (Goodwood, Canada: Richters, 2018).
54. Catherine Yronwode, *Hoodoo Herb and Root Magic: A Materia Magica of African-American Conjure* (Forestville, CA: Lucky Mojo Curio Company, 2002).

Chapter 15: Movement Building

1. Jess Gilbert et al., *The Decline (and Revival?) of Black Farmers and Rural Landowners: A Review of the Research Literature* (Madison, WI: Land Tenure Center at the University of Wisconsin, 2001), https://ageconsearch.umn.edu/bitstream/12810/1/ltcwp44.pdf.
2. "Food Sovereignty Prize Honors Grassroots Initiatives in Haiti, Brazil, Basque Country, Mali and India," US Food Sovereignty Alliance, August 13, 2013, http://foodsovereigntyprize.org/wp-content/uploads/2013/08/Food-Sov-Prize-Honorees-2013-Press-Release-8-13.pdf.
3. "Services," Black Belt Justice Center, 2012, http://blackbeltjustice.org/services.html.
4. Carol Estes, "Second Chance for Black Farmers: After Decades of Discrimination, Black Farmers Are Struggling for Justice," *Yes! Magazine*, June 30, 2001, http://www.yesmagazine.org/issues/reclaiming-the-commons/second-chance-for-black-farmers.
5. Pete Daniel, "African American Farmers and Civil Rights (*Pigford v. Glickman*)," *Journal of Southern History* (2007), http://www.federationsoutherncoop.com/pigford/AfricanAmericanfarmers.pdf.
6. More information about the cy pres distributions can be found at https://www.blackfarmercase.com.
7. Layla Eplett, "Organic Synthesis: Towards an Inclusion of African Americans in Organic Farming," *Scientific American*, November 5, 2013, https://blogs.scientificamerican.com/food-matters/organic-synthesis-towards-an-inclusion-of-african-americans-in-organic-farming.
8. Addie Louise and Joyner Butler, *The Distinctive Black College: Talladega, Tuskegee, and Morehouse* (Metuchen, NJ: Scarecrow Press, 1977); History of Tuskegee University, https://www.tuskegee.edu/about-us/history-and-mission.
9. "Clipped Wings," *Guardian*, July 26, 2000, https://www.theguardian.com/society/2000/jul/26/guardiansocietysupplement1.
10. "Afro-Descendant Women March in Defense of Life and the Ancestral Territories," Afro-Colombian Solidarity Network, November 16, 2014, https://afrocolombian.org/2014/11/16/afro-descendant-women-march-in-defense-of-life-and-the-ancestral-territories.
11. Kay Bailey, "Women Leaders Bring Indigenous and Black Issues to Colombia Peace Table," *Antyajaa: Indian Journal of Women and Social Change* 1 (2016): 10–18.
12. Wangari Maathai, "Speak Truth to Power (Speech)," Green Belt Movement, May 4, 2000, http://www.greenbeltmovement.org/wangari-maathai/key-speeches-and-articles/speak-truth-to-power.
13. Mara Gittleman, Lenny Librizzi, and Edie Stone, *Community Garden Survey New York City* (New York: GreenThumb and GrowNYC, 2010).
14. Claire Provost, "La Vía Campesina Celebrates 20 Years of Standing Up for Food Sovereignty," *Guardian*, June 17, 2013.
15. "Rationale & Strategy," National Black Food & Justice Alliance, http://www.blackfoodjustice.org/rationale-strategy.
16. Will Higgins, "Indiana's Pre–Civil War Black Farming Community a Smithsonian Surprise," *Indianapolis Star*, 2017.
17. *49th Anniversary 2015–2016 Report* (East Point, GA: Federation of Southern Cooperatives Land Assistance Fund, 2016).
18. Leah Douglas, "African Americans Have Lost Untold Acres of Land Over the Last Century: An Obscure Legal Loophole Is Often to Blame," *Nation*, June 26, 2017.
19. W. Terry Whalin, *Sojourner Truth: Liberated in Christ* (Uhrichsville, OH: Barbour Publishing, 1997).
20. Estes, "A Second Chance for Black Farmers."
21. "Racial Equity in the Farm Bill: Context and Foundations," National Sustainable Agriculture Coalition, December 1, 2017, http://sustainableagriculture.net/blog/racial-equity-in-the-farm-bill/?utm_source=roundup&utm_medium=email.
22. Leah Penniman, "4 Not-So-Easy Ways to Dismantle Racism in the Food System," *Yes! Magazine*, April 27, 2017, http://www.yesmagazine.org/people-power/4-not-so-easy-ways-to-dismantle-racism-in-the-food-system-20170427.
23. "Platform," Movement for Black Lives, https://policy.m4bl.org/platform.
24. Franky Abbott, "The United Farm Workers and the Delano Grape Strike," Digital Public Library of America, https://dp.la/primary-source-sets/the-united-farm-workers-and-the-delano-grape-strike.

25. William F. Holmes, "The Arkansas Cotton Pickers' Strike of 1891 and the Demise of the Colored Farmers' Alliance," *Arkansas Historical Quarterly* 32, no. 2 (1973): 107–19.
26. Sarafina Wright, "Black Farmers Org Still Stands 50 Years Later," *Washington Informer*, November 1, 2017, http://washingtoninformer.com/black-farmers-org-still-stands-50-years-later.

Chapter 16: White People Uprooting Racism

1. "Budget Highlights 2017," NYS Department of Corrections and Community, 2017, https://www.budget.ny.gov/pubs/archive/fy17archive/eBudget1617/agencyPresentations/appropData/CorrectionsandCommunitySupervisionDepartmentof.html.
2. Alex Friedmann, "Lowering Recidivism Through Family Communication," *Prison Legal News*, April 15, 2014, https://www.prisonlegalnews.org/news/2014/apr/15/lowering-recidivism-through-family-communication.
3. Audry Smedley, "The History of the Idea of Race . . . And Why It Matters," presented at the Race, Human Variation and Disease: Consensus and Frontiers Conference, March 14–17, 2007, http://www.understandingrace.org/resources/pdf/disease/smedley.pdf.
4. Nathan A. Rosenberg and Bryce Wilson Stucki, "Rural America's Trump Vote Was Decades in the Making. Democrats' Farm Policies—Starting with the New Deal—Are Partly to Blame," *The New Food Economy*, December 6, 2017, https://newfoodeconomy.org/rural-trump-vote-democrat-farm-policy.
5. Ellen Tuzzolo, personal communication to the author, 2016.
6. Tracy Dunn and Jeff Neumann, "40 Acres and a Mule Would Be at Least $6.4 Trillion Today—What the US Really Owes Black America," *Yes! Magazine*, May 14, 2015, http://www.yesmagazine.org/issues/make-it-right/infographic-40-acres-and-a-mule-would-be-at-least-64-trillion-today.
7. Laura Shin, "The Racial Wealth Gap: Why a Typical White Household Has 16 Times the Wealth of a Black One," *Forbes*, March 26, 2015, https://www.forbes.com/sites/laurashin/2015/03/26/the-racial-wealth-gap-why-a-typical-white-household-has-16-times-the-wealth-of-a-black-one.
8. Sarah Safransky, "Greening the Urban Frontier: Race, Property, and Resettlement in Detroit," *Geoforum* 56 (2014): 237–48.
9. "Role Models," White Noise Collective, https://www.conspireforchange.org/?page_id=7.
10. Amory Starr, *Global Revolt: A Guide to the Movements Against Globalization* (London: Zed Books, 2013).
11. Mel Mariposa, "A Practical Guide to Calling In," The Consent Crew, May 29, 2016, https://theconsentcrew.org/2016/05/29/calling-in.
12. Francis Lee, "Why I've Started to Fear My Fellow Social Justice Activists," *Yes! Magazine*, October 13, 2017, http://www.yesmagazine.org/people-power/why-ive-started-to-fear-my-fellow-social-justice-activists-20171013.
13. Katrina Brown, *Traces of the Trade: A Story from the Deep North* (Washington, DC: PBS, 2008), http://www.tracingcenter.org/library/discussion_guide.pdf.
14. "Truelove Seeds & Owen Taylor," *The Table Underground*, February 9, 2018, https://thetableunderground.com/the-table-underground/2018/2/5/truelove-seeds.
15. bell hooks, *Black Looks: Race and Representation* (Boston: South End Press, 1992), 31.
16. Robin Wall Kimmerer, *Braiding Sweetgrass: Indigenous Wisdom, Scientific Knowledge, and the Teachings of Plants* (Minneapolis: Milkweed Editions, 2013), 20–21.

INDEX

Abelmoschus esculentus. *See* okra
Abimbola, Wande, 56, 183
accommodations for overnight guests, 47
accountability and calling in process, 309–12
Achillea spp. (yarrow), 99, 182, 195, 196
Acmella oleracea, 189
Acorn community, 160
acorns as pig feed, 179
active listening, 277
Adams, Ezra, 119–20
Adams, Lewis, 286
Adams, Victoria, 183
Added Value Farm (NY), 157
adverse possession, 15
advertising on fast foods, 254–55
Africa. *See also specific countries*.
  Alliance for Food Sovereignty in, 150
  dark earth in, 94, 96
  diet in, traditional, 225–26, 233
  fermented grains in, 238
  guinea fowl raised in, 169
  irrigation systems in, 141
  medicine in, traditional, 181–82, 183
  polycultures in, 123, 126
  rotational grazing in, 164
  saving seeds in, 149, 150, 155
  soil testing in, 90
  swidden agriculture in, 127
  tools and equipment in, 129–30, 131, 139, 145
  traumatic events experienced by people from, 265
African Heritage Diet Pyramid, 226, 233
African mini bottle gourd, 108
African power cress, 189
African rice, 105, 116, 149
Afro-Colombian Solidarity Network, 257, 287
Afro-futuristic diasporic cuisine, 228
AfroSeder, 63
Agricultural Justice Project, 297
agricultural training, 16–20. *See also* education
agricultural universities, 19, 285
Agriculture and Land-Based Training Association, 16, 19
agroforestry, 78, 80–84
akara recipe, 230, 231
Akim Abuakwa people, 29
Alabama A&M University, 19
Alexander, Michelle, 282
Alexander, William, 119
alfalfa, 99, 215, 217
algae overgrowth, 92
Allen, Will, 257
alleopathic effects, 101–2, 124, 126
Alliance for Food Sovereignty, 150
Allium spp., 154, 156. *See also* garlic; leeks; onions
Allophylus cominia, 201
aloe, 183
Alpine pennygrass, 74
Alpinia speciosa, 201
aluminum, 91
Alvarez, Gabriela, 228
amaranth, 106, 116, 228
  in polycultures, 121, 125
  recipe for, 233
  saving seeds of, 149, 154, 159
  as traditional food, 225
"Amazing Grace" (song), 105
American Civil Liberties Union, 285
American Farmland Trust, 14
American Legislative Exchange Council, 291
American Medicinal Plants, 191
ancestors and elders
  connection with, in healing from trauma, 275–77
  food knowledge of, 223, 228
  legacy knowledge of, 30
  offerings to, 277
  as spirits of the land, 53–70
Angelou, Maya, 314
animals, 163–80
  cattle, 164, 180
  chickens, 163–76. *See also* chickens
  in Cuba, 216
  greenhouse gases from, 180
  pigs, 59, 164, 176–79
  rotational grazing of, 164
  and sustainability, 179–80
  in urban areas, 207, 209, 216
annual crops, 104–28, 153
anti-racism work, 299–315
  calling in process in, 309–12
  interracial alliances in, 304–6
  organizational transformation in, 306–9
  personal development in, 312–14
  recommended readings for, 315
  reparations in. *See* reparations
  training programs for, 315
  Uprooting Racism in, 59, 282, 306, 309, 311
Apium graveolens. *See* celery
apparel and clothing, 147
  in contaminated soil exposure, 73, 77, 78, 212
apples, 82, 236, 238, 243, 244
appreciation. *See* gratitude and appreciation
apprenticeship programs, 17, 28
appropriation, cultural, 69, 312–14
apricots, 82
Arachis hypogaea. *See* peanuts
arnica, 83
arsenic, 211
Artemisia spp.
  mugwort, 193
  wormwood, 182, 183, 189, 196, 199
Articum spp. (burdock), 192, 193
arugula, 106, 119, 159
asafoetida, 183–84
ash
  composting of, 96
  tomatoes stored in, 237
Ashbourne, Alexander P., 145
ashwagandha, 184
Asteraceae family, 118, 128
  dandelion, 118, 192, 233
  lettuce. *See* lettuce
ATTRA, 17
Azaka, 55, 59–60, 61

Bacon's Rebellion, 300
bacteria
  in fermentation, 238
  in soil, 96, 97, 113
  in water, 212, 214

bagging
  of chicken meat, 175–76
  of produce for delivery, 147
Baker Creek Heirloom Seeds, 139, 161
Balanta, Paulina, 287
Baldwin, James, 270
bambara groundnut, 114
bananas, 62, 123
Bandele, Owusu, 19, 257, 285
Barbosa, Flavio, 283
Barker, Dorathy, 43, 317
Barker, Phillip, 43
basil, 106, 181
  in container gardening, 218
  in crop rotation, 128
  dried, 237
  medicinal uses of, 184
  in polycultures, 124, 181
  saving seeds of, 149
  in square-foot gardening, 220
  starting seeds of, 138, 198
  varieties of, 184
baths, herbal. *See* herbal baths
Batista, Ysanet, 228
beans, 106, 113, 114
  bagged for delivery, 147
  bush, 113, 114
  in container gardening, 218
  in crop rotation, 126, 128
  green, 218, 238, 239
  pole, 113, 116, 125, 142–43, 220
  as power ingredient, 242
  saving seeds of, 152, 153, 154, 155–56, 159
  in square-foot gardening, 220
  in three sisters, 103, 116, 121, 124–26
Beckert, Sven, 271
bedding for chickens, 165, 166, 168
bed former tractor attachment, 132, 134
beds, raised. *See* raised beds
bee balm, 83, 149, 182, 196
beets, 106, 121
  in container gardening, 218
  in crop rotation, 128
  as microgreens, 217
  in pikliz, 239
  in polycultures, 124, 125
  saving seeds of, 153, 154, 159
  in square-foot gardening, 220
  starting seeds of, 121, 138
  storage of, 236
Bell, Herman, 298
Benjamin, Tavia, 257
Ben & Jerry's, 295, 297
Berry, Wendell, 129
Beta vulgaris, 154. *See also* beets; chard
biennials, saving seeds from, 153
bioaccumulators in lead phytoremediation, 74
biochar, 96
bird's-foot trefoil, 102
Black Belt Justice Center, 284
black cherry, 102, 192
The Black Child's Pledge (Williams), 253, 262
Black Codes, 6, 23, 266–267
black cohosh, 184–85
Black Consciousness Movement, 274
black-eyed peas, 113–14
  akara recipe for, 230, 231
  Hoppin' John recipe for, 232–33
  in polycultures, 125
  in urban farming, 208
Black Family Land Trust, 14, 259, 290
Black Farmers Conference, 30, 291
Black Farmers & Urban Gardeners, 2, 3, 286, 287
Black Is Beautiful movement, 274
Black Land & Liberation Initiative, 288
Black Latinx Farmers Immersion, 8, 9, 16
  business planning in, 49
  collards tended in, 119
  graduates of, 8, 290
  harvest tasks in, 17
  herbal baths in, 8, 53
  plant medicine in, 181
  seed planting in, 110
  soil workshops in, 87, 93
  spiritual practices in, 53
  tractor use in, 134
  trauma survivors in, 268
Black Latinx Youth Immersion, 58, 171
Black Left Unity blog, 150
Black Lives Matter, 5, 264
Black Lives Matter Toronto Freedom School, 97
black nightshade, 112, 228
The Black Panther newsletter, 253
Black Panther Party, 255, 262, 298
black plastic mulch, 108, 109
Black Power movement, 274
Black Rice (Carney), 105
Black Urban Growers, 212, 257
black walnut, 102
Black Women's Movement in Defense of Life and the Ancestral Territories, 287
Black Youth Project, 255
Blair, Henry, 145
blueberries, 82, 83, 84
Bobrow-Williams, Sarah, 294
body mass index, 225
bok choy, 106
Bolden-Newsome, Chris, 125–26, 151, 152, 194, 263
boneset, 190, 192, 198
boots, waterproof, 147
borage, 99, 123, 182, 198
boron, 91, 93
bottle gourd, 108
Boukman, Dutty, 59, 177
Bowen, Natasha, 17, 222
boycotts, 284, 295, 296
Boyd, Susie, 149
Braiding Sweetgrass (Kimmerer), 189–90
Brassicaceae family, 118–120. *See also specific brassicas.*
  in crop rotation, 128
  in polycultures, 118, 123, 124
  saving seeds of, 152, 153, 154, 155, 156
  in straw mulch, 111
breeds
  of chickens, 164, 168
  of pigs, 164, 176, 177
broccoli, 106, 118, 119
  in crop rotation, 128
  freezing of, 240
  harvest of, 119, 146
  saving seeds of, 152, 159
  sprouting seeds of, 215
  in square-foot gardening, 220
Bronx Green Up, 212
brooder, chicks in, 164–66, 168
Brown, Michael, 5, 272
Brown, Xavier, 297
brown crowder peas, 113–14
Browne, Katrina, 312
brush cutter tractor attachment, 132, 134
brussels sprouts, 106, 118, 152, 159, 236
buckwheat, 86, 99–100, 215, 217
budget, business plan on, 49–51
buena mulata pepper, 113
buildings
  chicken coops, 166, 167, 168–70, 209
  design of, 129–30
  greenhouses, 110–11, 135–36
  in urban areas, 209

burdock, 192, 193
Burgess, Yusuf, 247
Burkett, Ben, 43, 150, 257, 290
Burkina Faso, zai pits in, 81
burning
  ash from, 96, 237
  biochar from, 96
  in swidden agriculture, 127
Burr, John A., 145
bush beans, 113, 114
bush-fallow system, 127
business laws, 30–33, 285
business plans, 27–52
  budget in, 49–51
  communal labor in, 44–48
  farm-share model in, 38–42
  food hubs in, 42–44
  goals in, 49, 51
  lifestyle issues in, 49
  production decisions in, 51–52
  resources on, 48
  values and mission in, 48–49
  worker-owned cooperative model in, 29–38
  writing of, 48–52
butchering. *See* slaughter of animals
butter beans, 154
Bwa Kayiman, 59

cabbage, 106, 118, 119
  as chicken feed, 168
  in container gardening, 218
  in crop rotation, 128
  fermentation of, 238–39
  marigolds between, 123
  as microgreens, 217
  in pikliz, 239
  recipes for, 233, 235
  saving seeds of, 152, 159
  in square-foot gardening, 220
  storage of, 236
cabbage worms and moths, 118–19
cadmium, 211
calcitic limestone, 92
calcium, 91, 92, 93
calendula, 182
  medicinal uses of, 185, 203
  saving seeds of, 152
  starting seeds of, 198
California Blackeye Pea, 114
California FarmLink, 16
callaloo. *See* amaranth
calling in process, 309–12
canning, 240
cantaloupe, 158, 220
Capsicum spp., 112–13, 154. *See also* peppers
carbon, 57
  in African dark earth, 94
  in forest soil, 82
  in no-till methods, 84
  in swidden agriculture, 127
  in terraces, 78
carbon dioxide levels, atmospheric, 78, 87, 88, 96
Caribbean red pepper, 113
Carney, Judith, 105
Carrasquillo, Nelson, 257
carrots, 106, 117
  in container gardening, 218
  in crop rotation, 128
  dried, 237
  fermentation of, 238
  in pikliz, 239
  in polycultures, 123
  roasted, 235
  saving seeds of, 153, 154, 159
  in square-foot gardening, 220
  storage of, 236
Carter, Asha, 257
Carthan, Hattie, 206
Carver, George Washington, 7, 39, 257, 286
  on acorns as pig feed, 179
  on cover crops, 99, 101
  on cowpeas, 101, 113–14
  on erosion prevention, 78
  on herbal medicine, 183, 188, 192, 194
  on regenerative agriculture, 3
cassava, 60, 123, 237
castor bean, 126, 183
cation exchange capacity, 91, 93–94
cattle, 164, 180
cauliflower, 106, 118, 119
  freezing of, 240
  in pikliz, 239
  saving seeds of, 152, 159
  in square-foot gardening, 220
cayenne peppers, 113
celeriac, 159
celery, 106, 117, 154
  in crop rotation, 128
  dried, 237
  saving seeds of, 154, 159
  in square-foot gardening, 220
  starting seeds of, 117, 138
  storage of, 236
Census of Agriculture of USDA, 21, 293
centering, in calling in process, 310
Cesar (herbalist), 190–91
chamomile, 83, 182, 185–86, 198
Chaney, James, 305
chard, 106, 121
  in container gardening, 218
  in crop rotation, 128
  saving seeds of, 154, 159
  in square-foot gardening, 220
Chavez, Cesar, 130, 257, 296
Chenopodiaceae family, 121, 128. *See also* amaranth; beets; chard; spinach
cherries, 82
  black, 102, 192
chicken coops, 166, 167, 168–70
  in urban areas, 209
chickens, 163–76
  breeds of, 164, 168
  budget for costs related to, 50
  butchering of, 18, 53, 168, 169, 170–76, 179
  for eggs, 164–68, 169
  feed for, 164, 166–68, 170, 180
  floor space requirements for, 166
  illnesses of, 164, 168
  legacy in Black community, 169
  manure of, 74, 166, 168
  for meat, 164, 168–76
  and sustainability, 180
  in urban areas, 207, 209, 216
chicken tractors, 168
chickweed, 203
chicory, 118
children. *See* youth
chiogga beets, 121
chives, 82
chlorine, 93
cilantro, 106, 117–18
  in container gardening, 218
  harvest of, 263
  saving seeds of, 154
  in square-foot gardening, 220
Cimicifuga racemosa, 184–85
citron melon, 108
Citrullus caffer, 108
Citrullus lanatus, 108, 154. *See also* watermelons
civil rights laws, 285
civil rights movement, 7, 13, 264, 268–69, 281
  Black farmers in, 5, 281

USDA settlement in, 13, 20–21, 258–59, 268–69
clay soil, 89, 90, 91, 93, 117
Cleopatra, 97
Clientele Membership Club, 3, 39
climate change, 84
clothing and apparel, 147
in contaminated soil exposure, 73, 77, 78, 212
clover, 99, 100, 124
red, 100, 194–95
sprouting seeds of, 215
tick, 126
white, 100, 124
Coalition of Immokalee Workers, 295, 296, 297
Coates, Ta-Nehisi, 299
codonopsis, 182
coleslaw, 235
"Collage Biographies" activity, 256–57
collards, 104, 106, 116, 118–20
cooking with, 179, 180, 233
in crop rotation, 128
saving seeds of, 152, 159
in square-foot gardening, 220
in urban farming, 208
colon cancer, 225
Colorado potato beetle, 112, 126
Colored Farmers' National Alliance and Cooperative Union, 3, 297
The Color of Food (Bowens), 222
Color of Food Map, 17
color of soil, 90, 94
Combahee River Colony, 15
Comfort Farms, 179
comfrey, 83, 99
communal labor, 44–48
in irrigation, 141
in millet harvest, 62
in terracing, 80
community
dance rituals in, 278–79
marketing in, 39–40
training in, 39–40, 293–94
in urban farming, 222
community gardens, 206, 209, 212–13, 288
Community Seed Network, 161
community-supported agriculture, 3, 27
farm-share model in, 38–42
harvesting produce for, 145
home delivery in, 5, 27–28, 38, 42
loans for, 14
sliding scale in, 27, 41
Whatley on, 3, 19, 39, 286
companion plants, 123–26
for orchard, 82–83
three sisters as, 103, 116, 121, 124–26
compost, 94–96, 101
chicken offal in, 176
for clay soil, 89
forest model of, 95, 98
land-use agreement on, 210
in raised beds, 76–77
in remediation of contamination, 74, 210, 211
in sheet composting, 84–85, 95, 96
tea recipe, 98–99
in urban areas, 210, 216, 221
in vermicomposting, 221
Congress of Racial Equality, 305
consensus for decision making, 38
Conservation Stewardship Program, 22
consultative decision making, 37–38
consumer organizing, 295–97
container gardening, 218–19
contamination
of soil, 72–78, 94, 210–12
of water, 212–13
contractual agreements
on land use, 209–10
laws on, 285
on resource sharing and labor, 33
on temporary ventures, 31
Convolvulaceae family, 116, 121–22, 128. *See also* sweet potatoes
Cooke, Eugene, 15
cooking. *See* food preparation and preservation
CoolBot, 146, 147
cooling produce, 146–47
Cooper, Dara, 13, 291
cooperative economics, 4, 40
cooperatives, worked-owned, 29–38
copper, 91, 93, 211
Coral sorghum, 116
Corbin Hill Food Project, 44
Corchorus olitorius, 115
Coriandrum sativum. *See* cilantro
corn, 103–104, 107, 115–16
in crop rotation, 128
fermentation of, 238
freezing of, 240
in monoculture, 103–4
pollination of, 154, 155
in relay cropping, 123
saving seeds of, 149, 152, 153, 154, 155, 156, 159
spiritual mandate for, 115–16
in square-foot gardening, 220
in three sisters, 103, 116, 121, 124–26
as traditional food, 225
treated seeds of, 150
corn bread recipe, 233–34
corn earworms, 116, 125
Cornelius, Don, 274
Cornell Cooperative Extension, 171
Cornell Small Farms Program, 17
Cornerstone Farm Ventures, 171
cornmeal
offerings of, 53, 60, 65, 199
recipes using, 233–35
as traditional food, 44
cosmos, 123
costs
business plan on, 49–51
of farm share membership, 41–42
financing of. *See* financing
of food, 4–5, 27–28, 40–42, 241
of tools and equipment, 26, 50
of water supply, 214
cotton, 114, 123
picker strike, 297
saving seeds of, 150
and slavery, 114, 265, 271
counseling, peer-to-peer, 277
cover crops, 86, 99–102
in bed preparation, 132
for clay soil, 89, 91
in crop rotation cycle, 128
in terraces, 80
cowhorn okra, 114–15
cowpeas, 101, 113–114
black-eyed. *See* black-eyed peas
in polycultures, 123, 126
in relay cropping, 123
saving seeds of, 154
cows, 164, 180
manure of, 74, 164
Creole black pigs, 59, 164, 177
cress, 215
Crews, John and Elizabeth, 206
criminal justice system, 5–6, 249–52, 271, 300
incarceration in. *See* incarceration
police violence in, 5, 265–66, 272
racism in, 5, 282

crop planning, 103–28
crop rotation, 113, 126–28
peas in, 114, 128
potatoes in, 112, 126, 128
cross-pollination, 152–53
crowdfunding, 24–25
Cuba, 200–201, 216
cucumber beetle, 106
cucumbers, 106–7, 108
in container gardening, 218
in crop rotation, 128
fermentation of, 238
in pikliz, 239
in polycultures, 123
saving seeds of, 154, 156, 158, 159
in square-foot gardening, 220
trellising of, 106–7, 143
Cucumis spp., 108, 154. *See also* cucumbers; melons
Cucurbitaceae family, 105–8. *See also* cucumbers; melons; squash
in crop rotation, 128
gourds, 106, 108, 125, 154, 159
in polycultures, 124
pumpkins, 108, 159, 229
saving seeds of, 154, 159
starting seeds of, 138
transplanting seedlings of, 111
culinary justice movement, 228
Cullors, Patrisse, 292
cultivator for weeding, 144
culture
appropriation of, 69, 312–14
in interracial alliances, 304
and matrix of intersectionality, 314
organizational, transformation of, 307–8
Cureton, Rhyne, 176, 177, 178
curriculum in Youth Food Justice program, 252–62
Curtis, Zachari J., 257
cushaw pumpkin, green striped, 108

dairy products, 225–26
dairy workers, 295, 297
dance rituals, 278–79
dandelion, 118, 192, 233
Daniel, Pete, 269
Daniels, Jonathan, 310
dark earth, African, 94, 96
Datura stramonium, 193
Daucus carota, 154. *See also* carrots
Davis, Angela, 181
decision making
in cooperative business, 31, 35–38
organizational transformation of, 307
types of, 38
decoctions, herbal, 201–2
deer, 83
dehydrators, 238
delegation, 38
delivery of produce
boxes and containers for, 147
to homes, 5, 27–28, 38, 42
denial, in internalized racism, 275
Dennis, Benjamin "B. J.," 105
Derryck, Dennis, 44
DeSanctis, Cheryl, 132
Desmodium spp., 126
Dessalines, Jean Jacques, 150
Dessalines Brigade, 150, 283
Detroit Black Community Food Security Network, 208, 259
DeWitt, Sara, 25
diabetes, 4, 182, 183, 225, 272
and environmental racism, 273
Youth Food Justice curriculum on, 252–253, 259
Diaspora, 9
animals raised in, 164
dance rituals in, 278–79
food preparation in, 223, 228, 234
harvest rituals in, 62, 155
plant medicine in, 181, 183
spiritual practices in, 56, 60, 62
susus in, 20
sweet potatoes in, 121
diet
African food pyramid in, 225–26, 233
decolonization of, 225, 228
in Haitian konbit, 44
illness related to, 4, 224–25, 252–53, 259, 267, 272
meat in, 179–80, 225
vegetarian, 168, 180
Youth Food Justice program on, 252–56
digging sticks, 139
dill, 106, 220
Dioscorea spp., 116, 121. *See also* yams
direct action, 287–89
direct seeding, 46, 113, 139, 140, 148
discrimination
in criminal justice system, 271
in education, 266, 271
in employment, 270, 271, 285
in housing, 205–6, 269–70
in land ownership, 258–59, 266, 268–69
by USDA, 7, 12, 13, 20–21, 205, 258–59, 268–69, 284, 285, 289, 295, 317
Dispossession (Daniel), 269
divination, 56, 57, 69
herbs in, 189, 192, 195
spiritual instructions in, 59–60, 115
tools in, 54
Doctrine of Discovery, 265
dolomitic limestone, 92, 93
Domestic Fair Trade Association, 251, 297
Douglass, Frederick, 245
dragon's claw millet, 116
Dred Scott v. Sanford, 265
drip irrigation, 140–42
drying
of fruits, 238
of herbs, 200–201, 202, 237–38
of vegetables, 237–38
D-Town Farm (MI), 208, 214, 222
DuPont, 150
Dutch white clover, 100, 124
dynamic meditation, 278
Dysphania ambrosioides, 186, 201, 279

"each one teach one" model, 8
Earth, as Sacred Mother, 1
earthworms, 97, 221
Earvin, Eddie, 267
East African kale, 120
EBT (Electronic Benefits Transfer), 27, 41, 42, 293
echinacea, 83, 182, 186
propagation of, 196, 198, 199
ecology, soil, 96–99
edamame, 106, 113, 240
education, 16–20, 285–87
in agricultural universities, 19, 285
anti-racism programs in, 315
in Black Latinx Farmers Immersion. *See* Black Latinx Farmers Immersion
Black Panther Party on, 255
community-based, 39–40, 293–94
discrimination in, 266, 271
"each one teach one" model in, 8
on food preparation and preservation, 223
in incubator programs, 16
on organic and sustainable practices, 285
people of color leading programs in, 17, 18
on plant medicine, 197

Pygmalion Effect in, 255
racial justice commitment of programs in, 19
racism in, 8, 16
scholarships for, 17
Youth Food Bill of Rights on, 250, 251
in Youth Food Justice program, 5–6
eggplant, 106, 109, 112
in container gardening, 218
in crop rotation, 128
saving seeds of, 154, 156, 158, 159
in square-foot gardening, 220
eggs
as power ingredient, 242
preserved in vinegar, 239
raising chickens for, 164–68, 169
Egypt
earthworms in, 97
fowl in, 169
herbal medicine in, 182
hoes in, 130
irrigation in, 141
sesame in, 122
soil in, 90, 97
Egyptian spinach, 115
Egyptian walking onion, 121
Ehrlichman, John, 271
elderberries, 82
elderflower, 182
electric fencing
for chickens, 166, 167
for pigs, 176
Electronic Benefits Transfer (EBT), 27, 41, 42, 293
Eleusine coracana, 116. *See also* millet
Elfe, Willie, 191
Emanuel, James A., 87
empathy, in calling in process, 311
employment. *See also* labor
discrimination in, 270, 271, 285
interracial alliances in, 304–5
Environmental Protection Agency, 72, 73, 213
Environmental Quality Incentives Program, 22
environmental racism, 210, 272–73
epazote, 186, 201, 279
epigenetics, 273
epis recipe, 230–32
Epsom salts, 93
Equal Employment Opportunity Commission, 285
equipment. *See* tools and equipment
Equity Trust, 14, 16
Eritrean basil, 181
erosion, 130–31
estimating risk for, 90
Haitian techniques in, 72
soil mounds in, 75
terracing in, 78–80
Eupatorium perfoliatum (boneset), 190, 192, 198
evisceration of chickens, 175
expenses. *See* costs
Ezili Danto, 53, 59, 177, 181, 182

Fabaceae family, 113–114, 128. *See also* beans; peanuts; peas
Facing the Black Shadow (Watson), 273
Fair Food Program, 295
Fair Labor Standards Act, 270, 293
fair trade, 251, 288, 297
fallow or rest periods, 127, 128
fanya-juu practice, 78, 80
Farm Bill, 291
Farmers Home Administration, 13, 258
farmers markets, 42
farm incubator programs, 16
farmland access. *See* land access
Farmland Access website, 16
Farm Service Agency, 21
farm-share model, 4–5, 27–28, 38–42
chicken eggs in, 164
sliding scale for costs in, 4, 41
sprouts in, 215
farm skills training, 16–20. *See also* education
farm work. *See* labor
fast food advertisements, 254–55
Fatiman, Cecile, 59
Federal Housing Administration, 7, 205–6, 269–70
Federation of Southern Cooperatives, 7, 11, 13, 43, 257, 259, 290, 297
Land Assistance Fund, 289, 297–98
feed for livestock
for chickens, 164, 166–68, 170, 180
for pigs, 176, 177, 179, 180
and sustainability, 180
fencing
for chickens, 166, 167
for pigs, 176
repair of, in konbit, 46
fennel, 106, 239
fenugreek, 215
fermentation
in food preservation, 238–39
in seed ripening, 156–57, 158
Ferula assa-foetida, 183–84
field peas, 86, 100–101
Fields, Tanya, 257
financing, 20–26
of agricultural studies, 17
of cooperative businesses, 30, 35
of Corbin Hill Food Project, 44
crowdfunding in, 24–25
government sources in, 13, 20–23
of land, 7, 16, 21–22, 258, 268–69
lending society in, 25
of operating expenses, 22
of Soul Fire Farm, 4, 20, 24–25, 26, 30, 33
susu groups in, 9, 20, 25
Finding Farmland: A Farmer's Guide to Working with Land Trusts, 14
finger millet, 116
Finke, Jens, 163
firewood tasks in konbit, 46
fish, preservation of, 237
fish emulsion, 139
fish pepper, 112–13
Fitzgerald, Oleta Garrett, 294
flame-weeding, 117
flaxseeds, 217
flea beetles, 109, 118
Flint, Michigan, 212–13
floating row covers, 109, 118, 137
floor space requirements for chickens, 166
flowers
in crop rotation, 128
as offering, 199
in polycultures, 123
saving seeds of, 155
foggara irrigation systems, 141
food access, 8, 271–72
cost affecting, 4–5, 27–28, 40–42, 241
and diet-related illnesses, 4, 5, 224–25
in food apartheid and food deserts. *See* food apartheid and food deserts
food hubs in, 42–44
home delivery in, 4–5, 27–28, 38, 42
policy demands on, 293
racial disparity in, 271–72
scarcity in, 27
transportation affecting, 5, 27–28
wealth redistribution in, 40–41
Youth Food Bill of Rights on, 250–51

food access (*continued*)
Youth Food Justice program on, 252–56, 259
Food and Agriculture Organization, 125, 127, 152, 180
food apartheid and food deserts, 3–5, 39, 224–25, 259, 272
home delivery of foods in, 4–5, 27–28, 38, 42
for immigrants, 41
Youth Food Justice program on, 253
The Food Hub Collaboration, 42
food hubs, 42–44
food justice program for youth, 5–7, 245, 248
curriculum in, 252–262
food preparation and preservation, 223–44
ancestral legacy in, 228
canning in, 240
drying in, 237–38
fermentation in, 238–39
freezing in, 240
in Haiti, 44
konbit tasks in, 44, 45, 46
learning from elders about, 223
pickling in, 239
recipes for, 226–36
safety regulations in, 209
soil storage in, 236–37
in time and cost limitations, 241–44
in West Africa, 237
youth involved in, 226–28
The Food Project (MA), 1–2, 19, 129, 263
Food Safety Modernization Act, 145
food scraps, vermicomposting of, 221
food sharing in communal labor, 44, 47
food stamps, 27, 41
forests
and agroforestry, 78, 80–84
decomposition in, 95, 98
interactions of trees in, 80–82, 246–48
pigs in, 178–179
spiritual experiences in, 248
in swidden agriculture, 127
"40 acres and a mule," 267, 291, 301
Foster, W. F., 286
Frazier, Garrison, 267, 291
Freedom Farm, 3, 29, 257, 297
Freedom Food Alliance, 36, 282, 298
freezing for food preservation, 240, 241
Fruition Seeds, 139, 161
fruits
in agroforestry, 82, 83
cobbler recipe for, 236
dried, 238
freezing of, 240
in lead contaminated soil, 77
pest protection for, 83
in polycultures, 82–83, 123
spacing between, 82
spring planting of, 81
in terraces, 80, 82
fungi in soil, 96, 97

garden egg (white eggplant), 112
Gardening the Community (MA), 219, 222
Garden of Happiness (NY), 212–13, 222
garlic, 106, 120–21, 152
dried, 237
plant spacing, 106, 220
in square-foot gardening, 220
Garner, Eric, 5, 103
Garza, Alicia, 292
Gay, Ross, 103
gbogname collard eggplant, 112
Geechee red pea, 114, 151–52
George, Harrison, 300
Georgia Rattlesnake watermelon, 108
geranium, scented, in lead phytoremediation, 74
Ghana, 53, 129
Akim Abuakwa people of, 29
contractual agreements in, 33
dark earth in, 94
food preservation in, 237
millet harvest festival in, 62
nature-based religion in, 54
traditional diet in, 225, 226
trees as cover crops in, 102
gherkins, 108
GI Bill, 205–6, 270
Gilliard Farm, 30
Giovanni, Nikki, 48
Giuliani, Rudolph, 206, 213, 288
Glickman, Dan, 284
Global Village Farms (MA), 30
GMO crops, 104, 160, 250, 294
goals of business, 49, 51
goats, 180
goober nuts, 114
Good Food Jobs, 17
good food movement, 39, 44
Goodman, Andrew, 305
gooseberries, 82
Gossypium spp., 114. *See also* cotton
gourds, 106, 108, 125, 154, 159
government funding and loan programs, 13, 20–23, 258–59, 284, 291, 295
Gran Bwa (Great Forest), 64–65
Grant, Ulysses S., 291
grapes, 82, 296
gratitude and appreciation
for harvest, 53–54, 60, 62–63
in herbal baths, 65
for konbit volunteers, 48
for medicinal herbs, 199
for new team members, 35–36
for spirits of the land, 53–54, 60
for water, 199
grazing, rotational, 164, 177–79
Great Migration, 205
green beans, 218, 238, 239
Green Belt Movement, 257, 287
Green Glaze collards, 119–20
Green Guerillas, 213
greenhouse gases, 180, 251
greenhouses, 110–11, 135–36
Solanaceae family in, 109
in urban areas, 209
green striped cushaw pumpkin, 108
Green Thumb, 213
Greenwood Food Blockade, 268
grits recipes, 234–35, 244
groundnuts
bambara, 114
in relay cropping, 123
Groundnut Stew, 230
Groundswell Center for Local Food & Farming, 306, 308
Group of Four, 150
Growing Food and Justice, 17
Grow Where You Are, 15–16
Guillen, Rosalinda, 257
guinea fowl, 169
gumbo, 114–15

hairy vetch, 100
Haiti, 8, 9, 10
agroforestry in, 82
Azaka as Spirit of Agriculture in, 59–60
Creole black pigs in, 59, 164, 177
Dessalines Brigade in, 283
harvest festival in, 62–63
konbit in, 44
mushrooms in, 220

reforestation in, 71
sacred herbalism in, 200–201
seeding saving in, 151
soil conservation in, 72
stone art in, 276
Haitian grits with spinach, 234–35
Haitian Peasant Movement, 150, 287
Haitian pumpkin, 108
Haitian Revolution, 59, 60, 65, 108, 112, 164
pig sacrifice at time of, 59, 177
and Soup Joumou, 108, 229
Halfkenny, Babalawo Enroue, 48
Hamelia patens, 201
Hamer, Fannie Lou, 3, 29, 236, 257
hand pollination, 153–55
hand seeding, 139
hand weeding, 144
Hargrove, Tasha, 19
Harmony Homestead & Wholeness Center (NY), 23
Harriet's Apothecary, 196, 197, 278, 304
Harriet Tubman Freedom Farm (NC), 23, 301
harvest, 17
of chicken eggs, 168
of herbs, 65, 199–201
konbit tasks in, 46
of microgreens, 217
offerings at time of, 199, 201
of rainwater, 140, 215, 216
of seeds for keeping, 155–59
spiritual practices in, 53–54, 55, 57, 58, 62–63, 65, 199, 201
tools and equipment in, 133, 145–47, 148
of wild plants, 64, 190–91
harvest knives, 119, 121, 146
Haudenosaunee people, 103
H2A visa program, 293
Hayes, Cynthia, 257, 285
Haymarket Fund, 307
hazelnuts, 82
Hazzard, Dominique, 257
Heads Together Small Producers of Haiti, 150
HEAL Food Alliance, 12, 294
healing, 275–80
caucus spaces in, 278
dance in, 278–79
dynamic meditation in, 278
nature in, 246
partners and circles in, 277–78
plant medicine in, 181–204, 279–80
"rock therapy" in, 251–52
in Youth Food Justice program, 252
healthy food choices, 223
cost as barrier to, 4–5
food deserts and food apartheid affecting, 4, 224–25
meat in, 179, 180
policy demands on, 293
recipes for, 226–36
in time and cost limitations, 241–44
in traditional African diet, 225–26
transportation as barrier to, 5
Ujamaa Farm Share providing, 41
Youth Food Bill of rights on, 250–51
Youth Food Justice program on, 252–56
heart disease, 4, 225, 272
and environmental racism, 273
Youth Food Justice curriculum on, 252–53, 259
heat
in chicken brooder, 165, 168
in greenhouses, 111, 135, 136
heavy metal contamination
of soil, 72–78, 91, 94, 210–12
of water, 213, 214
Hebrew prayers, 172
heirs property, 268, 289–90, 295
Helianthus annuus, 74, 107. *See also* sunflowers
herbal baths, 8, 55, 64–65, 204
after animal slaughter, 8, 53, 176
plants used in, 65
in Vodou, 64–65, 201
herbalists, Black, 190–91, 197
herbs, 54, 181–204, 279–80
in agroforestry, 82–83
classes offered on, 197
in crop rotation, 128
in Cuba and Haiti, 200–201
cultivation of, 196–99
drying of, 200–201, 202, 237–38
harvest of, 65, 199–201
as microgreens, 217
in polycultures, 82–83, 123, 124, 181
preparations of, 201–4
safety of. *See* safety of plants
saving seeds of, 155
starting seeds of, 198–99
hibiscus (sorrel), 114, 115, 189, 228
medicinal uses of, 189
planting charts on, 106, 154
red, 115, 189
Roselle, 154, 159
saving seeds of, 149, 154, 159
white, 115
high blood pressure, 225
high tunnels, 136–37
cost of, 26
funding for, 21, 22
polycultures in, 125
starting seeds in, 111
tomatoes in, 107, 109, 124, 136, 144
in urban areas, 209
Hill, Walter, 19
hoes, 145
African style, 129, 130, 131
back position during use of, 131
short-handled, 130
The Holistic Orchard (Phillips), 84
Homecoming: The Story of African-American Farmers, 258
home delivery of foods, 4–5, 27–28, 38, 42
Homestead Act (1862), 266
honeybees, 99, 100
honey berry, 82
hoop houses, 26, 111, 135–36
"Hope Tree" activity, 261–62
Hoppin' John, 114, 152, 232–33
horehound, 187, 190–91
Horne, Savi, 290
hot peppers, 113, 156, 240
housing discrimination, 205–6, 269–70
Howell, Donna, 255
Hudson Valley Seed Library, 139, 161
Huerta, Dolores, 257, 296
Huggins, Erika, 255
humility in offerings to Azaka, 60
Hurricane Sandy, 130
Hurston, Zora Neale, 67
hybrid varieties, 152, 159
hydrocooling of produce, 146–47
hyssop, 187, 196, 198

Ifa, 55
appropriate use of practices, 69
divination in, 54, 56, 60
financial practices in, 35
individualized instructions in, 115
Orisa Oko in, 55, 60, 61
plant medicine in, 183
sacred literature in, 56–58
Igbo traditions, 63
Ika Irosun, 57

illness
of chickens, 164, 168
diet-related, 4, 224–25, 252–53, 259, 267, 272
in diminished contact with nature, 273
in environmental racism, 273
in lead exposure, 273
of pigs, 176, 177
plant medicine in, 181–204
immigrants, 41, 207, 213
education programs for, 285
employment of, 31, 287
policy platform on, 292
immobilization of lead contamination, 75–78
incarceration, 299–300
in Black Codes, 266–67
labor of prisoners in, 5, 267, 293
racial disparities in, 270–71, 300
and Victory Bus Project, 282, 298
Youth Food Justice program as alternative to, 5–6, 249–52
income
business plan on, 49–51
and wealth gap, 40–41, 271, 301
incubator programs, 16
Indigenous Knowledge and Peoples Foundation, 127
Indigenous Seed Initiative, 161
infusions, herbal, 201, 202–4
inhaling steam of herbal infusions, 203–4
inheritance laws, 285
and heirs property, 268, 289–90, 295
insects, beneficial, 123, 124
intensive growing, 219–20
intercropping. *See* polycultures
internalized racism, 5, 273–75
International Coalition to Commemorate African Ancestors of the Middle Passage, 64
interracial alliances, 304–6
intersectionality, matrix of, 314
Intervale Center (VT), 16
Ipomoea batatas. *See* sweet potatoes
iron, 91, 93
irrigation systems, 140–42, 148
cost of, 26
rainwater in, 140, 215
isolation distances for seed saving, 152–53, 154
"It Is Our Duty" activity, 262
Itliong, Larry, 296
Iwa, offerings to, 200, 201

Jackson, Andrew, 291
Jackson, Jon, 179
Jacobson, Larisa, 143, 313
jaden lakou, 82, 83
jail time. *See* incarceration
Jefferson, Sajo, 219
Jefferson, Thomas, 169
Jim Crow laws, 258, 267–68
jimsonweed, 193
Johnson, Andrew, 267
Johnson, C. J., 66
Johnson, JoVonna, 15
jostaberries, 82
joumou pumpkin, 108
Juglans nigra, 102
juglone, 102
Julien, Leonard, 145
juneberries, 82
justice
criminal justice system. *See* criminal justice system
culinary justice movement, 228
racial, 5, 19, 315
youth food justice program, 5–7, 245, 248, 252–62

Kabui, Njathi, 228
Kaepernick, Colin, 255
kale, 106, 118
in crop rotation, 128
in low tunnels, 137
as microgreens, 217
saving seeds of, 159
in square-foot gardening, 220
kaolin clay, 106
Karenga, Maulana, 4
kebarika bush bean, 114
Kenya, 78, 80, 126
Kimmerer, Robin Wall, 65, 189–90, 313
King, Martin Luther, Jr., 264, 288, 296
kiwi, hardy, 82, 84
knives
harvest, 119, 121, 146
planting, 139
konbit, 44–48, 80
krump dance, 279
Ku Klux Klan, 258, 268, 305, 317
Kwanzaa, 4
lablab, 114, 123
labor
agreements on, 33
and Black Codes, 266–67
budget on, 50
clothing for, 147
communal, 44–48, 62, 80, 141
dignity in, 293, 297
of migrant workers, 5, 270, 293
policy demands on, 293
of prisoners, 5, 267, 293
rights of workers in, 270, 285, 293, 295–97
spiritual teachings on, 57–58
strikes of, 296, 297
labor laws, 270, 285, 293
lactic acid fermentation, 238
Lactuca sativa, 154. *See also* lettuce
lady's mantle, 196
Lagenaria siceraria, 108
Lahens, Forrest, 93
Lamar, Kendrick, 8
land access, 11–16
in land trusts, 3, 8, 13–14, 209
in sharecropping, 267
in urban areas, 209
Youth Food Justice program on, 253–54, 257–59
Land Contract Guarantee Program, 22
land defense, 289–91
Land Equivalency Ratio in three sisters, 124
Land Link projects, 16
"Land Loss and Resistance" activity, 257–59
Land Loss Prevention Project, 289, 290
land ownership, 7, 8, 12
and civil rights movement, 281
cost of, 26
defense of, 289–91
discrimination in, 258–59, 266, 268–69
financing for, 7, 16, 21–22, 258, 268–69
in heirs property, 268, 289–90, 295
in New Communities, 13
and racial disparity in wealth, 271
in reparations, 8, 12, 23–24
at Soul Fire Farm, 4, 11–12, 29
and theft of Black-owned land, 268
Youth Food Justice program on, 257–59
land restoration. *See* soil restoration
Land Trust Alliance, 14
land trusts, 3, 8, 13–14, 209
land-use agreements, 209–10
lavender, 182, 187, 196, 198
laws and regulations

Black Codes in, 6, 23, 266–67
business laws in, 30–33, 285
changing policies in, 291–95
direct action against, 287–89
on heirs property, 268, 289–90, 295
and historical trauma, 265–71
Jim Crow laws in, 258, 267–68
labor laws in, 270, 285, 293
litigation on, 284–85
in slavery, 265–66
in urban farming, 207–10
lead, 273
in soil, 72–78, 91, 94, 211
in water, 213, 214
leafhoppers, 112
leafminers, 121
leasing land, 14, 15–16
land-use agreements in, 209–10
Lee, Joseph, 145
leeks, 106, 120
plant spacing, 106, 220
saving seeds of, 153, 154, 155, 159
in square-foot gardening, 220
legal issues. *See* laws and regulations
legumes, 113–14. *See also specific plants.*
as cover crops, 99, 100–101
in nitrogen fixation, 99, 101, 113
sprouting seeds of, 215, 217
lemon balm
as companion plants, 82–83
medicinal uses of, 182, 187
starting seeds of, 198
lemon drop marigolds, 123
lending society, 25
lentils, 215, 217, 238
Lentinula edodes, 220–21
lettuce, 118
bagged for delivery, 147
in container gardening, 218
in crop rotation, 128
in high tunnel, 136
as microgreens, 217
planting chart on, 106–7
in polycultures, 118, 124
saving seeds of, 152, 153, 154, 155, 156, 159
in square-foot gardening, 220
starting seeds of, 118, 138
Levant cotton, 114
Lewis, Michael, 13
liability protection
in communal labor, 48
in cooperative business, 31
land-use agreement on, 210
Liberia, dark earth in, 94
lifestyle, business plan on, 49
lighting for seed starting, 110–11, 135
Liliaceae family, 120–21, 128. *See also* garlic; leeks; onions
chives, 82
scallions, 107, 120, 124
limestone, 92, 93
limited liability corporations, 31, 33
literature, sacred, 56–58
litigation, 284–85
against USDA, 13, 20–21, 258–59, 269, 284, 285
Live Power Farm (CA), 4
livestock, 163–180. *See also* animals
living mulch, 124
loam, 89, 90
loans
for housing, residential security maps in, 205–6, 269–70
from USDA, 21–22, 258–59, 268–69, 284, 291, 295
logistics meetings, 36
Louisiana long green eggplant, 112
Lovejoy, Elijah P., 305
low tunnels, 137
Lukumi, 189, 200, 201
Lumumba, Chokwe Antar, 294
Lunsford, Lindsey, 291
Lycopersicon spp. *See* tomatoes

Maathai, Wangari, 257, 287
macronutrients in soil, 92–93
Madison, Stanley, 289
magnesium, 91, 92, 93
maize. *See* corn
majority vote, 38
malabar spinach, 116
Malcolm X, 1, 7, 63, 257, 258
Malcolm X Grassroots Movement, 255
Malvaceae family, 114–15, 128. *See also* cotton; hibiscus; okra
manganese, 91, 93
Manje Yam festival, 62–63
manure
chicken, 74, 166, 168
cow, 74, 164
pig, 176, 179–80
Manye (Queen Mothers), 54–55
Many Hands Organic Farm (MA), 4, 176, 251
Marek's disease, 164
marigolds, 118, 123, 185
Mariposa, Mel, 309
marketing, community-based, 39–40
Márquez, Francia, 257
Marrubium vulgare (horehound), 187, 190–91
Martin, Trayvon, 292
Matricaria recutita, 185–86
Mayi Moulin ak Fèy, 234–35
Mboga pepper, 113
Mbombo green, 114
meals eaten with forks or hands, 59–60
meat in diet, 179–180, 225
chickens raised for, 164, 168–76
pigs raised for, 176–79
"Media Does Not Have My Mind" activity, 254–56
medicinal plants, 181–204, 279–80. *See also* herbs
meditation, dynamic, 278
meetings in cooperative business, 32, 35–38
Megbezorli, 186, 279
Melissa officinalis. *See* lemon balm
melons, 107–108
in crop rotation, 128
in polycultures, 123
saving seeds of, 154, 158, 159
membership
in cooperatives, 33, 34
in CSA, 27, 41
Mentha spp. (mints), 82–83, 187–88, 196, 198
mentoring, 16, 17, 197
Merikin Moruga Hill Rice, 105
mesclun, 107
microgreens, 217
microloans, 22
micronutrients in soil, 93
microorganisms in soil, 96–99, 113
migrant workers, 5, 270, 293
milk, 225–26
and dairy workers, 295, 297
Milk With Dignity, 297
millet, 107, 116
as cover crop, 86
harvest festival, 62
in raised beds, 75
saving seeds of, 149
soil type for, 90
as traditional food, 44, 228
in zai pits, 81
milpa polyculture (three sisters), 103, 116, 121, 124–26
Minaj, Nicki, 8

Minnesota Food Association, 16, 19
mint, 82–83, 188, 196, 198
Mission Accomplished Transition Services, 249
mission statement, 48–49, 50
Mississippi Association of Cooperatives, 43
Mississippi Freedom Democratic Party, 29
Mohican people, 12
Moi, Daniel arap, 287
molokhia, 107, 115
molybdenum, 93
Momordica charantia, 201
Monsanto, 150–51, 283–84, 317
moringa, 228
Morrill Act, 266
Morrison, Toni, 8–9, 10
mortgages, discrimination in, 205–6, 269–70
movement building, 281–98
  consumer organizing in, 295–97
  direct action in, 287–89
  education in, 285–87
  land defense in, 289–91
  litigation in, 284–85
  mutual aid and survival programs in, 297–98
  policy change in, 291–95
Movement for Black Lives, 259, 292
"Move Your Butt" activity, 252–54
Moyamensing tomato, 112
mugwort, 193
Muhammad, Baba Curtis Hayes, 5, 281
mulch
  in agroforestry, 82
  black plastic, 108, 109
  konbit tasks in, 46
  on lead contaminated soil, 77
  living, 124
  overwintering crops in, 137
mullein, 193–194, 199
mung beans, 215
Murray, George W., 145
mushrooms, 220–21
mustard greens, 119, 154
  in lead phytoremediation, 74
  as microgreens, 217
  saving seeds of, 154, 159
  in square-foot gardening, 220
mutual aid and survival programs, 297–98
mycorrhizal fungi, 97

NAACP, 7
Nana Buruku, 54, 55
napa cabbage, 107
nasturtiums, 123, 182
National Black Farmers Association Scholarship Program, 17
National Black Food & Justice Alliance, 12, 259, 288, 291
National Community Land Trust Network, 14
National Congress of Papaye Peasant Movement, 150
National Coordination of Peasant Organizations of Mali, 150
National Farm Workers Association, 257, 296
National Housing Act, 205, 269
National Incubator Farm Training Initiative, 16
National Labor Relations Act, 270, 293
National Scholars Program, 17
National Sustainable Agriculture Coalition, 308
National Young Farmers Coalition, 14, 16, 294, 306, 308
Natural Resources and Conservation Service, 21, 89
nature
  healing power of, 246, 247
  illness in reduced contact with, 273
  observation and learning from, 260
  in urban areas, 246, 247, 273
nature-deficit disorder, 246
"Nature Is My Teacher" activity, 260
Ndeup dance ritual, 279
Nduwimana, Jean, 237
Negro Cooperative Guild, 297
Nelson, Wallie and Juanita, 257
New Communities, 3, 13, 257, 297
New Entry Food Hub (MA), 41
New Entry Sustainable Farm Project (MA), 16, 19
New Farmers Grant Fund (NY), 23
The New Jim Crow (Alexander), 282
Newsome, Demalda and Rufus, 98, 206
Newsome Community Farms, 98
Newton, Huey P., 255
Newton, John, 105
New York Botanical Garden, 213
New York State Prisoner Justice Network, 6
ngai ngai, 228
Ngmayem festival, 62
nickel, 211
Nicotiana tabacum (tobacco), 189, 199, 265
Niger, 90
Nigeria, 35, 123, 220
nightshade, black, 112, 228
nitrate levels in water, 214
nitrogen, 92
  fixation of, 97, 99, 100, 101, 102, 113, 125
Nixon, Richard, 270–71
noise in urban farming, 209
Noninsured Crop Disaster Assistance Program, 22
nonprofit organizations, 30, 32–33, 44
North East Community Center, 41
Northeast Farmers of Color Network, 14, 282
  policy demands of, 293–95
  reparations map of, 282, 302
Northeast Organic Farming Association, 2, 285–87
Northeast Sustainable Agriculture Working Group, 308
no-till methods, 84–86, 97, 123, 133
  tarping in, 85–86, 117, 133
nuisance ordinances, 207, 209
nuts, 82, 243
nzinzingrolo, 187, 279

Oakland Community School, 255
oats, 86, 99, 100–101
obesity, 4, 224, 225, 272
  in environmental racism, 273
  Youth Food Justice curriculum on, 252–53, 259
obi obata, 54
Ocimum spp., 184. *See also* basil
Odi Ogunda, 57
Odu Ifa, 56–58
offerings, 59–62, 172, 180
  to ancestors, 277
  to Gran Bwa, 65
  before herb harvest, 199, 201
  pig sacrifice in, 59, 177
Ofori, Nana, 29
Ofun Irete, 57
Ofun Oturupon, 57
Ogbe Iwori, 57
Ogou, 53
oils, herbal, 202–3
okra, 107, 114–15, 154
  in crop rotation, 128
  in polycultures, 123
  saving seeds of, 149, 152, 153, 154, 155, 156, 159, 161
  in square-foot gardening, 220
  as traditional food, 225

in urban farming, 208
Oldways Preservation Trust, 226
Olugbala, Amani, 69, 262
Olympia Food Co-op, 38
On-Farm Poultry Slaughter Guidelines, 171
onions, 107, 120, 121, 154
in crop rotation, 128
dried, 237
Egyptian walking onion, 121
in high tunnel, 125, 136
in pikliz, 239
in polycultures, 123, 124, 125
saving seeds of, 153, 154, 159
sprouting seeds of, 215
in square-foot gardening, 220
open-pollinated varieties, 152, 153
Operation Spring Plant, 43
oregano, 182
organic farming, 3, 285
organic matter, 87–88, 89, 101
composting of, 84–85, 94–96
in raised beds, 75, 76–77
soil testing for, 91
in terraces, 78, 80
Organic Seed Alliance, 160
organizational transformation, 306–9
orisa, 56, 57, 172, 201
Orisa Oko, 55, 60, 61
Orisa Osum, 201
Ortiz, Merelis Catalina, 228
Oryza spp. *See also* rice
O. glaberrima, 105, 116, 149
O. sativa, 105
Ose Otura, 58
Ovambo people, raised beds of, 75
Owonrin Obara, 58

Paige, Ralph, 7, 11, 257, 258, 317
parsley, 107, 117
in container gardening, 218
in crop rotation, 128
in high tunnel, 125, 136
in polycultures, 124, 125
saving seeds of, 154
in square-foot gardening, 220
starting seeds of, 117, 138
parsnips, 107, 117
in crop rotation, 128
roasted, 235
in square-foot gardening, 220
storage of, 236
Parson, Louis, 25
partnerships
businesses organized as, 31
institutional, in farm-share CSA, 41
pastures
chickens raised on, 164, 165, 166, 168, 170
rotational grazing of, 164, 177–79
patents of Black farmers, 145
pathways between beds, 75, 77, 131
in sheet composting, 85
pattypan squash, 108
Paul Robeson tomato, 112
pawpaws, 82, 84
payment plans, 41–42
peaches, 82, 236, 238
peanut butter, 230, 242
peanuts, 113, 114, 154
nitrogen fixation by, 99, 101
planting chart on, 107
saving seeds of, 154, 159
pears, 82, 123
peas, 113, 114, 151–52
bagged for delivery, 147
as cover crop, 86, 92, 99, 100–101
cowpeas. *See* cowpeas
in crop rotation, 114, 128
plant spacing, 107, 220
as power ingredient, 242
preservation of, 239, 240
saving seeds of, 151–52, 153, 154, 155–56, 159
sprouting seeds of, 215, 217
in square-foot gardening, 220
trellising of, 113, 142–43
Peasant Movement of Papaye, 150
Pedaliaceae family, 122
Pelargonium spp., 74
Penniman, Naima, 318–19
pennyroyal, 187–188
People's Institute for Survival and Beyond, 307
peppermint, 188
peppers, 107, 109, 112–13
in container gardening, 218
in crop rotation, 128
dried, 237
hot, 113, 156, 240
in pikliz, 239
in polycultures, 123, 181
saving seeds of, 149, 153, 154, 155, 156, 159
in square-foot gardening, 220
perennials
in herb garden, 196–99
saving seeds of, 153
perfluorooctanoic acid, 213
persimmons, American, 82
personal development, 312–14
pest protection
crop rotation in, 126–28
dogs in, 126
for fruit trees, 83
kaolin clay in, 105, 106
polycultures in, 123–24, 125
row covers in, 109, 118, 137
Petroselinum crispum. *See* parsley
Phaseolus spp., 114, 154. *See also* beans
Phillips, Michael, 84
pH of soil, 92, 93
for blueberries, 84
in contamination remediation, 73, 74, 212
soil testing for, 91
taste and smell as indicators of, 90
phosphorus, 74, 91, 92
phytochemicals, 183
phytoremediation of soil contamination, 73–75
pickling, 239
Pigford v. Glickman, 13, 20–21, 284, 317
pig manure, 176, 179–80
pigs, 164, 176–79, 180
Creole black, 59, 164, 177
in industrial agriculture, 179–80
pikliz, 239
pile fèy ceremony, 201
pine, 194
Pippin, Horace, 113
Pisum sativum. *See* peas
planning
of business. *See* business plans
of crops, 103–28
plantain, 190–91, 194, 225
planting. *See* seed planting and starting
planting knives, 139
plant medicine, 181–204, 279–80. *See also* herbs
plant spacing, 106–7
in intensive growing, 219–20
for seed saving, 152–53, 154
plastic greenhouse covering, double layer of, 111, 135–36
plastic mulch, black, 108, 109
Plat de Haiti tomato, 112
Plessy v. Ferguson, 258, 268
plow and harrows, 131–32
plucking chickens, 174–75

plums, 82
Poaceae family, 115–16, 128. *See also* corn; millet; rice
oats, 86, 99, 100–101
rye, 99, 100, 124
poison sumac, 195
pole beans, 113
in square-foot gardening, 220
in three sisters, 116, 125
trellising of, 113, 142–43
police presence, spiritual concerns about, 59–60
police violence, 5, 265–66, 272
policy change, 291–95, 302
pollination, 152–55
attraction of pollinators for, 99, 100, 123
pollution
and environmental racism, 273
soil contamination in, 72–78, 94, 210–12
water contamination in, 212–13
polycultures, 123–26
herbs in, 82–83, 123, 124, 181
legumes in, 113
three sisters in, 103, 116, 121, 124–26
ponds, 54
potassium, 91, 92
potato beetles, 112, 126
potatoes, 107, 109, 112, 152
in crop rotation, 112, 126, 128
roasted, 235
in square-foot gardening, 220
storage of, 236
sweet. *See* sweet potatoes
poultices, 203
power structure, 272, 307
prayer, 58, 61
in chicken slaughter, 172
in herb harvest, 199
in "Hope Tree" activity, 261–62
preserving food. *See* food preparation and preservation
Price, Richard, 149
prison time. *See* incarceration
privacy, in calling in process, 310
profits, 31, 32–33, 35
Project Growth, 250
propagation
of herbs, 196–99
tools and equipment for, 135–38, 148
property laws, 285
Prunus serotina (black cherry), 102, 192
pumpkins, 108, 159, 229
purins, 99
Pygmalion Effect, 255

quality of life, business plan on, 49

Race Forward, 8
racial justice, 5
readings on, 315
training programs with commitment to, 19
racism, 7, 205
in agricultural training, 8, 16
anti-racism work in, 299–315. *See also* anti-racism work
in criminal justice system, 5, 282
environmental, 210, 272–73
in food system, 5
internalized, 5, 273–75
radishes, 107, 119
in container gardening, 218
in crop rotation cycle, 128
fermentation of, 238
in high tunnel, 136
marking carrot rows with, 117
as microgreens, 217
in pikliz, 239
saving seeds of, 159
sprouting seeds of, 215
in square-foot gardening, 220
Raiford, Matthew, 30
rainwater
erosion from, 130–31
prayer on, 58
water supply from, 140, 215, 216
raised beds, 130–33
in clay soil, 89
of Ovambo people, 75
pathways between, 75, 77, 85, 131
in sheet composting, 84–85
in soil contamination, 75–78, 211
tools for, 130–33, 134, 148
Rambo, Salim, 287
Rapp Road community (NY), 25
raspberries, 82, 83, 84
Reagan, Ronald, 269
Reagon, Toshi, 68
recipes, 226–36
reciprocity, 53
in communal labor, 48
in spiritual practices, 57, 58
red clover, 100, 194–95
redlining, 205–206, 269–70, 288
red peas, 114, 151–52
red sorrel, 115, 189
red wiggler worms, 221
reflection meetings, 36, 37
refuge, farms as place of, 58
regenerative agriculture, 3, 286
Regional Coordination of the Southeast, 150
Regional Food Hub Resource Guide (USDA), 44
Reich, Lee, 84
relay cropping, 123
religion and spiritual practices, 53–70. *See also* spiritual practices
reparations, 8, 12, 23–24, 283, 288, 301–4, 317
"40 acres and a mule" plan in, 267, 291, 301
map on, 23, 282, 301, 302
policy changes in, 292, 295, 302
for USDA discrimination, 13, 20–21, 289
residential security maps, 205–6, 269–70
Resora, 13
Resource Generation, 23–24
resources, 303, 321–24
on anti-racism programs, 315
on business planning, 48
on farm incubator programs, 16
on farm training programs, 18–19
on herbal classes, 197
on racial justice, 315
on reparations, 302
on scholarship opportunities, 17
on seed suppliers, 138–39
on tools and equipment, 139
on urban farming, 222
Resources of the Southern Fields and Forests, 190
responsibilities in cooperative business, 33–35
"returning generation," 263, 264
Rhus spp., 195
rice
African, 105, 116, 149
fermentation of, 238
in polycultures, 123
as power ingredient, 242
in rotational grazing, 164
saving seeds of, 149, 155
starting seeds of, 110
Richardson, William H., 145
Rillieux, Norbert, 145

rituals
  in harvest, 62–63
  in meetings, 35–36
  in planting, 62, 63–64
roasted roots recipe, 235
Robeson, Paul, 112
Robinson, Stepheney, 191
Rockman, Seth, 271
rock phosphate, 74–75
rocks and stones
  balancing and carving of, 276, 278
  throwing as therapy, 251–52
Rock Steady Farm (NY), 41
rodent damage, 83
Roderick, Libby, 66
Rojas, Charo Minas, 257
role-playing, 310, 311
roles in cooperative business, 33–35
rollover protection of tractors, 134
Rooftop Garden Project, 218
rooftop gardens, 209, 216, 218–19
Roosevelt, Franklin D., 267
roosters, 163–64, 172
Rooted in Community, 249
  Youth Food Bill of Rights of, 249, 250–51
root knot nematodes, 102
Roselle, 154, 159
rosemary, 182, 196, 198
rotary tiller tractor attachment, 132–33, 134
rotational grazing, 164, 177–79
rotation of crops, 112, 113, 114, 126–28
Roundup Ready seeds, 150
row covers, 109, 118, 137
rue, 188, 196
Rural Training & Research Center (AL), 298
Ruta graveolens (rue), 188, 196
rye, 99, 100, 124

Sabur, Jalal, 282, 300
sacred herbalism, 200–201
sacred literature, 56–58
safety
  in communal labor, 46, 47–48
  in food processing, 209
  in harvesting, 145
  in ponds, 54
  in soil contamination, 72–78, 210–12
  in sprouting seeds, 217
  in tractor use, 134
  in water supply, 212–15
safety of plants, 184
  of black cherry, 192
  of black cohosh, 185
  of boneset, 192
  of hyssop, 187
  of jimsonweed, 193
  of pennyroyal, 188
  of pine, 194
  of red clover, 195
  of rue, 188
  of sage, 189
  of sumac, 195
  of wormwood, 189
sage, 82, 182, 188–89, 196, 198
Sahel restoration, 81
Salanova, 118
Sales, Ruby Nell, 264, 310
salves, herbal, 203
Salvia officinalis (sage), 82, 188–89, 196, 198
sandy soil, 89, 90
sassafras, 195
Sawadogo, Yacouba, 81
Scafidi, Susan, 312
scallions, 107, 120, 124
scallop squash, 108
scarification process, 199
"Scavenger Hunt" activity, 259–60
Scotch bonnet pepper, 113
Sea Island red peas, 114, 151
Seal, Bobby, 255
season extension, 136–37
seaweed emulsion, 139
seeders, 139
seed exchange, 159–61
The Seed Farm (PA), 16
Seed Keepers, 160
seedling tables, 138
seed planting and starting
  in field, 46, 113, 139, 140, 148
  hand seeding in, 139
  of herbs, 198–99
  indoors, 110–11, 135–36, 137–38, 148
  konbit tasks in, 46
  for microgreens, 217
  scarification in, 199
  spacing of seeds in, 139
  spiritual practices in, 53, 55, 62, 63–64
  for sprouts, 215–17
  stratification in, 198–99
  vernalization in, 198, 199
  watering in, 138, 140
Seed Savers Exchange, 139, 160–61
seed saving and keeping, 149–61
  braided into hair, 149, 161
  exchange of seeds in, 159–61
  growing seeds for, 152–55
  harvest of seeds for, 155–59
  of hybrid varieties, 152, 159
  importance of, 150–52
  isolation in, 152–53, 154
  longevity in, 159
  pollination in, 152–55
  population size in, 154, 155
  storage conditions in, 137–38, 157, 159
  tools and supplies in, 157
  Youth Food Bill of Rights on, 251
Seeds for Change UK, 38
seed storage, 137–38, 157, 159
seed suppliers, 138–39
segregation, racial, 258, 267–68
selenium, 211
self-healing, 275–80
self-image, 274–75
self-pollination, 152
sesame, 122
seu, 62
Shaffer, Lloyd, 268, 284
Shakur, Assata, 8, 69
sharecropping, 3, 25, 29, 267
  and USDA, 205
  Youth Food Justice program activities on, 253, 258
sharpening of tools, 133
sheep, 180
sheet composting, 84–85, 95, 96
shelter for animals
  for chickens, 166, 167, 168–70, 209
  for pigs, 176, 178
  in urban areas, 209
Sherman, William T., 267, 291
Sherman Park Community Garden, 76
Sherrod, Charles, 13
Sherrod, Shirley, 13, 257, 317
shiitake mushrooms, 220–21
Shwerner, Michael, 305
silage tarps, 85–86, 133
silt in soil, 89
Six National Indigenous reserve, 97
skills transfer in interracial alliances, 304
skullcap, 83, 182, 198, 199
Sky Woman, 103
Slade, Clifton, 160
slaughter of animals
  of chickens, 18, 53, 168, 169, 170–76, 179
  of pigs, 179
  spiritual practices in, 8, 53, 172, 176

slavery, 3
and Black Codes, 266–67
and chickens, 169
and cotton, 114, 265, 271
diet and health in, 225
and Haitian Revolution, 59, 108
and herbal medicine, 182, 183, 190–91
honoring ancestors lost to, 63
and internalized racism, 273–75
and joumou pumpkin, 108
and millet, 116
and patents, 145
reparations for, 23–24, 267, 301–4
and rice, 105
seed keeping in, 149
trauma of, 263, 265
and wealth gap, 271, 301
Youth Food Justice program on, 253, 258
Slavery's Capitalism (Beckert & Rockman), 271
sliding scale for farm share membership, 4, 41
slopes
measurement of, 79
terracing of, 78–80, 82
small spaces, growing in, 215–21
smell of soil, 87, 90
Smith, Barbara, 15
Smith, Joseph, 140, 145
Smith, Peter, 145
Smith-Penniman, Adele, 242
SNAP (Supplemental Nutrition Assistance Program), 5, 27, 41–42, 241, 282, 293
soil, 71–102
cation exchange capacity of, 91, 93–94
color of, 90, 94
compaction of, 133
contamination of, 72–78, 94, 210–12
ecology of, 96–99
erosion of. *See* erosion
macronutrients in, 92–93
micronutrients in, 93
microorganisms in, 96–99
pH of. *See* pH of soil
restoration of, 71–86, 210–12
smell and taste of, 87, 90, 91
at Soul Fire Farm, 4, 71, 78, 89, 91
storage of vegetables in, 236–237
texture of, 88–89, 90
in urban areas, 210–12
soil amendments in lead contamination, 73, 74–75
Soilful City, 297
soil preparation, 130–33
cover crops in, 99, 132
for herbs, 199
no-till, 84–86, 97, 123, 133
pigs in, 176
tarping in, 85–86, 117, 133
tillage in, 84, 87–88, 97, 132–33, 134
tools in, 130–33, 148
soil restoration, 71–86
agroforestry in, 78, 80–84
Carver on, 101, 114
in contamination, 72–78, 210, 212
in erosion problems, 72, 78–80
no-till and biological tillage in, 84–86
of Ovambo people, 75
at Soul Fire Farm, 4, 71, 78, 89, 91
in urban areas, 210–212
zai pits in, 81
soil tests, 88–94
in Africa, 90
for heavy metals, 73, 74, 91, 94, 210, 211
texture in, 88–89
Solanaceae family, 109–113, 124, 128. *See also* eggplant; peppers; potatoes; tomatoes
tobacco, 189, 199, 265
Solanum spp., 112, 154, 228. *See also* eggplant; tomatoes
solar energy, 13, 14, 33, 130, 166
Solenostemon monostachyus (nzinzing-rolo), 187, 279
sole proprietorships, 30, 31
songs and chants, 8, 58, 66–69, 260–61, 262
"Amazing Grace," 105
on basil, 181
in dynamic meditation, 278
in harvesting, 62, 104
in herbal baths, 53, 65
in meetings, 36–37
in pile fèy ceremony, 201
in planting, 63
sorghum, 107, 116
fermentation of, 238
in polycultures, 126
saving seeds of, 154, 159
sorghum sudangrass, 99, 101–2
sorrel. *See* hibiscus (sorrel)
Soul on Fire Hot Sauce, 240
Soul Train (television show), 274
Soup Joumou, 108, 229, 241
South Africa, 225
South Carolina Gazette, 191
Southeastern African American Farmers' Organic Network, 257, 285
Southern Exposure Seed Exchange, 138, 160, 161
Southern peas, 154. *See also* cowpeas
Southern Tenant Farmers Union, 258, 300
Southern University Agricultural Research and Extension Center, 19
southernwood, 196
soybeans, 86, 99, 150
space requirements
for chickens, 166
for pigs, 176
spacing between plants, 106–7
in direct seeding, 139
of fruits, 82
in intensive growing, 219–20
for seed saving, 152–53, 154
spider plant, 228
spilanthes, 189
spinach, 107, 116, 121
in container gardening, 218
in crop rotation, 128
Egyptian, 115
in Haitian grits recipe, 234–35
in high tunnel, 136
malabar, 116
as microgreens, 217
saving seeds of, 154, 155, 159
in square-foot gardening, 220
spiritual practices, 53–70
in animal slaughter, 8, 53, 172, 176
appropriation of, 69, 313
baths in. *See* herbal baths
in forests, 248
in Haitian Revolution, 59, 177
in harvest, 53–54, 55, 57, 58, 62–63, 65, 199, 201
herbs and plants in, 8, 53, 54, 64–65, 181–204
offerings in. *See* offerings
in planting, 53, 55, 62, 63–64
sacred literature in, 56–58
songs and chants in, 8, 65, 66–69
teachers of, 70
sprinklers for watering, 140, 142
sprouts, 215–17
square-foot gardening, 219–20
squash, 106–8
in crop rotation, 128
saving seeds of, 149, 152, 154, 155, 156, 158, 159

in Soup Joumou, 108, 229
in square-foot gardening, 220
summer. *See* summer squash
in three sisters, 103, 116, 121, 124–26
winter. *See* winter squash
squash bug, 106
squatting on land, 14–15
staghorn sumac, 195
steam inhalation of herbs, 203–4
Still, James, 191
stones and rocks
balancing and carving of, 276, 278
throwing as therapy, 251–52
storage
of eggs, 168
of seeds, 137–38, 157, 159
of tools, 133
of vegetables, 147, 236–37
strategic discussion meetings, 36–37
strategic goals, 49, 51
stratification of seeds, 198–99
strawberries, 82, 83–84, 124, 238
Strengthening Our Rural Roots (Fitzgerald & Bobrow-Williams), 294
strikes of farmworkers, 296, 297
subordination in internalized racism, 273–74
Sugar Drip Sorghum, 116
sugar plantations, 265
sugar snap peas, 113
sulfur, 93
acidifying soil with, 73, 74, 92
soil testing for, 91
sumac
poison, 195
staghorn, 195
winged, 195
summer squash, 106–7
in crop rotation, 128
saving seeds of, 159
in square-foot gardening, 220
Sun, Jun, 278
Sunday Soup, 241–42, 244
sunflowers, 107
in lead phytoremediation, 74, 75
as microgreens, 217
in polycultures, 123, 125
sprouting seeds of, 215, 217
sunn hemp, 99, 102
supermajority vote, 38
Supplemental Nutrition Assistance Program (SNAP), 5, 27, 41–42, 241, 282, 293
suppliers, 138–39
supplies. *See* tools and equipment
sustainability, 179–80, 250
Sustainable Agriculture Consortium for Historically Disadvantaged Farmers Program, 19
susus, 9, 20, 25
sweet potatoes, 107, 121–22
in crop rotation, 128
leafy greens from, 122, 233
in polycultures, 123, 126
as power ingredient, 242
roasted, 235
selling slips of, 160
in square-foot gardening, 220
storage of, 236
in urban farming, 208
sweet potato pie pumpkin, 108
swidden agriculture, 127
swine flu, 177

Taco Bell, 295
Tagetes spp. (marigolds), 118, 123, 185
Taraxacum officinale (dandelion), 118, 192, 233
taro, 225
tarping, 85–86, 117, 133
taste of soil, 87, 90, 91
Tate, Terressa, 104
taxes, 31, 32
Taylor, Gail, 257
Taylor, Owen, 151, 312
terracing, 78–80, 82
terra preta, 96
Terror Campaign, 267
Terry, Bryant, 228
texture of soil, 88–89, 90
Their Eyes Were Watching God (Hurston), 67
Thlaspi caerulescens, 74
Thompson, Joe, 14
three sisters, 103, 116, 121, 124–26
Thurman, Howard, 26, 256
thyme, 182, 189, 198
tick clover, 126
Tierra Negra Farm (NC), 160
tillage, 84, 87, 132–33, 134
and soil ecology, 97
The Timber Press Guide to Vegetable Gardening in the Southeast (Wallace), 160
tinctures, 202
tobacco, 189, 199, 265
tomatillos, 107, 124
tomatoes, 107, 109, 112
canning of, 240
in container gardening, 218
in crop rotation, 128
dried, 237
freezing of, 240
harvest of, 109, 146, 295
in high tunnels, 107, 109, 124, 136, 144
in polycultures, 123, 124
saving seeds of, 149, 152, 153, 154, 156, 158, 159
in square-foot gardening, 220
storage in ash, 237
trellising of, 109, 124, 136, 143–44
Tometi, Opal, 292
tools and equipment, 129–48
African hoes, 129, 130, 131
in bed preparation, 130–33, 148
in canning, 240
care of, 133, 148
checklist on, 148
in chicken processing, 171–72, 173
clothing, 73, 77, 78, 147, 212
cost of, 26, 50
digging sticks, 139
in harvesting, 133, 145–47, 148
in irrigation, 140–42, 148
patents by Black farmers on, 145
in propagation, 135–38, 148
in seeding, 139, 148
in seed saving, 157
storage of, 133
suppliers of, 138–39
tractors, 26, 129, 131–32, 134
in transplanting, 139–40, 148
in trellising, 142–44, 148
in weed control, 144–45, 148
toothache plant, 189
Top Leaf Farms (CA), 219
topsoil, 71, 78, 90. *See also* soil
Toxicodendron vernix, 195
Traces of the Trade: A Story from the Deep South, 312
tractors, 129, 134
attachments for, 134
in bed preparation, 131–32
cost of, 26
safe use of, 134
Tradescantia zebrina, 201
transplanting, 110–11, 139–40, 148
konbit tasks in, 46
watering at time of, 139–40, 142

transportation barrier to food access, 5, 27–28, 42
home delivery of foods in, 4–5, 27–28, 38, 42
trauma, 263–280
and agricultural training, 16
and connection to land, 8, 263
and cotton, 114
epigenetics in, 273
healing in, 275–80. *See also* healing
historical events causing, 265–73
imagination limitations in, 48
internalized racism in, 273–75
restorative justice program in, 252
of slavery, 263, 265
spiritual practices in, 59
of violence exposure, 58
trees
in agroforestry, 78, 80–84
as cover crops, 102
at edges of fields, 102
interactions between, 80–82, 246–48
in lead contaminated soil, 77
pest protection for, 83
planting of, 81, 82, 83
in polycultures, 82–83, 123
in swidden agriculture, 127
in zai pits, 81
trellising, 142–44, 148
of beans and peas, 113, 142–43
of cucumbers, 106–7, 143
in square-foot gardening, 220
of tomatoes, 109, 124, 136, 143–44
Trichilia glabra, 201
Trifolium pratense (red clover), 100, 194–95
triticale, 99
Truelove Seeds, 138, 151, 161
True Reformers Bank, 25
Trust for Public Land, 209
Truth, Sojourner, 291
Tubman, Harriet, 15, 63, 182, 190, 197
turbidity of water, 214
turnips, 107, 118, 119, 154
in crop rotation, 128
fermentation of, 238
leafy greens from, 233
in pikliz, 239
roasted, 235
saving seeds of, 154, 159
in square-foot gardening, 220
storage of, 236
Tuskegee University, 3, 17, 19, 39, 257, 286
Twitty, Michael W., 104, 119, 223, 228
Tyrrell County Training School, 25

Ujamaa Farm Share, 4–5, 27–28, 38–42
chicken eggs in, 164
sliding scale for costs in, 4, 41
sprouts in, 215
Umbelliferae family, 117–118, 128. *See also* carrots; celery; cilantro; parsley; parsnips
Uncommon Fruits for Every Garden (Reich), 84
Underground Railroad, 182, 190
undocumented people, 31
United Farm Workers, 296
United Movement of Small Peasants in the Artibonite, 44
United States Department of Agriculture (USDA)
census of, 21, 293
county committees, 294
discrimination by, 7, 12, 13, 20–21, 205, 258–59, 268–69, 284, 285, 289, 295, 317
on earthworms, 97
Farmers Home Administration, 13, 258
on food hubs, 44
funding and loan programs, 13, 20–22, 258–59, 284, 291, 295
National Scholars Program, 17
Office of Rural Development, 32
and *Pigford v. Glickman* case, 13, 20–21, 284, 317
small farms affected by, 299
SNAP program, 5, 27, 41–42, 241, 282, 293
universities, agricultural, 19, 285
unused land, cultivation of
right to, 251
in urban areas, 206, 209, 212–13, 288
Uprooting Racism training, 59, 282, 306, 309, 311
Urban Agricultural Legal Resource Library, 209
urban areas, 205–22
chickens in, 207, 209, 216
community gardens in, 206, 209, 212–13, 288
community in, 222
compost in, 210, 216, 221
in Cuba, 216
and growing in small spaces, 215–21
land trusts in, 14, 209
laws and land access in, 207–10
nature in, 246, 247, 273
renewal programs in, 270, 273, 288
rooftop gardens in, 209, 216, 218–19
soil in, 210–12
water in, 210, 212–15
urban renewal, 270, 273, 288
Utah, Adaku, 197

vacant lots, cultivation of, 209
right to, 251
in urban areas, 206, 209, 212–13, 288
Value-Added Producer Grants, 22
value-added products, 22, 44
values and mission of business, 48–49
vegetarian diet, 168, 180
venting of greenhouse, 136
Verbascum thapsus (mullein), 193–94, 199
vermicomposting, 221
vernalization, 198, 199
vervain, 182
vetch, 99, 100
Via Campesina, 150, 287–88
victim identity in internalized racism, 274
Victory Bus Project, 282, 298
Vigna spp., 114. *See also* cowpeas
vinegar, preserving in, 239
violence
farm as refuge from, 58
in internalized racism, 275
police-related, 5, 265–66, 272
Virginia Baker sweet potato, 160
visa program H2A, 293
"Vision for Black Lives: Policy Demands for Black Power, Freedom and Justice," 292
Vitex trifolia, 201
Vodou, 54–55, 59, 60, 69
herbal baths in, 53, 64–65, 201
pig sacrifice in, 59, 177
plant medicine in, 64–65, 184, 200–201, 279–80
prayer to Lwa of death in, 172
traditional song in, 68
Vodun, 54, 55, 69, 115
volunteers in communal labor, 48
Voodoo, 55
plant medicine in, 184, 185, 188

wages and salaries
in cooperative business, 31, 32, 35
financial agreements on, 35

in interracial alliances, 304–5
in organizational transformation, 307
Walker, David, 268
Walker, Lloyd T., 19
Walker, Margaret, 149
walking onion, 121
Wallace, Ira, 160
wall gardens, 209
wandering Jew, 201
Washington, Booker T., 19, 286, 310
Washington, Karen, 3, 212–13, 224–25, 257, 287
water
for chickens, 166, 168, 170
contamination of, 212–13
daily intake of, 243
for pigs, 176, 177
rainfall as source of, 140, 215, 216
sustainable use of, 180
testing quality of, 213, 214
in urban areas, 210, 212–15
watercress, 217
watering
in container gardening, 218–19
irrigation systems in, 26, 140–42, 148, 215
in seed starting, 138, 140
in transplanting, 139–40, 142
watermelons, 108
in polycultures, 123
saving seeds of, 149, 154, 156, 158, 159
in square-foot gardening, 220
in urban farming, 208
Watson, Marlene, 273
Watt, James, 164
wealth gap, 271, 301
redistribution of wealth in, 40–41
reparations in. *See* reparations
weed control, 144–45, 148
for carrots and parsnips, 117
cover crops in, 100, 102
flame-weeding in, 117
konbit tasks in, 46
tarping in, 85–86, 117, 133
"We Gonna Be All Right" activity, 260–61
welcome ritual in meetings, 35–36
West Africa, 105, 129, 237
West India burr gherkin, 108
Whatley, Booker T., 3, 19, 39, 257, 286
wheat, 124
White, John T., 145
white bush scalloped squash, 108
white clover, 100, 124
white eggplant, 112
White Noise Collective, 304
white sorrel, 115
white supremacy views, 8, 299, 300
anti-racism work in, 299–315
behavior patterns in, 308–9
dismantling of, 300
internalized racism in, 273
self-healing in, 275
statutory exclusions in, 270
and USDA, 12, 205
Whitney, Eli, 145
Who Owns Culture? (Scafidi), 312
WIC program, 27, 241, 293
wild plants
harvest of, 64, 190–91
identification of, 191
in spiritual baths, 201
Wildseed Community Farm & Healing Village (NY), 23, 34
Wilkes, H. Garrison, 125
William Alexander heading collards, 119
Williams, Shirley, 253
Wilson, August, 205
Wilson, Darren, 272
winged sumac, 195
winter squash, 107–8
in crop rotation, 128
living mulch under, 124
saving seeds of, 159
in square-foot gardening, 220
in three sisters, 116
wire worms, 124
Wise, Eddie and Dorothy, 288
Withania somnifera, 184
Women Infants and Children (WIC) program, 27, 241, 293
Woodbourne Correctional Facility, 299
Worcester Roots Project, 72, 74
worker-owned cooperatives, 29–38
The Working World, 35
work songs, 66, 104
worktables, 138
World Health Organization, 183
World Wide Opportunities on Organic Farms, 16
worms, composting with, 221
wormwood, 182, 183, 189, 196, 199

Yakini, Malik, 60, 208, 210, 257
yams, 116, 121
harvest festival for, 62–63
in polycultures, 123
as traditional food, 44, 225
yarrow, 99, 182, 195, 196
Yoruba people
Ifa spiritual practices of. *See* Ifa
prayer for chicken sacrifice in, 172
soil classification by, 90
youth
in criminal justice system, 5–6, 249–52, 271
diet-related illnesses of, 4, 224, 225
fatalism of, 5, 248
school discipline of, 271
Youth Ed-Venture & Nature Network, 247
Youth Food Bill of Rights, 249, 250–51
Youth Food Justice Pledge, 262
Youth Food Justice program, 5–6, 245, 248, 252–62
Youth Grow program (MA), 211
youth programs, 245–62
best practices in, 248–52
in community-based marketing, 39
curriculum in, 252–62
food preparation in, 226–28
principles in, 248–49
Youth Food Justice program, 5–6, 245, 248, 252–62

zai pits, 81
Zea mays, 154. *See also* corn
Zimmerman, George, 292
zinc, 91, 93, 211
Zinn, Howard, 300
zinnias, 123
zoning laws in urban areas, 209
zucchini, 106, 158, 220

ABOUT THE AUTHOR

Jamel Mosely | Mel eMedia

Leah Penniman is a Black Kreyol farmer who has been tending the soil for 22 years and organizing for an anti-racist food system for 16 years. She began with The Food Project in Boston, Massachusetts, and went on to work at Farm School in Athol, Massachusetts, and Many Hands Organic Farm in Barre, Massachusetts. She co-founded Youth Grow urban farm in Worcester, Massachusetts. She currently serves as founding co-executive director of Soul Fire Farm in Grafton, New York, a people-of-color-led project that works to dismantle racism in the food system through a low-cost fresh-food delivery service for people living under food apartheid, training programs for Black, Latinx, and Indigenous aspiring farmer-activists, Uprooting Racism training for food justice leaders, and regional-national-international coalition building between farmers of color advocating for policy shifts and reparations. She has dedicated her life's work to racial justice in the food system and has been recognized by the Soros Equality Fellowship, NYSHealth Emerging Innovator Awards, The Andrew Goodman Foundation Hidden Heroes Award, Fulbright Distinguished Awards in Teaching Program, New Tech Network National Teaching Award, Presidential Award for Excellence in Teaching (New York finalist), among others. She has contributed to two published volumes, authored numerous online articles, and given dozens of public talks on the subject.